AF788180

RECENT ADVANCES OF PLANT ROOT STRUCTURE AND FUNCTION

Developments in Plant and Soil Sciences

VOLUME 90

RECENT ADVANCES OF PLANT ROOT STRUCTURE AND FUNCTION

PROCEEDINGS OF THE 5TH INTERNATIONAL SYMPOSIUM ON STRUCTURE AND FUNCTION OF ROOTS

Stará Lensná, Slovakia, 30 August – 4 September, 1998

Guest Editors:

Otília Gašparíková,
Milada Čiamporová,
Igor Mistrík and
František Baluška

SPRINGER-SCIENCE+BUSINESS MEDIA, B.V.

A C.I.P. Catalogue record for this book is available from the library of Congress.

ISBN 978-90-481-5598-9 ISBN 978-94-017-2858-4 (eBook)
DOI 10.1007/978-94-017-2858-4

Printed on acid-free paper

All Rights reserved
© 2001 Springer Science+Business Media Dordrecht
Originally published by Kluwer Academic Publishers in 2001
Softcover reprint of the hardcover 1st edition 2001
No part of the material protected by this copyright notice may be reproduced or utilized in any form or by any means, electronic or mechanical, including photocopying, recording or by any information storage and retrieval system, without written permission from the copyright owner.

CONTENTS

*Reprinted from *Plant and Soil* Volume 226, No. 1 (2000).

Recent Advances of Plant Root Structure and Function, 1–2
Edited by O. Gašparíková, M. Čiamporová, I. Mistrík and F. Baluška
© 2001 *Kluwer Academic Publishers.*

Preface

It has been claimed that many of the most important chemical and physical processes which drive change in the material and biological world take place at interfaces. It is here that processes such as adsorption and de-absorption, uptake and efflux, attraction and repulsion, and slip and slide re-arrange molecules and elements in ways that alter their properties radically by provoking new interactions. It is difficult to find a functional interface of greater significance for terrestrial life than the one that exists between plant roots and the soil they exploit. Only when this complex interface between bio-, geo- and hydrophase became established in relatively recent biological and geological time (400 million years ago) could extensive land colonization and evolution of terrestrial life begin in earnest. Likely reasons for this dependency on roots and their interactions with mineral- and water-bearing substrates are not hard to find. At their heart are the gravity-guided penetrative and branching properties of roots that enable soil and more rocky substrates to be explored thoroughly and in a steadily expanding way that keeps in proportion to the aerial shoot system it supports both physically (for stability) and chemically. This presents an enormous absorptive interface to the soil (several hundred square meters even in a modest herbaceous plant) that selectively takes-in the twenty four or so essential elements from highly diluted solutions (μM concentrations) plus a few others, even gold. This impressive permeation elevates metabolically driven nanoscale absorption processes in to a vast mining operation that sustains the mineral requirements not only of plants but almost all other forms of terrestrial life. Without the action of roots the minerals would remain locked away and mostly unavailable to non-aquatic life. The root/soil interface is also essential for water uptake in amounts that maintains cell hydration and expansive growth, and creates the medium for cellular biochemistry. Of even more fundamental significance perhaps is that water supplied by roots serves as a co-substrate in photosynthesis along with carbon dioxide. This means that hydrogen atoms split-off by photolysis from water supplied by roots comprise roughly half the atoms of each photosynthetically assembled glucose molecule and thus the many substances derived from it that build and maintain living organisms of all kinds. If we add to this the importance for respiration of the oxygen generated from this same water we are left in little doubt about the key role played by roots and their interface with the soil in the original colonisation of the land and its healthy continuance.

Thus placing roots in their true perspective makes a stirring story that contrasts strongly with their modest outward appearance, at least as individual roots axes. That roots are largely hidden from our view adds a further piquancy since, for the majority of plant scientists, out of sight has meant out of mind despite the overwhelming importance of the activities of roots to living things. There can be little doubt that the excitement created by the opening up this hitherto secret but vital world of plant roots through experimental and observational studies inspired Professor Jozef Kolek and his Slovak colleagues to arrange the first International Symposium on Structure and Function of Roots at Tatranská Lomnica in 1971. This sparked a successful and influential series of international symposia devoted to the investigation of plant roots and attended by root devotees from many countries, almost as a necessary pilgrimage. The most recent of these meetings (the fifth) held at Stará Lesná, Slovak Republic in September 1998 was attended by root scientist from 17 countries. The symposium was organized by Dr. Otília Ga paríková and her colleagues of the Institute of Botany and the Commenius University in Bratislava. They have sustained and nourished a fine tradition for which a grateful botanical community benefits much. Some chauvinists may claim that plant science has reached a stage where focusing on different organs, species, even phyla is a distraction as we increasingly recognize the extensive homeology between the genomes of disparate plant species and also their closely similar regulatory mechanisms such as those of signal sensing and transduction. Gene sequence databases contain no taxonomic or morphological boundaries and it can be argued that roots should be studied only where they offer advantages for understanding basic processes that apply more widely. However, as we have seen, roots possess unique characteristics that played a crucial role in establishing life on the land and continue to sustain it. The human race depends on them and this alone is reason enough to study roots in their own right and for us to look forward expectantly to the next international conference about plant root systems. This book presents selected contributions to the Fifth International Symposium on Structure and Function of Roots.

Michael B. Jackson
IACR-Long Ashton Research Station

Recent Advances of Plant Root Structure and Function, 3–12
Edited by O. Gašparíková, M. Čiamporová, I. Mistrík and F. Baluška
© 2001 *Kluwer Academic Publishers.*

Histochemical analysis of root meristem activity in *Arabidopsis thaliana* using a cyclin:GUS (β-glucuronidase) marker line

Marie-Theres Hauser and Elisabeth Bauer
Center of Applied Genetics, University of Agricultural Sciences Vienna, Muthgasse 18, A-1190 Vienna, Austria. E-mail: hauser@mail.boku.ac.at

Key words: Arabidopsis thaliana, cell cycle, root meristem, mitotic cyclin, postembryonic development

Abstract

We used a transgenic *Arabidopsis* line expressing a translational fusion between a mitotic cyclin and the reporter gene β-glucuronidase (GUS) to investigate cell divisions in postembryonic root meristems. The fusion protein contains the cyclin destruction box (CDB) and this leads to a rapid degradation of the chimeric GUS-protein after mitosis. Hence, the staining pattern of the meristem marks dividing cells. We observed that upon germination the first cell divisions occur in epidermis cells at the junction with the hypocotyl. Moreover, the accelerated root growth on media supplemented with sucrose correlates with an increased number of dividing cells and an enlargement of the root meristematic zone. The conditional root expansion mutants *pom pom1* and *procuste1* (*quill*) suppress this sugar effect leading to a smaller meristematic zone. Simultaneous visualisation of the nucleus revealed that the CYCAT1:CDB:GUS expression is subcellularly localised around the nucleus. This particular staining starts at prophase and disappears after the completion of the new cell wall. In metaphase the staining invades the cytoplasm whereas in the telophase it concentrates again around the nucleus. This cell cycle-dependent distribution was used to characterise the two root specific cytokinesis mutants *pleiade1* and *hyade1*. In both mutants, cells which fail to develop a complete cell wall during cytokinesis divide synchronously in further cell divisions leading to multinucleate cells. These experiments demonstrate the usefulness of the CYCAT1:CDB:GUS marker line for studying cell division of wild-type and mutants. Furthermore, this line can be used to analyse the influence of biotic and abiotic signals on the rate and spatial distribution of cell divisions.

Introduction

Most of the plant organs are generated by postembryonic cell divisions realised by meristems at the end of the apical-basal axis of the plant. The meristem activity, i.e. the rate at which daughter cells are produced is developmentally controlled and modified by environmental signals such as light or nutrient availability. Classical experimental approaches to asses the meristematic activity are ^{3}H-thymidine and bromodeoxyuridine feeding which specifically labels cells in S-phase (Dolan et al., 1993; Scheres et al., 1994). Other studies counted mitotic figures in serial sections of root meristems. An extensive overview of the microscopic and mathematical methods used to determine the relationship between cell proliferation and elongation is given by Ivanov (1997). Recently, Beemster and Baskin (1998) calculated from kinematic and cell length measurements, rates of local cell production (cells per mm and h) and cell division (cells per cell and h) for the root cortex and the two epidermal cells types, trichoblasts and atrichoblasts in *Arabidopsis* root meristems. All these methods were applied on already established postembryonic root meristems. We wanted to analyse cell division frequencies during germination and under different growth conditions as well as in different root morphogenesis mutants.

As scoring mitotic frequencies in sections is difficult, we decided to use a mitosis specific marker line to test their usefulness for studying and measuring meristem activity.

The cell cycle machinery is regulated by the cyclins which together with cyclin-dependent kinases (CDKs) are necessary for cell cycle progression (Evans et al., 1983). While plant CDKs are expressed throughout the cell cycle in division competent cells

(Hemerly et al., 1993; Martinez et al., 1992), expression of cyclins oscillates in these cells (Doerner et al., 1996; Fobert et al., 1994; Hemerly et al., 1992). Different classes of cyclins regulate similar cell cycle controls and checkpoints in animals and plants. For example, A type cyclins are expressed during S-phase until the G2 phase and are destroyed during M phase whereas the B type cyclins accumulate only late during G2 phase and start also to be destroyed parallel to the end of mitosis (Ferreira et al., 1994; Gentschik, et al., 1998; Hirt et al., 1992; Kouchi et al., 1995; Meskiene et al., 1995; Setiady et al., 1995; Shaul et al., 1996). The A and B type cyclins contain a conserved motif of nine amino acids at their N-terminus which resembles the destruction box of mitotic cyclins of animals and fungi. Minor structural differences of this motif probably reflect the functional distinctions in the mode and timing of cyclin degradation (Renaudin et al., 1996).

The cyclin Arath:CycB1;1 (former name: At-cyc1; Renaudin et al., 1996) which was used for the CYCAT1:CDB:GUS marker line belongs to the B type cylins and is known to be expressed during mitosis in root meristematic cells (Doerner et al., 1996). For the translational fusion the first 150 amino acids of the cyclin protein were combined with the normally stable β-glucuronidase gene (John Celenza, personal communication). The expressed chimeric protein therefore contains the cyclin B type destruction box (CDB) which resides between amino acid 40 and 50 of the N-terminus of the Arath:CycB1;1. Hence, the activity of the chimeric GUS protein detected by histochemical analysis marks cells in late G2 and mitosis. Moreover, GUS staining of transgenic *Arabidopsis* plants expressing this construct have a spotty pattern indicating that the chimeric protein is degraded rapidly. In this paper we describe the difference between embryonic and postembryonic root meristems. We use the CYCAT1:CDB:GUS marker line to visualise and quantify the temporal and spatial distribution of dividing cells within the root meristem of wild type and root morphogensis mutants. Furthermore, the marker line facilitates the quantification of the stimulatory effect of sucrose on the activity of the root meristem. We report some preliminary observations on the subcellular distribution of the cyclin destruction box containing reporter gene during cell division.

Materials and methods

Plant strains and growth conditions

The CYCAT1:CDB:GUS line of ecotype Columbia was kindly provided by John Celenza. The mutants *pom1-5, pom1-21, prc1-11* (former name: *qui1-3*), *hya1-1* and *ple1-3* were isolated in the lab during several mutant screens (Hauser and Benfey, 1994; Hauser et al., 1995; Hauser, unpublished). Genetic crosses between these mutants and the CYCAT1:CDB:GUS line were performed in both directions and F1 plants were allowed to self-fertilise. F2 seeds were plated as described below and wild type as well as mutant plants were separately stained for GUS activity (see below).

To observe the root phenotype and to analyse young seedlings, seeds were sterilised with 5% sodium hypochlorite, 0.5% Tween-20 solution for about 3 min, washed twice with sterile water and plated on agar-solidified Murashige & Skoog (1962) medium without or with 4.5% ($\sim$130 mM) sucrose. After 2 days of stratification at 4 °C 12$\times$12 cm square tissue culture plates were vertically placed in a growth chamber under 16 h light (80–90 μmol m^{-2} s^{-1}) and 8 h dark cycles at 22 °C.

Histological staining

Seedlings were stained from 30 min to several hours in 250 μL freshly prepared staining solution by incubation at 37 °C on microscope slides. A 1–2 mm high barrier of vaseline on the microscope slide avoided pressure damage of the plant organs and reduced evaporation. The staining solution contained 100 mM Tris/HCl pH 7.5, 50 mM NaCl, 2mM $K_3Fe(CN)_6$, 20% MeOH and 0.5–1 mg/ml *X*-Gluc (5-bromo-4-chloro-3-indolyl-β-D-glucuronic acid cyclohexylammonium salt, Biomol). Pictures were taken from cleared preparations. Clearing was done in microtiter plates after the following protocol. After staining, seedlings were treated for 15 min at 57 °C in 0.24 N HCl in 20% MeOH, then transferred to 7% NaOH in 60% EtOH for 15 min at room temperature. Seedlings were rehydrated in 40%, 20% and 10% EtOH each 5 min and finally 5% EtOH, 25% glycerol for 15 min. For visualising the nuclei, the GUS stained seedlings were incubated with the following solutions for 15 min each: 50% EtOH, 75% EtOH, 1 μg/ml DAPI (4,6,-diamino-2-phenylindole, Sigma) in H_2O, 30% glycerol and 50% glycerol and mounted in 70% glycerol on microscope slides. All samples were observed in a Zeiss Axiovert microscope using DIC (differential interference contrast). DAPI stained nuclei were analysed under UV with 365 nm exitation and 420 nm emission filters. Pictures were taken with a Contax 167mt camera using Konica R-100 films. The slides were scanned and arranged for publication using Adobe Photoshop.

The number of cells in mitosis were counted under the microscope. The size of the meristem was analysed on cleared preparations under the microscope using a measurement optic and defined as the region between the initials surrounding the quiescent centre (QC) and the last cell before rapid cell elongation and vacuolisation takes place. The SPSS 6.1 program was used for statistical analysis.

Confocal microscopy

For confocal laser scanning microscope (CLSM) analysis of mature embryos dry seeds were fixed over night in MeOH: acetic acid=3:1. After two washes with water for at least one hour, seeds were stained in a 1:2000 dilution of YO-PRO^{TM} −1 iodide (491/509, Molecular Probes, Leiden) for about two hours. After several washes with water and a glycerol series of 30%, 50% and 70%, each for about 10 min, the embryos were isolated with fine tweezers under a dissecting microscope on a wet filter paper and transferred to a microscope slide in a drop of mounting media (Citifluor, Plano). Seedlings were stained similarly.

The embryos were analysed under 488 nm exitation with the Argon/Krypton laser of the Biorad MSE 600 mounted on an Olympus microscope. PIC images were converted to TIFF files with PaintShop Pro and manipulated in Adobe Photoshop.

Results and discussion

Size and structure of the embryonic and postembryonic root meristem

The primary root of *Arabidopsis* has a simple architecture consisting of a defined number of epidermal, cortical, endodermal and pericycle cell files each arranged in single radial layers around the diarch organized vasculature (Dolan et al., 1993). This organisation is established in the embryo (Dolan et al., 1994; Scheres et al., 1994) and perpetuated after germination via anticlinal divisions of the postembryonic root meristem to increase cell number along the root axis.

We have performed extensive microscopic analysis to compare the sizes of the embryonic and postembryonic root meristem in *Arabidopsis*.

The embryonic root meristem (number of root cells per cell file) has been defined in several slightly different ways. Clowes (1961) proposed that the minimal promeristem of the root apex consists of a quiescent centre surrounded by initials for all cell types. According to this definition and the data of Dolan et al. (1993), the *Arabidopsis* promeristem would be about three cells long. A different definition was chosen by Cheng et al. (1995). Here, the number of cells were determined by subtracting the hypocotyl cell number from that of the embryonic axis. This calculation was reasoned because of a constant number of hypocotyl cells which did not divide during germination. However, fate map analysis indicated that cells within the root/hypocotyl boundary did not originate from divisions of the root meristem (Scheres et al., 1994). This particular region bears characteristics of the root – root hairs – as well as of the hypocotyl – the second cortex layer.

The later definitions were framed in relation to already germinated seedlings which possess root hairs. Since our aim was to compare the embryonic with the adult root meristem, we extend the root related definition of the promeristem of Clowes (1961) and define the embryonic root meristem as the zone between the cortex/endodermal initial cells (Dolan et al., 1993) and the hypocotyl which can by distinguished by the occurrence of a second cortex layer (Figure 1). According to this definition, the root meristem of mature *Arabidopsis* embryos is 8.4±1.3 endodermis, 8.9±1.4 cortex and 8.4±1.4 epidermis cells long with an absolute length of 60 μm±12.7 (mean and ±SD, $n\sim 50$). Furthermore, our extensive microscopic analysis revealed that the position of the uppermost lateral root cap cells correlates with the end of the embryonic root meristem and the basal end of the intermediate root/hypoctoyl zone with two cortex cell layers. Thus, the lateral root cap cells completely cover the meristem. Similar observations concerning the position of the lateral root cap could be detected in the adult root meristem. Here, the meristematic zone was defined as the region between the initials and the last cell before rapid cell elongation and vacuolarisation takes place (Figure 1). Hence, depending on the availability of sucrose in the nutrient agar, the postembryonic root meristem could reach up to 300 μm (see below). In the presence of sucrose the number of cortical cells was, on average, 20.6±3.2 but could bear as much as

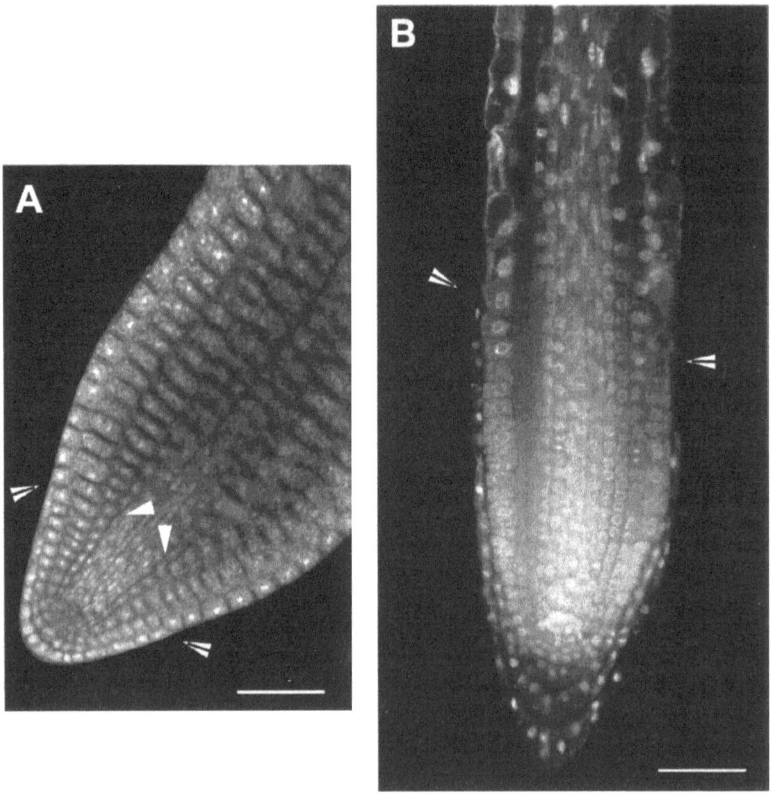

Figure 1. *Arabidopsis thaliana* root meristems stained with YO-PRO™ and visualised by confocal laser scanning microscopy. The open arrow heads indicate the apical end of the lateral root cap. The closed arrow heads point to the junction between the embryonic root meristem and the hypocotyl. (**A**) Root meristem of a mature embryo, (**B**) root meristem of a 12 days old seedling grown on nutrient agar medium supplemented with sucrose. Bar, 50 μm.

29 cells (n~80). This shows that the root meristem enlarges dramatically upon germination.

The correlation between the uppermost lateral root cap cells and the end of the meristem may point to their important role in the regulation of size and activity of the root meristem. On the other hand the position of the root cap cells may depend on the meristem. It is generally accepted that the root cap is protecting the meristem. In addition, the root cap cells produce mucilage which also protects the root tip from toxic soil products, prevents desiccation, helps absorbing ions and dissolving nutrients. Although a wealth of information is available for the responsibility of the root cap and particularly the columella cells in graviresponse (Blancaflor et al., 1998) the function of the lateral root cap in maintaining or regulating the root meristem or vice versa remains unclear. Recently, Blancaflor et al. (1998) performed laser ablation studies of root cap cells to analyse the root's response to gravistimulation. Ablation of the peripheral cap cells did not influence the growth rate and curvature of roots of 3 day old *Arabidopsis thaliana* seedlings. However, in their study only lateral cap cells were deleted close to the apex and ablation data are still missing for the uppermost lateral cap cells. It remains to be seen whether cell wall components or soluble signals or other signalling mechanism will provide evidence on the significance between the positional correlation between lateral root cap and the end of the meristem.

Onset of cell divisions upon germination

Similar to other plants, the meristematic zone of the primary root of *Arabidopsis* has to increase in size

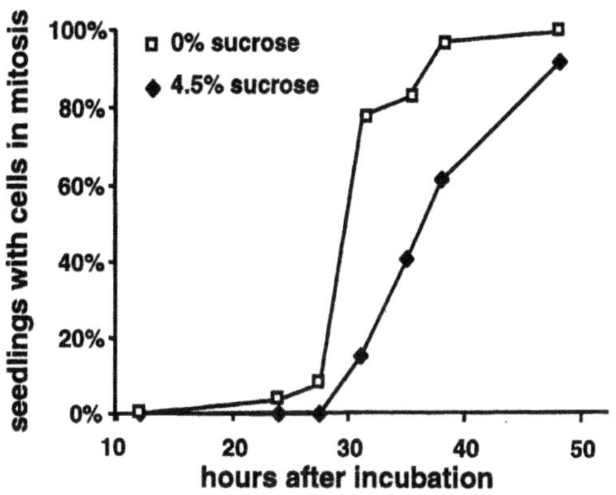

Figure 2. Time course of the progress of cell divisions in the root meristem upon germination. Cell division was monitored by activity staining of the CYCAT1:CDB:GUS maker line. The percentage of seedlings with cells in mitosis was plotted against the time after transferring the seeds to light and 22 °C. Seeds on nutrient agar media without sucrose started to germinate approximately 5 hours earlier than seeds on sucrose containing media. The data were gathered in two different experiments and at least 25 to 100 germinating seeds were analysed for each time point.

(Figure 1). The question we asked was when and where in the meristem do the first cell divisions occur. The availability of a simple histochemical analysis for the visualisation of cells in late G2 and mitosis with the CYCAT1:CDB:GUS marker lines allowed us to determine the position of the first cell division upon germination. Simultaneously, the CYCAT1:CDB:GUS marker line permitted an analysis of the temporal distribution of the first mitosis upon germination. First cell divisions were detected 24 h after transferring the seeds on nutrient agar plate to 22 °C and illumination. As shown in Figure 2 the presence of sucrose in the media delayed the germination for approximately 5 h.

The first cell division was always seen in epidermis cells close to the root/hypocotyl junction (Figure 3). The number of mitotic cells increases rapidly and at 38 h the meristem reached a size of 162 μm$\pm$15.5 and maintains in average 8.2$\pm$2.9 cells in mitosis. The meristem reached an adult size in two or three days after incubation on a media supplemented without or with sucrose, respectively.

Effect of sucrose on the activity and size of the root meristem

It was previously shown that 4.5% sucrose can promote root growth (Hauser and Benfey, 1994). Our experiments indicate that, although 4.5% sucrose delays germination slightly, the roots are longer and already develop lateral roots following five days of incubation whereas without sucrose lateral root development is about two days behind (Figure 4). Hence, the roots growing on sucrose supplemented media overcompensate the delayed germination by enhanced growth.

This effect could be explained by either an enhanced cell elongation or by an increased cell production. Since previous experiments showed that wild type cell lengths did not change in response to sucrose availability we used the CYCAT1:CDB:GUS marker line to analyse the activity of the root meristem. The results are summarised in Figure 5 and reveal that the meristems of roots grown on media supplemented with sucrose maintain more dividing cells. Thirty five percent of these roots had more than 15 cells in mitosis (n=44) whereas without sucrose (n=80) this percentage dropped to 3.5% and the majority of the meristems had less than 10 cells in mitosis.

To further quantify this effect we compared the meristem size with the number of dividing cells and observed a highly significant correlation (r=0.8179, p<0.0001, n=135) between these two meristematic parameters (Figure 6). This result is in agreement with the data of a kinematic analysis by Beemster and Baskin (1998). The authors highlight from their extensive studies on 6 to 10 day old *Arabidopsis* seedlings the importance of the control of the number of dividing cells to regulate the rate of organ enlargement. Hence, sucrose containing media enhances root growth by enlarging the meristematic zone which leads to an increased cell production.

In conclusion, we find that the root meristem of *Arabidopsis* can change its size which correlates with a higher number of dividing cells as well as with an increased growth rate.

Cell elongation versus cell division – analysis of root morphogenesis mutants

The CYCAT1:CDB:GUS marker line provides the unique opportunity to analyse the meristematic activity of root morphogenesis mutants. Hence, we crossed the marker line into the two conditional root expansion mutants *pom1* and *prc1* (previous name *qui1*) (Hauser and Benfey, 1994; Hauser et al., 1995; Desnos et al., 1996). Both mutants exhibit short expanded roots under conditions that favour higher growth rates. Furthermore, these mutants also have smaller root meristems. The size of the mutant root meristem was reduced to 108.8 μm$\pm$37.2 for *prc1–11* (n=23) and

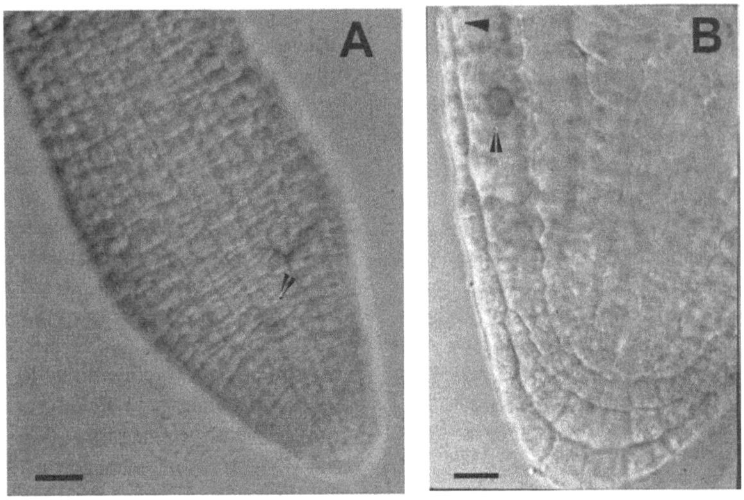

Figure 3. Detection of the first cell division upon germination by activity staining of the CYCAT1:CDB:GUS marker line. Open arrow heads mark cells in mitosis and the closed arrow head marks the apical end of the lateral root cap. (**A**) Complete root and hypocotyl of a germinating *Arabidopsis* seedling. Bar, 25 μm. (**B**) Magnified region of a root meristem with an epidermis cell in prophase. Bar, 10 μm. In both pictures the first cell division occurs at the junction to the hypocotyl region about 60 μm apical from the quiescent centre.

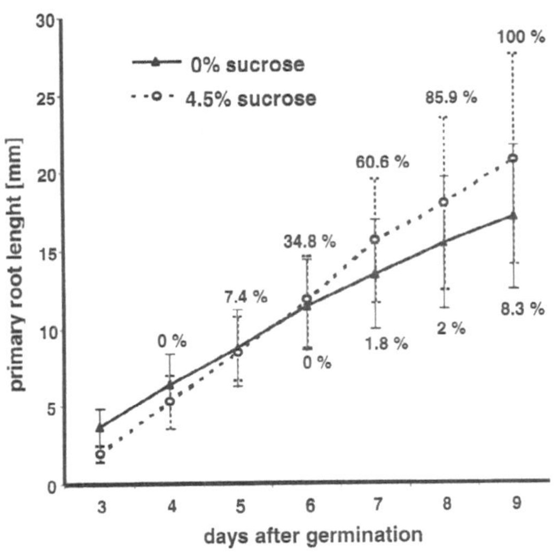

Figure 4. Effect of sucrose in the nutrient agar medium on the root growth and lateral root formation of *Arabidopsis* seedlings. Primary root length (mm) was measured daily starting 3 days after germination (DAG, $n>55$, the error bars correspond to SDs). The percentage of seedlings with lateral roots is indicated above for the seedlings on media supplemented with sucrose and below for the seedlings on media without sucrose. Only lateral roots were counted that emerged and could be detected under a dissecting microscope.

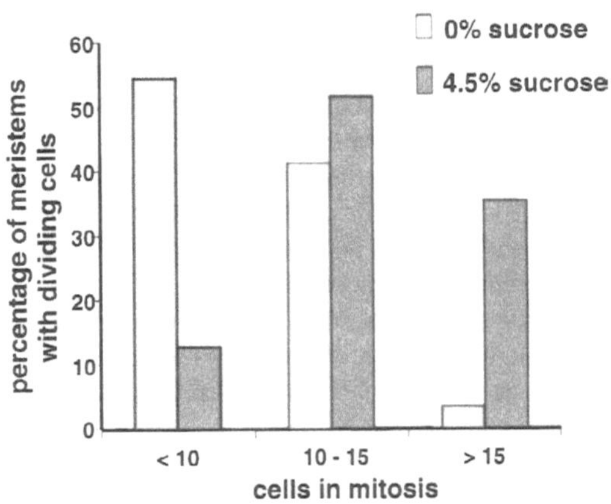

Figure 5. Comparison of the root meristematic activity of *Arabidopsis* seedlings grown with and without sucrose. The meristematic activity was measured by counting cells in mitosis of any cell type in the primary root meristem which were visualised by GUS staining. Meristems were ordered in groups with less than 10 cells, between 10 and 15 cells and with more than 15 cells in mitosis ($n>40$).

100 μm±36.2 for *pom1–5* and *pom1–21* ($n=18$). Consequently, approximately 6 cells of the *prc1–11* and 5.5 cells of the *pom1–5* and *pom1–21* root meristem expressed the reporter gene and where in mitosis (Figure 8B, C). These results partly confirm previous data

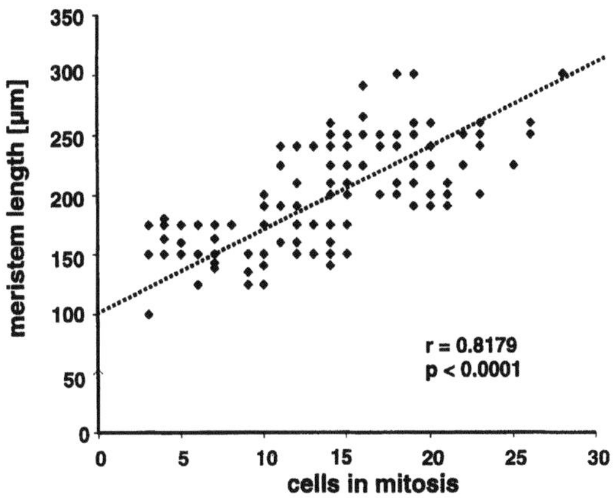

Figure 6. Correlation between cells in mitosis and the size of the root meristem. Root meristematic cells in mitosis were visualised by GUS staining of seedlings containing the CYCAT1:CDB:GUS construct. The correlation is highly significant with r=0.8179 and p<0.0001. (n=135). The meristem size of mature embryos is 60 μm±12.7 (n~16) and contains about 8.3±1.4 endodermis, 8.6±1.3 cortex and 8.4±1.4 epidermis cells which are dormant.

were the division rate was calculated by dividing the root growth rate per day with the average cell length (Hauser et al., 1995). While for *pom1* the reduction in root growth appeared to be due to a decreased cell division rate as well reduced cell elongation, the shorter *prc1* root was interpreted to be primarily a defect in cell elongation (Hauser et al., 1995). In contrast, the data with the CYCAT1:CDB:GUS line indicate that both cell division and elongation are affected in *prc1*. Our interpretation of the discrepancy between the old and new *prc1* data are: i) The earlier root growth data were measured between day 4 and 6 after germination just before the majority of lateral roots emerge. At this developmental stage the roots of *prc1* do not exhibit the full expansion phenotype and are therefore made up of longer cells. ii) The growth rate information was combined with the cell dimension data which were obtained from older roots. These roots already reached the maximum cell expansion defect. Hence, the calculated cell division rates were misleading and resulted in an apparently higher meristematic activity. The new data for the conditional root expansion mutants fit to the correlation between meristem size and number of cells in mitosis (Figure 6). Thus, under conditions that favour higher growth rates for wild type roots, the two conditional expansion mutants *pom1* and *prc1* exhibited not only reduced cell elongation but reduced meristem size which results in a reduced cell production. In other words, the root expansion mutants *pom1* and *prc1* repress the sugar effect on the size of the root meristem.

In conclusion, the CYCAT1:CDB:GUS marker line can be used to elucidate the influence of cell elongation versus cell division on the defects of root morphogenesis mutants.

Subcellular localisation of the CYCAT1:CDB:GUS reporter gene in root meristems

The GUS reporter gene expresses a soluble protein and unless the reporter protein is targeted to other compartments of the cell, the activity staining normally yields a strong blue water-insoluble indigo dye in the cytoplasm. By monitoring the time course of the activity staining of the chimeric GUS protein, we observed a distinct subcellular staining pattern which changes during mitosis. However, this subcellular pattern is lost by overstaining. Double staining of the GUS reporter and the DNA in the nucleus indicates that in prophase the dye precipitates around the nucleus (Figure 7A). In meta- and anaphase the dye is distributed in the whole cell (Figure 7B, C, D). In telophase the staining gathers once again around the nucleus probably concomitantly with the development of the nuclear membrane (Figure 7E, F). The staining disappears gradually after the completion of cytokinesis.

In conclusion, the presence of the mitotic cyclin destruction box or other adjacent part of the first 150 amino acids of the cyclin B translational fusion construct seem to direct the GUS protein i.e. its reaction product toward distinct loci within the cells during mitosis. Further analysis is needed to decide upon the composition of these distinct loci. The cell cycle dependent subcellular localisation of the GUS staining in this marker line was further used to analyse the spatial and temporal distribution of cell divisions of two potential root specific cytokinesis mutants, *pleiade1* (*ple1*) and *hyade1* (*hya1*). Previous electron microscopic studies of *ple1* roots exposed large cells with incomplete cell walls (Ovecka et al., 1996). Hence, nuclei inadequately separated by imperfect cell walls should divide synchronously. In both mutants, we observed root cells which contained more than one nucleus with identical subcellular staining pattern. In some of these cells, cell walls were visible (Figure 8D, E). Histochemical analysis of the F2 progeny of a genetic cross between the cytokinesis mutants and the marker line revealed up to 32 synchronised nuclei in the mutant *ple1* whereas in *hya1* the maximum number was four. Although, in wild type root meristems it is

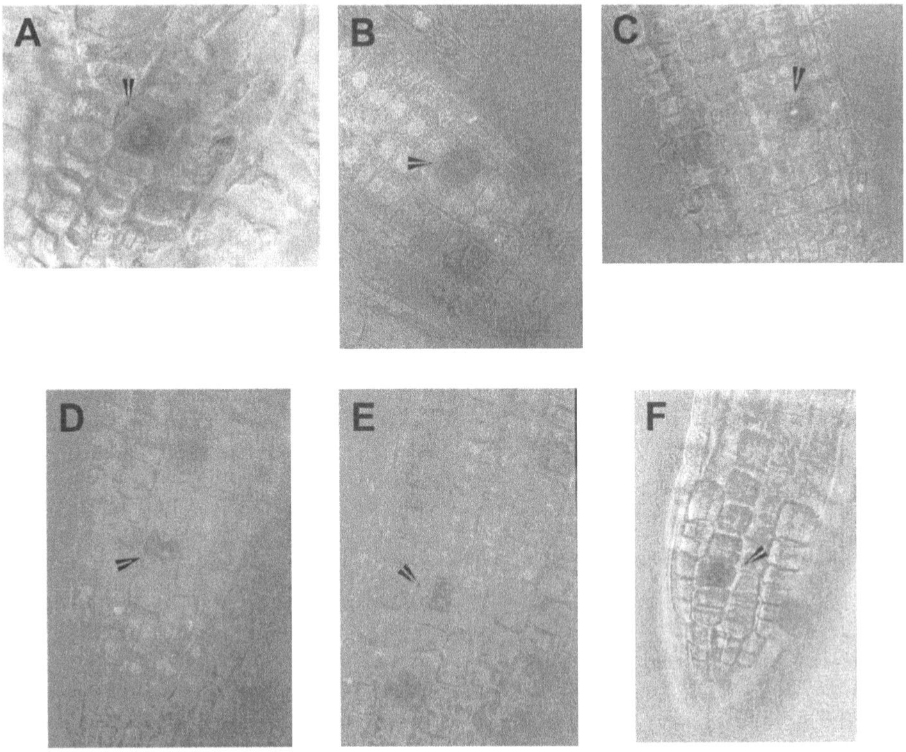

Figure 7. Subcellular GUS staining patterns in the root meristem of CYCAT1:CDB:GUS marker lines. The DNA of the nuclei were visualised with DAPI. Open arrow heads mark cells in (**A**) prophase: the GUS staining is distributed around the nucleus and the cell is large, (**B**) prometaphase: the nucleus staining is quenched by the GUS staining of the cytoplasm, (**C**) metaphase: the condensed strongly DAPI stained chromosomes are arranged at the equatorial plane and the GUS staining is distributed in the cytoplasm, (**D**) anaphase: the separated chromosomes are visible with DAPI and the GUS staining is still distributed in the cytoplasm, (**E, F**) telophase: the GUS staining occurs again around the two nuclei and the new cell wall is not visible yet.

possible that neighbouring cells are synchronised, the high amount of synchronisation was only detectable in the mutants. These observations add to the anatomical observations that the mutants *ple1* and *hya1* develop incomplete cell walls and that the corresponding gene products are probably involved in the cell wall formation during cytokinesis.

Concluding remarks

The quantification of root meristem activity is of fundamental importance for the exploration of the mechanisms that coordinate the processes of cell division and cell elongation. We have introduced a GUS marker line of *Arabidopsis thaliana* which allows to identify and quantify cells in different stages of mitosis under the light microscope. Hence with this marker line, the analysis of cells during mitosis avoids sectioning and can be performed on whole roots. The staining appears during G2 and stops after cytokinesis and follows the expression pattern of mitotic cyclin B (Genschik et al., 1998). Thus the time window is larger than when counting cells in metaphase and subsequently more cells can be identified which are just before and short after mitosis. This is a major advantage over counting metaphase bands which are only visible by staining with fluorescent dyes as DAPI and which have to be analysed in either a fluorescence microscope or CLSM.

One major limitation of all these staining procedures is that the roots have to be fixed and therefore *in vivo* analysis is impossible. This obstacle can probably be overcome by using a transgenic marker line with the same translational fusion of the mitotic cyclin B promoter fused to the green fluorescent protein (GFP).

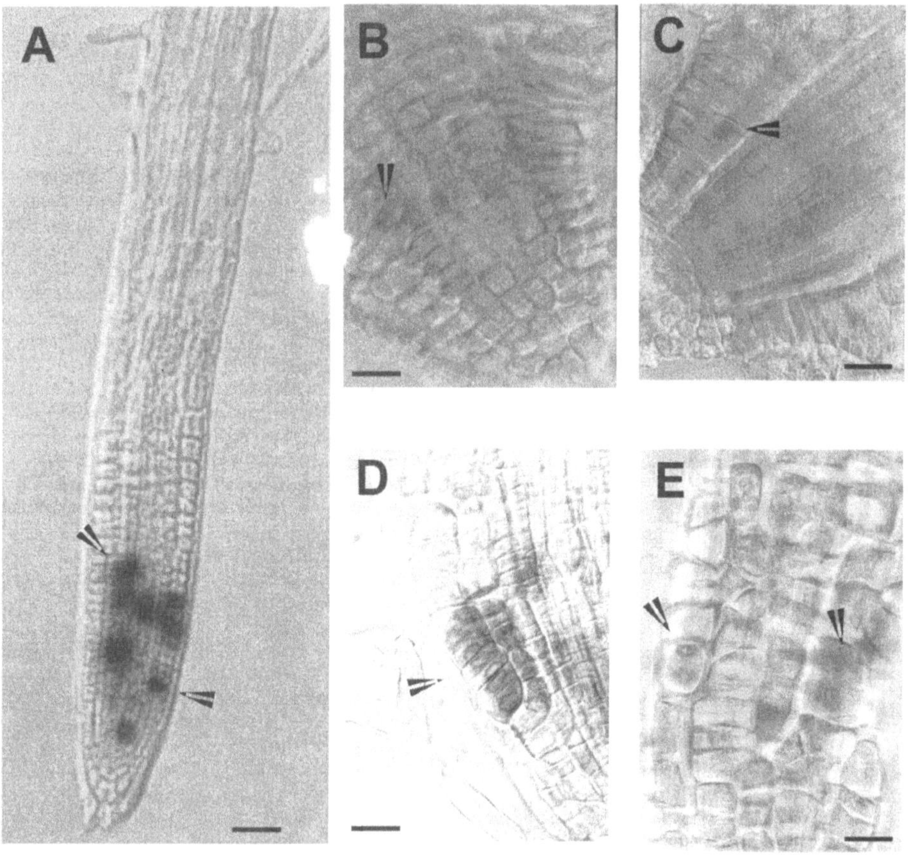

Figure 8. CYCAT1:CDB:GUS expression in wild type and root morphogenesis mutants. (**A**) Wild type, (**B**) *pom1-5*, (**C**) *prc1-11*, (**D**) *ple1-3* and (**E**) *hya1-1* homozygotes. The open arrow heads mark cells in mitosis. In B, C and D the open arrow heads indicate cells in telophase and in E in prophase. Bar, 50 μm.

In case this chimeric GFP will behave and oscillate in the same way as the GUS protein than this marker line would allow the *in vivo* observation of cell division in *Arabidopsis* roots under a CLSM.

Nevertheless by using the CYCAT1:CDB:GUS marker line we could show which cells in the embryonic root meristem start dividing upon germination. We further could measure the influence of sucrose on the meristem size and activity. Finally, the availability of this reporter line improved the analysis of root morphogenesis mutants.

Hence, the CYCAT1:CDB:GUS marker line is a valuable tool to identify different stages around and during mitosis and to analyse mutants which affect the activity and size of the root meristem.

Acknowledgements

This work was funded by the FWF (P11001-GEN) and the EU-project PL960217. We are obliged to John Celenza for the CYCAT1:CDB:GUS marker line. We like to thank Dr. Josef Glössl, head of the Center of Applied Genetics, for his attentive support. Special thanks is given to Christian Schlötterer, Christian Luschnig and Martin Hülskamp for their valuable comments on the manuscript.

References

Blancaflor E B, Fasano J M and Gilroy S 1998 Mapping the functional roles of cap cells in the response of *Arabidopsis* primary root to gravity. Plant Physiol. 116, 213–222.

Beemster G T S and Baskin T I 1998 Analysis of cell division and elongation underlying the developmental acceleration of root growth in *Arabidopsis thaliana*. Plant Physiol. 116, 1515–1526.

Cheng J-C, Seeley K A and Sung Z R 1995 *RML1* and *RML2*, *Arabidopsis* genes required for cell proliferation at the root tip. Plant Physiol. 107, 365–376.

Clowes F A L 1961 Apical Meristems. Blackwell, Oxford.

Desnos T, Orbovic V, Bellini C, Kronenberger J, Caboche M, Traas J and Höfte H 1996 *Procuste1* mutants identify two distinct genetic pathways controlling hypocotyl cell elongation, respectively in dark- and light-grown *Arabidopsis* seedlings. Development 122, 683–693.

Doerner P W, Jorgensen J-E, You R, Steppuhn J and Lamb C J 1996 Control of root growth and development by cyclin expression. Nature 380, 520–523.

Dolan L, Janmaat K, Willemsen V, Linstead P, Poethig S, Roberts K and Scheres B 1993 Cellular organisation of the *Arabidopsis* root. Development 119, 71–84.

Dolan L, Duckett C M, Grierson C, Linstead P, Schneider K, Lawson E, Dean C, Poethig S and Roberts K 1994 Clonal relationships and cell patterning in the root epidermis of *Arabidopsis*. Development 120, 2465–2474.

Evans T, Rosenthal E T, Youngblom J, Distel D and Hunt T 1983 Cyclin: A protein specified by maternal mRNA in sea urchin eggs that is destroyed at each cleavage division. Cell 33, 389–396.

Ferreira P, Hemerly A, Engler J D, Bergounioux C, Burssens S, Van Montagu M, Engler G and Inze D 1994 Three discrete classes of *Arabidopsis* cyclins are expressed during different intervals of the cell cycle. Proc Natl Acad Sci USA 91, 11313–11317.

Fobert P R, Coen E S, Murphy G J and Doonan J H 1994 Patterns of cell division revealed by transcriptional regulation of genes during the cell cycle in plants. EMBO J 13, 616–24.

Genschik P, Criqui M C, Parmentier Y, Derevier A and Fleck J 1998 Cell cycle-dependent proteolysis in plants: identification of the destruction box pathway and metaphase arrest produced by the proteasome inhibitor MG132. Plant Cell 10, 2063–2075.

Hauser M-T and Benfey P N 1994 Genetic regulation of root expansion in *Arabidopsis thaliana*. *In* NATO-ASI Plant Molecular Biology Series. Eds P Puigdomenech and G Coruzzi. pp 31–40. Springer-Verlag, New York.

Hauser M-T, Morikami A and Benfey P N 1995 Conditional root expansion mutants of *Arabidopsis*. Development 121, 1237–1252.

Hemerly A S, Ferreira P, Engler J, Van Montagu M, Engler G and Inze D 1993. Cdc2a expression in *Arabidopsis* is linked with competence for cell division. Plant Cell 5, 1722–1723.

Hemerly A S, Bergounioux C, Van Montagu M, Inze D and Ferreira F 1992 Genes regulating the plant cell cycle: isolation of a mitotic like cyclin from *Arabidopsis thaliana*. Proc Natl Acad Sci USA 89, 3295–3299.

Hirt H, Mink M, Bögre L, Györgyey J, Jonak C, Gartner A, Dudits D and Heberle-Bors E 1992 Alfalfa cyclins: differential expression during the cell cycle and in plant organs. Plant Cell 4, 1531–1538.

Ivanov V B 1997 Relationship between cell proliferation and transition to elongation in plant roots. Int. J. Dev. Biol. 41, 907–915.

Kouchi H, Sekine M and Hata S 1995 Distinct classes of mitotic cyclins are differentially expressed in the soybean shoot apex during the cell cycle. Plant Cell 7, 1143–1155.

Martinez M C, Jorgensen J-E, Lawton M A, Lamb C J and Doerner P W 1992 Spatial pattern of *cdc2* expression in relation to meristem activity and cell proliferation during plant development. Proc Natl Acad Sci USA 80, 7360–7364.

Meskiene I, Bögre L, Dahl M, Pieck M, Ha D T C, Swoboda I, Heberle-Bors E, Ammerer G and Hirt H 1995 CycMs3, a novel B-type alfalfa cyclin gene, is induced in the G(0)-to-G(1) transition of the cell cycle. Plant Cell 7, 759–771.

Murashige T and Skoog F 1962 A revised medium for rapid growth and bioassays with tobacco tissue cultures. Physiol. Plant 15, 473–479.

Ovecka M, Ciamporova M and Hauser M-T 1996 Root-specific mutation of *Arabidopsis thaliana* affecting cytokinesis: production and fate of multinucleate cells. J. Computer-Assisted Microscopy 8, 277–278.

Renaudin J-P, Doonan J H, Freeman D, Hashimoto J, Hirt H, Inze D, Jacobs T, Kouchi H, Rouze P, Sauter M, Savoure A, Sorrell D A, Sundaresan V and Murray J A H 1996 Plant cyclins: a unified nomenclature for plant A-, B- and D-type cyclins based on sequence organization. Plant Mol Biol 32, 1003–1018.

Scheres B, Wolkenfeldt H, Willemsen V, Terlouw M, Lawson E, Dean C and Weisbeek P 1994 Embryonic origin of *Arabidopsis* primary root and root meristem initials. Development 120, 2475–2487.

Setiady Y Y, Sekine M, Hariguchi N, Yamamoto T, Kouchi H and Shinmyo A 1995 Tobacco mitotic cyclins: cloning, characterization, gene expression and functional assay. Plant J 8, 949–957.

Shaul O, Mironov V, Burssens S, Van Montagu M and Inze D 1996 Two *Arabidopsis* cyclin promoters mediate distinctive transcriptional oscillation in synchronized tabocco BY-2 cells. Proc Natl Acad Sci USA 93, 4868–4872.

Section editor: H. Lambers

Recent Advances of Plant Root Structure and Function, 13–21
Edited by O. Gašparíková, M. Čiamporová, I. Mistrík and F. Baluška
© 2001 *Kluwer Academic Publishers.*

Ultrastructural effects of the herbicide chlorpropham (CIPC) in root tip cells of wheat

E.P. Eleftheriou* and E. Bekiari
Department of Botany, School of Biology, Aristotle University of Thessaloniki, GR-540 06, Thessaloniki, Greece

Key words: Chlorpropham (CIPC), microtubules, nuclei, recovery, roots, *Triticum aestivum*

Abstract

The present ultrastructural investigation on the effects of 50 μM chlorpropham (previously called CIPC) on growing roots of wheat (*Triticum aestivum* (L.) Thell cv. Vergina) was undertaken to clarify the mechanism of a carbamate herbicide action in plant cells, since the wide range of responses of plant cells to carbamate herbicides is based mainly on immunofluorescence studies. Cells of control roots contained abundant microtubules both in interphase and mitotic arrays. In chlorpropham-treated roots, however, no microtubules could be detected at all, neither in dividing nor in differentiating cells. Cycling cells became binucleate, polyploid or contained incomplete cell walls, the result of inhibition of cytokinesis. In long-term drug treatments (24 h or more) the affected cells entered a new cycle, which, however, did not progress beyond mid-metaphase. The nuclei of binucleate cells initiated prophase synchronously. Small vacuoles and Golgi vesicles were trapped within the nucleoplasm of the multilobed nuclei. In roots recovering from 8 h chlorpropham treatment, cells continued to exhibit polyploid nuclei, intranuclear vacuoles and incomplete walls. Microtubules reappeared but they were sparse and lacked a definite orientation. Preprophase cells did not form normal preprophase bands of microtubules, while mitotic cells occasionally contained microtubules bound to chromosomes and converged to minipoles. It is concluded that chlorpropham disorganized directly microtubules in addition to irreversibly affecting microtubule organizing centres, which failed to further support microtubule arrays.

Abbreviations: c – chlorpropham treated; CIPC – isopropyl N-(3-chlorophenyl)-carbamate; MT – microtubule; MTOC – microtubule organising centre; PPB – preprophase band

Introduction

Microtubules (MTs) are highly dynamic cell components involved in numerous functions, including cell motility, cell division, growth and differentiation. In dividing plant cells, they rearrange into four successive major arrays: the interphase cortical array, the preprophase band (PPB), the mitotic spindle apparatus and the phragmoplast MTs. Besides, additional MTs are generated at the nuclear envelope of the reforming nuclei in late telophase, to be finally replaced by the cortical array (Cyr and Palevitz, 1995; Mizuno, 1993).

Based on the frequently observed parallelism between the cortical MTs and the most recently synthesized cellulose microfibrils in interphase or differentiating cells, it was widely postulated that MTs mediate the orientation of cellulose microfibrils and affect the pattern of wall development, thus playing a crucial role in plant cytomorphogenesis (see Giddings and Staehelin, 1991). The MT-microfibril parallelism rule has, however, been disputed due to observations of ordered microfibril deposition in the absence of MTs (see Emons et al., 1992).

The experimental depolymerization of MT arrays by specific MT-disrupting agents has been a frequent approach to identify various MT-dependent processes in plant cells. Colchicine is the most widely used

* Corresponding author. Tel: ++30-31-99 8385. FAX No: ++30-31-99 8389. E-mail: eelefth@bio.auth.gr

anti-MT drug, but many other compounds have been employed, including dinitroaniline, phosphoric amide and N-phenyl carbamate herbicides (Morejohn, 1991).

N-Phenyl carbamate herbicides, such as propham, chlorpropham (discontinued name CIPC) and pronamide, are used as pre-emergence compounds with post-emergence activity on several weed species in broad-leaf crops. They are also used as inhibitors applied to suppress potato sprouting during winter storage (Kleinkopf et al., 1997). When carbamate herbicides are used in the high micromolar concentration they produce root tip swelling similar to that caused by colchicine (Nooden, 1971), though not so marked, and arrest dividing cells at metaphase or anaphase (Morejohn, 1991). Based mainly on immunofluorescence studies, it was found that they induce a range of responses in plant cells, depending upon the particular compound used, its concentration and the type of cells and plant species treated. In some cases, MTs remain long, apparently unaffected by the herbicide treatment, while in other cases they are short, disoriented, or lost completely, depending upon the stage in the cell cycle (Clayton and Lloyd, 1984; Gunning and Wick, 1985; Lloyd and Traas, 1988; Lehnen et al., 1990; Marc and Palevitz, 1990). These experimental results, in combination with the common observation of multipolar mitotic spindles and multiple MT converging foci, were taken to mean that the effects of carbamates were exerted on plant MT organizing centres (MTOCs) rather than MTs themselves. On the other hand, based on polymerization assays, Morejohn and Fosket (1986) indicated that tubulin was the subcellular target of chlorpropham, and Akashi et al. (1988) provided biochemical evidence that the carbamate herbicides do bind to plant tubulin and directly affect microtubule polymerization in cells.

Thus, results are mixed: many reports based on immunofluorescence microscopy indicate that carbamate herbicides affect MTOCs, while biochemical studies maintain that they cause MT depolymerization by binding directly to plant tubulin. To the best of our knowledge, fine structural studies are missing. For these reasons, the present ultrastructural investigation was undertaken in an attempt to contribute towards clarifying the mechanism of a carbamate herbicide effects on plant cells.

Materials and methods

Seedlings of the cultivated variety of the hexaploid wheat *Triticum aestivum* (L.) Thell cv. Vergina (kindly provided by the Institute of Cereals, Thessaloniki) were grown for two days on moist filter paper in Petri dishes in a growth chamber at 21±0.5 °C in the darkness. They were then transferred in other dishes containing working solutions of chlorpropham [isoprophyl N-(3-chlorophenyl)-carbamate] prepared by dilution with double distilled water of a 1 M stock prepared in 100% DMSO (dimethyl sulfoxide). All working solutions contained 0.5% (v/v) DMSO, a level that does not appear to affect the cytoskeleton of plant cells (Eleftheriou and Palevitz, 1992). For controls, other seedlings were exposed to distilled water or 0.5% (v/v) DMSO. To ensure continuous and uniform exposure to the media, root tips were covered with moistened cotton.

Preliminary experiments were carried out by exposing seedlings to 50, 100 or 200 μM chlorpropham in order to estimate the growth rate and appreciate the most appropriate drug concentration, which was decided to be the 50 μM. In the main experiment, roots were exposed to 50 μM chlorpropham for 8, 16, 24, 32, 48 or 72 h. In a recovery experiment, other roots exposed to the same drug concentration for 8 h were then thoroughly washed and transferred to recover on distilled water for 8, 16, 24 or 48 h. Root tips of the treated seedlings were processed for electron microscopy as previously described in detail (Eleftheriou, 1993). In brief, they were fixed in a mixture of 3% (v/v) glutaraldehyde and 3% (w/v) paraformaldehyde, in 50 mM sodium cacodylate buffer, pH 7.2 for 3 h at room temperature (~21 °C), post-fixed in 1% (w/v) OsO_4 similarly buffered, dehydrated in a graded cold ethanol series and propylene oxide, and finally embedded in Durcupan epoxy resin ACM (Fluka Chemie AG, Buchs, Switzerland). Ultrathin sections, cut parallel to the long axis of the root on a Reichert Om U-2 ultramicrotome, were examined with a Zeiss EM 9 S-2 electron microscope after double staining with uranyl acetate and lead citrate.

Results

In control roots of wheat, MTs constituted a conspicuous cytoplasmic component. In interphase and differentiating cells they abounded in the cortical cytoplasm being parallel arranged and evenly distributed along the lateral wall, having a transverse orientation relative to the long axis of the root, an orientation mirrored by that of cellulose microfibrils in the cell wall (Figures 1 and 2). Tangential sections to the cell surface further revealed the presence of smooth and coated vesicles amongst the MTs. Preprophase and prophase cells contained well defined PPBs of MTs (Figure 3), encircling symmetrically the premitotic nucleus.

In the chlorpropham-treated roots (c-roots) MTs disappeared completely even in the shortest (8 h) drug treatment. Longer exposures to the drug resulted in binucleate (Figure 4) or polyploid cells. Polyploid nuclei of short treatments had conspicuous nucleoli and deep cytoplasmic invaginations at sites where the division plane of normal nuclei would be expected (Figure 5). Many nuclei of c-cells, both polyploid or binucleate, entered a new division cycle, but only preprophase, prophase and prometaphase stages could be observed (Figure 6), without, however, forming a typical metaphase plate. In binucleate cells only synchronous prophases could be found.

Another feature of c-cells was the occurrence of incomplete cell walls formed during cytokinesis in the presence of the drug. Occasionally they were united with the lateral parental wall and projected inwards between the two nuclei (Figure 7), while independent wall portions (islands) between the nuclei or within deep cytoplasmic invaginations intruding in the polyploid nuclei were also found (Figure 8).

Long-term treated cells contained multilobed nuclei (Figure 9) and became highly vacuolated. Usually there were two large vacuoles on either side of the polyploid nucleus or groups of numerous smaller vacuoles. Small vacuoles also occurred near the nuclei or within the cytoplasmic invaginations of the polyploid nuclei (Figure 10). Strangely, structures similar to vacuoles could frequently be encountered within the multiploid nuclei, especially of the long-term c-root cells (Figures 9 and 11). These structures were surrounded by a continuous membrane (tonoplast), sometimes bearing electron opaque material on the internal side. A similar highly contrasting tonoplast was also found in cytoplasmic vacuoles. Moreover, intranuclear vesicles similar to Golgi-derived ones could be seen scattered throughout the nucleoplasm (Figure 11). In mitotic cells, Golgi bodies also occurred among the prometaphase chromosomes (Figure 12).

In roots recovering from an 8 h chlorpropham treatment, sparse MTs could be found along the lateral cell walls even in the shortest recovery time, having a prevailing transverse (Figure 13) or oblique (Figure 14) orientation relative to the long axis of the root. The nascent cellulose microfibrils could not be resolved clearly enough to determine their relationship in orientation with that of cortical MTs. Coated, dense or large smooth vesicles occurred amongst MTs. Preprophase cells contained only a small number of MTs along the midway of the lateral walls where a PPB in control cells would be expected to form (Figure 15). The density of MTs in the supposed PPBs did not increase with increasing time of recovery (Figure 16). Recovering cells continued to bear large multiploid interphase nuclei with deep cytoplasmic invaginations and intranuclear vacuole-like structures. Occasionally, spindle MTs were found connecting chromosomes with converging centres or minipoles located at sites

figures 1–6. 1. Control. Electron micrograph of medial longitudinal section of a lateral wall separating two calyptrogen interphase cells overlaid on both sides by many cross-sectioned MTs. ×73 500. 2. Control. Tangential section of an interphase cell revealing MTs transversely aligned and cellulose microfibrils in the cell wall (double-headed arrows) reflecting the orientation of MTs. Arrow points to a coated vesicle. ×50 000. 3. Control. Medial longitudinal section of a lateral wall separating two premitotic calyptrogen cells each containing a well-developed PPB of MTs, the one (on the left) being transversely aligned and the other (on the right) longitudinally. ×63 000. 4. A 16 h c-cell containing two nuclei (N). ×3200. 5. A 16 h c-cell containing one large polyploid nucleus with a deep cytoplasmic invagination in a mid-position (arrow) and two sizeable nucleoli. ×3200. 6. A 32 h c-cell with two nuclei in synchronous prophase. ×3400.

figures 7–11. 7. An 8 h c-cell containing two nuclei (N) (the one entirely out of view) and an incomplete cell wall united with the lateral parental wall. ×12 300. 8. An 'island' cytokinetic wall close to the polyploid nucleus (N) of a 16 h c-cell. ×10 000. 9. A 72 h c-multi-invaginated nucleus containing small vacuoles (arrows) in the nucleoplasm. V=vacuole. ×3400. 10. Portion of a 72 h multi-lobed c-nucleus bearing small vacuoles (V) in deep cytoplasmic invaginations. ×11 000. 11. A 72 h c-nucleus containing a vacuole (V) and several vesicles (arrows) in the nucleoplasm. ×28 500.

figures 12–17. 12. A 16 h c-cell showing Golgi bodies (G) among the prometaphase chromosomes (Chr). ×45 000. 13. An 8 h recovery. Reappearance of a few mostly transversely oriented MTs in the cortical cytoplasm of a tangentially sectioned cell. Arrow points to a coated vesicle. ×62 000. 14. A 16 h recovery. The MTs have an oblique orientation. Arrows point to dense or coated vesicles. ×50 000. 15. An 8 h recovery. Medial longitudinal section of a lateral wall of a preprophase cell containing one nucleus (out of view) showing the reappearance of a few MTs (arrowheads) at the site of the expected PPB. ×63 000. 16. A 48 h recovery. Portion of the lateral wall of a preprophase cell containing two nuclei (out of view) at the site where normally a PPB of MTs would appear. Only a few MTs are visible (arrowheads). ×57 000. 17. A 32 h recovery. Portion of a metaphase cell illustrating spindle MTs connecting chromosomes (Chr) with a minipole (asterisk), to which they converge (double-headed arrows). ×42 000.

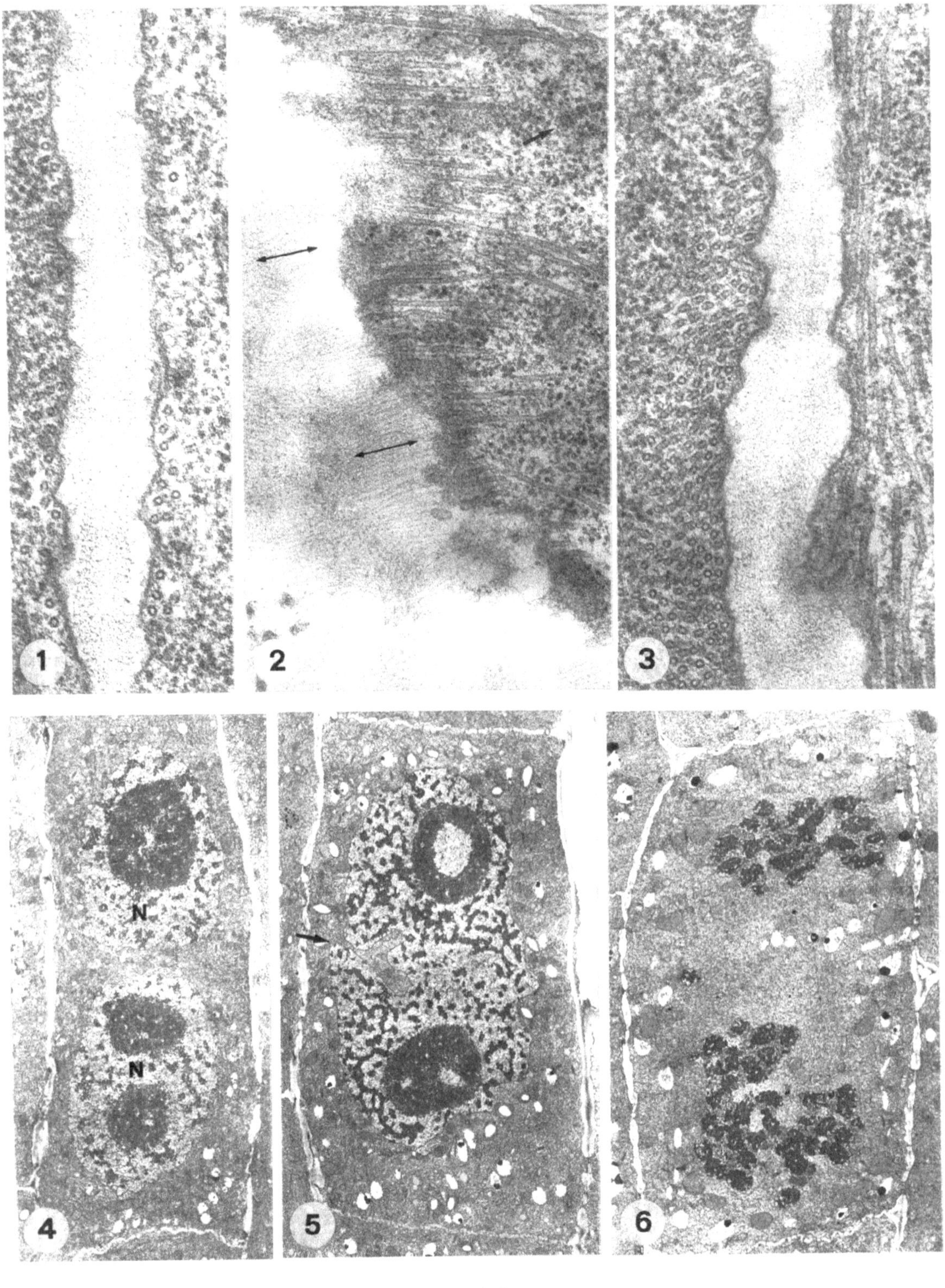

figures 1–6.

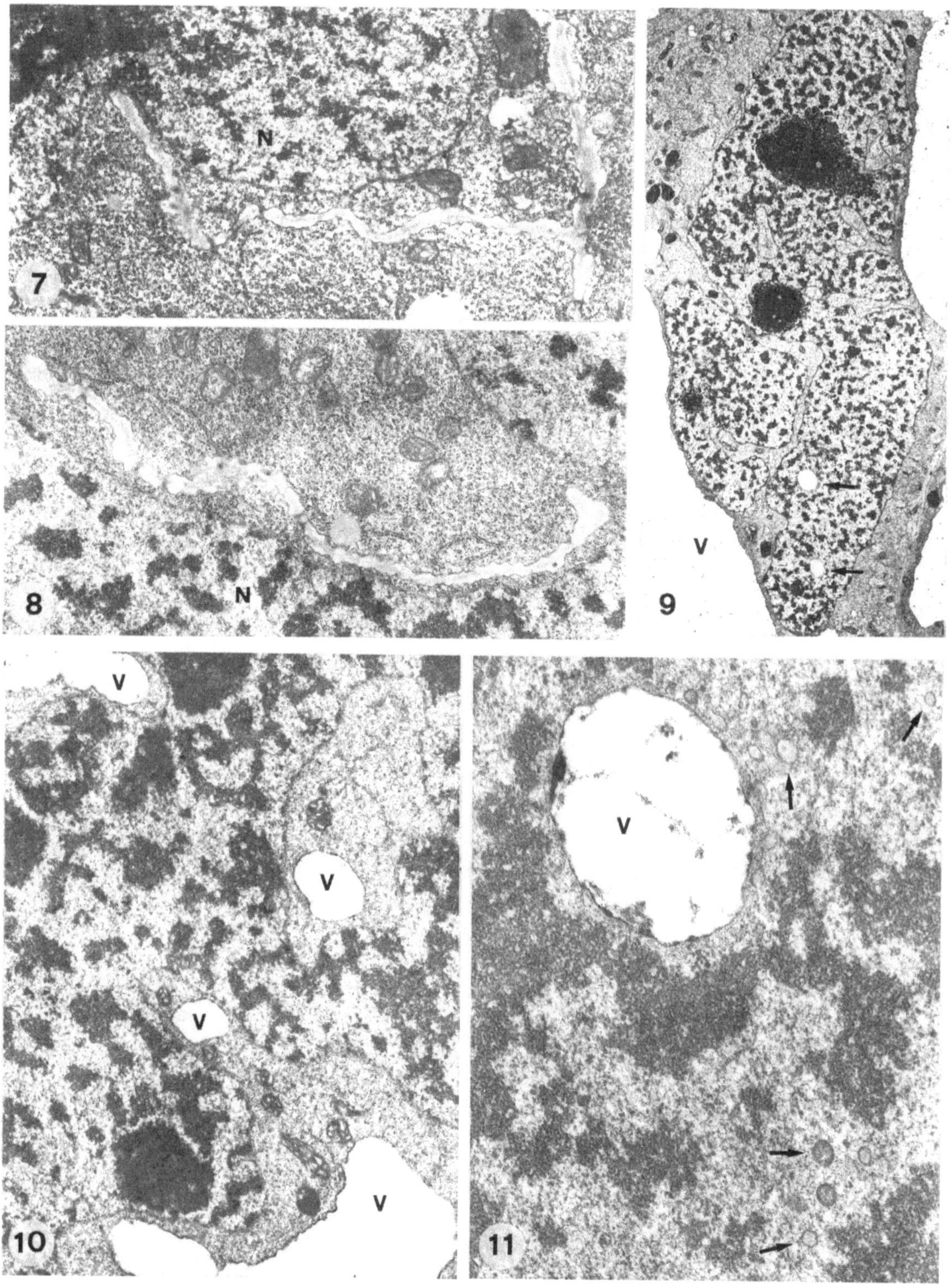

figures 7–11.

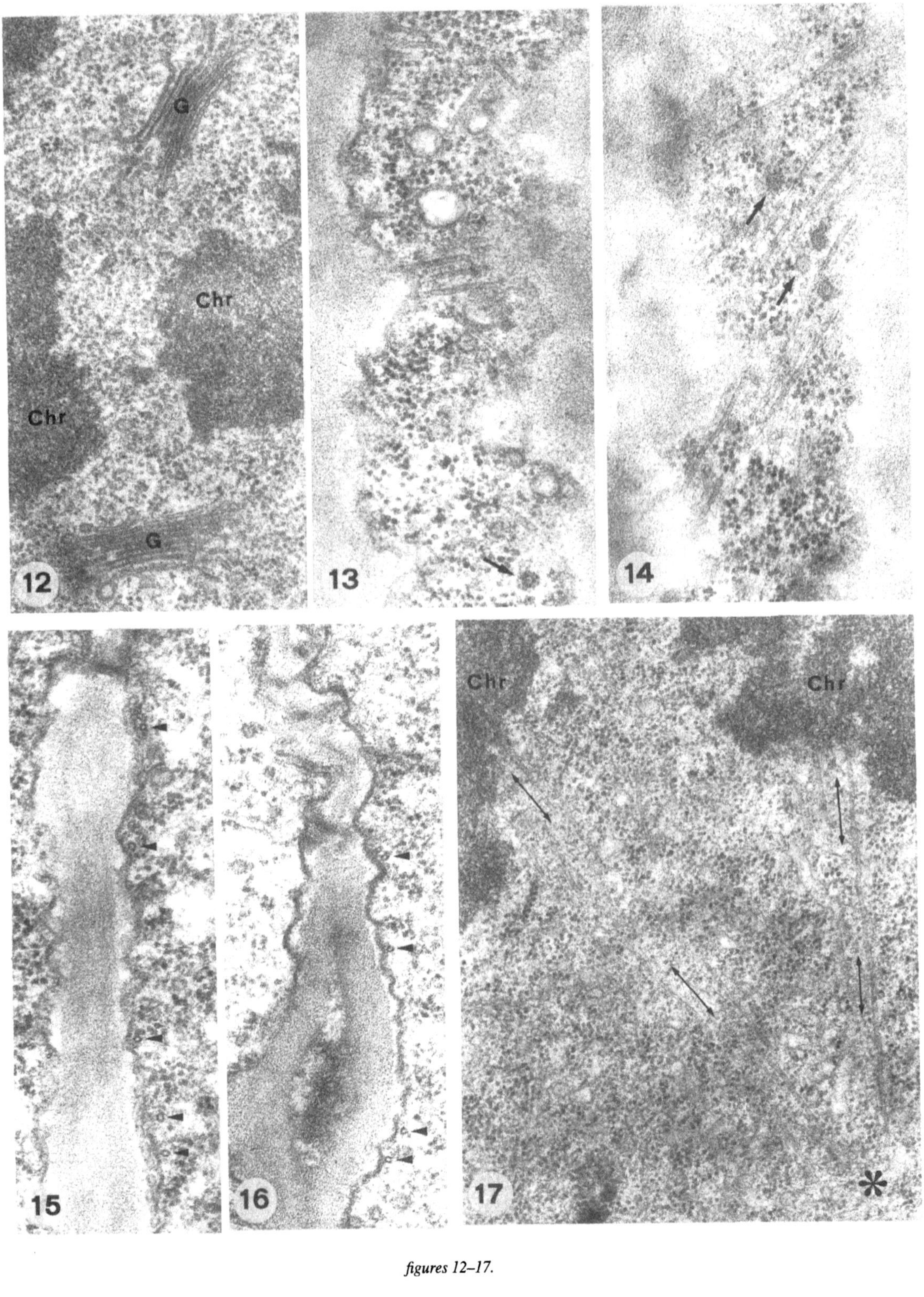

figures 12–17.

where the normal poles of the spindle would occur (Figure 17).

Normal arrays of either mitotic or interphase MTs could not be found.

Discussion

The investigation by electron microscopy of wheat roots exposed to 50 μM of chlorpropham (CIPC), a member of N-phenyl carbamates, showed a complete disappearance of all MTs of dividing and differentiating cells. Although the present study did not identify the primary target of chlorpropham action, which then leads to disintegration of MT cytoskeleton, it provides an ultrastructural support to the biochemical evidence that chlorpropham binds to plant tubulin in cells and may result in the formation of tubulin-carbamate complexes leading to MT depolymerization (Akashi et al., 1988; Morejohn and Fosket, 1986). Other anti-MT drugs such as colchicine (Apostolakos et al., 1990; Bennett and Smith, 1979; Eleftheriou, 1993) and the Vinca alkaloids vinblastine and vincristine (Takanari et al., 1990), induce the formation of extensive cytoplasmic paracrystals, which in the case of colchicine were shown to consist of tubulin-colchicine complexes. In the present study, no paracrystals could be detected, denoting the presumed formation of a different kind of tubulin-carbamate complex that may not crystalize.

On the other hand, there are several reports claiming that cytoplasmic MTs were not disassembled by CIPC. In onion root tip cells (Clayton and Lloyd, 1984) and in carrot suspension cells (Lloyd and Traas, 1988), treatment with 100 μM CIPC for as long as 23 h induced the formation of tripolar spindles creating three nuclei separated by three-way (Y-shaped) actin-containing phragmoplasts. Less commonly, multipolar spindles followed by branched phragmoplasts separating equal number of nuclei were found. Treatment of onion root tips with even higher CIPC concentrations (1 mM) for 2 h led to the formation of highly branched and contorted cell plates that separated multible nuclei (Gunning and Wick, 1985). These observations favour the notion that the primary effect of CIPC is upon the replication or splitting of the spindle poles. However, branched cell plates are not typical of carbamate herbicides as they have been observed in recovering cells after 2 mM colchicine treatment for 6 h (Galatis and Apostolakos, 1991) or after 5 mM caffeine treatment for 2 h (Manandhar et al. 1996). In our study, neither multible phragmoplasts nor multi-nucleate cells that might denote the formation of aberrant spindles could be found, even in prolonged and continuous treatment with the drug. Instead, we observed incomplete cell walls of c-cytokinesis and multilobed or two nuclei in each cell, similar to effects usually caused by colchicine (Apostolakos et al., 1990), caffeine (Manandhar et al., 1996; Samuels and Staehelin, 1996), or oryzalin (Wernicke and Jung, 1992). This discrepancy might be attributed to the fact that chlorpropham in wheat roots had completely disassembled MTs, which are responsible, along with actin filaments, for the organization of the spindle.

Structures resembling vacuoles and Golgi vesicles were seen trapped in the nucleoplasm of polyploid nuclei of long-term c-cells of wheat. To the best of our knowledge, such intranuclear enclosures have not been previously reported in carbamate-treated plant cells. Presumably, they originated from small vacuoles and Golgi vesicles that intermingled with chromosomes in a retarded c-prometaphase and, since chromosomes could not be moved to a normal metaphase plate neither did they separate at anaphase, structures found in the vicinity of or amongst chromosomes might have been accidentally enclosed in the polyploid nuclei by the reorganizing nuclear envelope in the supposed telophasic cells. Small vacuoles (and paracrystals) among chromosomes also occurred in tapetal cells of wheat anthers treated with colchicine for as long as 93 h (Bennett and Smith, 1979). The existence of highly lobed polyploid nuclei with deep cytoplasmic invaginations bearing small vacuoles (Figures 9 and 10) provides structural evidence indicating that the nuclear envelope at c-telophase 'struggles' to exclude components other than chromosomes from the reforming nucleus, but unsuccessfully in prolonged drug treatment. As a result, Golgi vesicles and small vacuoles were presumably trapped in the c-nuclei.

A considerable amount of research work has been accumulated on reinstallation of MTs following their experimental depolymerization in plant cells (Galatis and Apostolakos, 1991; Wasteneys and Williamson, 1989; and the literature cited therein). The general pattern of MT reappearance involves repolymerization at sites where they would be expected for unaffected cells, such as the cell surface, initially having a random orientation and then gradually adopting a transverse orientation. These results indicate that a MT-orienting factor(s) is associated with MTs. Our results, however, deviate from the general pattern in that recovery to the normal transverse dense arrays could not be documented, implying a failure of the mechanism regulating MT orientation to resume full activity. Similarly, incomplete recovery of cortical MTs has also been recorded in wheat mesophyll cells after oryzalin treatment (Wernicke and Jung, 1992).

The converging MTs to minipoles observed in the present study may manifest a MT nucleation site. Similar 'V-like' MT configurations have also been reported by other authors (Falconer et al., 1988; Wasteneys and Willamson, 1989). These findings are consistent with the view that functioning MTOCs are reestablished in recovering plant cells. On the other hand, however, this view is contradicted by our observations: 1. The already mentioned sparsity of cortical MTs and their lack of a definite orientation even in the long-term recovery times, and 2. The failure of a PPB, which is typical for plant cells and manifests the site of an active MTOC, to reorganize at the predicted plane. If MTOCs are to influence the number, structure, polarity and spatial positioning of MTs (Marc, 1997), our observations are rather indicative that putative MTOCs, under the conditions of our experiment, were irreversibly affected by the drug failing to further support MT arrays. Given the ability of assembled MTs to self-organize into bipolar spindles and the disperse nature of plant MTOCs (Marc, 1997), the often observed multiple foci may be the result of a combination of real MTOCs and secondary, nonspecific nucleation sites.

Concluding remarks

The present study revealed that chlorpropham (CIPC) caused the complete depolymerization of all MT arrays in root tip cells of wheat by presumably binding to plant tubulin and affected irreversibly the MTOCs, a specificity attributed to carbamate herbicides. The structural abnormalities and aberrant divisions obtained in the absence of MTs are in harmony with the view regarding MTs as highly morphogenetic cell organelles. However, further experimentation is needed to more precisely estimate the behaviour of meristematic cells and appreciate the recovery rates of plants after removal of the anti-MT compound. Although the drug concentration applied in the present study is found towards the minimum of the broad scale of concentrations used in the literature, experimentation with even less concentrated and/or shorter exposure time is needed to obtain more viable plants after treatment, a practice that might be helpful in estimating effective herbicide dose with less residual impact to the environment.

Notice

All pictures have been obtained from longitudinal sections to the root. Special care was exercised so that they were positioned vertically in the plates in correspondence with the long axis of the root. All orientations are then referred to relative to this axis.

Acknowledgements

This work was supported in part by a NATO grant (LST.CLG 975775).

References

Akashi T, Izumi K, Nagano E, Enomot M, Mizuno K and Shibaoka H 1988 Effects of propyzamide on tobacco cell microtubules *in vivo* and *in vitro*. Plant Cell Physiol. 29, 1053–1062.

Apostolakos P, Galatis B, Katsaros C and Schnepf E 1990. Tubulin conformation in microtubule-free cells of *Vigna sinensis*. An immunofluorescent and electron microscope study. Protoplasma 154, 132–143.

Bennett M D and Smith J B 1979 Colchicine-induced paracrystals in the tapetum of wheat anthers. J. Cell Sci. 38, 23–32.

Clayton L and Lloyd C W 1984 The relationship between the division plane and spindle geometry in *Allium* cells treated with CIPC and griseofulvin: An anti-tubulin study. Eur. J. Cell Biol. 34, 248–253.

Cyr R J and Palevitz B A 1995 Organization of cortical microtubules in plant cells. Curr. Opin. Cell Biol. 7, 65–71.

Eleftheriou E P 1993 Differentiation of abnormal sieve elements in roots of wheat (*Triticum aestivum* L.) affected by colchicine. New Phytol. 125, 813–827.

Eleftheriou E P and Palevitz B A 1992 The effect of cytochalasin D on preprophase band organization in root tip cells of *Allium*. J. Cell Sci. 103, 989–998.

Emons A M C, Derksen J and Sassen M M A 1992 Do microtubules orient plant cell wall microfibrils? Physiol. Plant. 84, 486–493.

Falconer M M, Donaldson G and Seagull R W 1988 MTOCs in higher plant cells: An immunofluorescent study of microtubule assembly sites following depolymerization by APM. Protoplasma 144, 46–55.

Galatis B and Apostolakos P 1991 Patterns of microtubule reappearance in root cells of *Vigna sinensis* recovering from a colchicine treatment. Protoplasma 160, 131–143.

Giddings T H Fr and Staehelin L A 1991 Microtubule-mediated control of microfibril deposition: A re-examination of the hypothesis. *In* The Cytoskeletal Basis of Plant Growth and Form. Ed. C W Lloyd. pp 85–99. Academic Press, London.

Gunning B E S and Wick S M 1985 Preprophase bands, phragmoplasts, and spatial control of cytokinesis. J. Cell Sci. Suppl. 2, 157–179.

Kleinkopf G E, Brandt T L, Frazier M J and Moller G 1997 CIPC residues on stored russet burbank potatoes. 1. Maximum label application. Am. Potato J. 74, 107–117.

Lehnen L P, Vaughn M A and Vaughn K C 1990 Terbutol affects spindle microtubule organizing centres. J. Exp. Bot. 41, 537–547.

Lloyd C W and Traas J A 1988 The role of F-actin in determining the division plane of carrot suspension cells. Drug studies. Development 102, 211–221.

Manandhar G, Apostolakos P and Galatis B 1996 Cell division of binuclear cells induced by caffeine: spindle organization and determination of division plane. J. Plant Res. 109, 265–275.

Marc J 1997 Microtubule-organizing centres in plants. Trends Plant Sci. 2, 223–230.

Marc J and Palevitz B A 1990 Regulation of the spatial order of cortical microtubules in developing guard cells of *Allium*. Planta 182, 625–634.

Mizuno K 1993 Microtubule-nucleation sites on nuclei of higher plant cells. Protoplasma 173, 77–85.

Morejohn L C 1991 The molecular pharmacology of plant tubulin and microtubules. *In* The Cytoskeletal Basis of Plant Growth and Form. Ed. C W Lloyd. pp 29–43. Academic Press, London.

Morejohn L C and Fosket D E 1986 Tubulins from plants, animals and protists. *In* Cell and Molecular Biology of the Cytoskeleton. Ed. J W Sharp. pp. 257–329. Plenum Press, New York.

Nooden L D 1971 Physiological and developmental effects of colchicine. Plant Cell Physiol. 12, 759–770.

Samuels A L and Staehelin L A 1996 Caffeine inhibits cell plate formation by disrupting membrane reorganization just after the vesicle fusion step. Protoplasma 195, 144–155.

Takanari H, Yosida T, Morita J, Izutsu K and Ito T 1990 Instability of pleomorphic tubulin paracrystals artificially induced by Vinca alkaloids in tissue-cultured cells. Biol. Cell 70, 83–90.

Wasteneys G O and Williamson R E 1989 Reassembly of microtubules in *Nitella tasmanica*: quantitative analysis of assembly and orientation. Eur. J. Cell Biol. 50, 76–83.

Wernicke W and Jung G 1992 Role of cytoskeleton in cell shaping of developing mesophyll of wheat (*Triticum aestivum* L.). Eur. J. Cell Biol. 57, 88–94.

Section editor: H Lambers

Recent Advances of Plant Root Structure and Function, 23–30
Edited by O. Gašparíková, M. Čiamporová, I. Mistrík and F. Baluška
© 2001 Kluwer Academic Publishers.

Measurement of viscoelastic properties of root cell walls affected by low pH in lateral roots of *Pisum sativum* L.

Eiichi Tanimoto[1,*], Shuhei Fujii[2], Ryoichi Yamamoto[2] and Shinobu Inanaga[3]
*[1] Division of Biomolecular Science, Institute of Natural Sciences, Nagoya City University, Mizuho, Nagoya 467-8501, Japan; [2] Biology and Chemistry Laboratory, Tezukayama College, Gakuen-minami 3, Nara 631-0034, Japan and [3]Arid Land Research Center, Tottori University, Tottori 680-0001, Japan *Fax No: +81 52 872 5867. E-mail: tanimoto@nsc.nagoya-cu.ac.jp*

Key words: acid pH, cell wall, creep extension, lateral roots, *Pisum sativum*, viscoelastic parameters

Abstract

Mechanical extensibility of the cell wall limits the elongation growth of roots. Low pH, ranging from pH 3–4.5, induces rapid elongation of excised roots, a phenomenon known as acid growth. The creep-extension analysis was carried out to measure and elucidate the viscoelastic properties of root cell walls in the acidic environment *in vitro*. The viscoelastic properties were determined at the elongation zone of the lateral roots of pea (*Pisum sativum* L. cv. Alaska) and described by the physical parameters of three elastic (E_0, E_1, E_2) and three viscosity (η_0, η_1, η_2) parameters using a Kelvin–Voigt–Burgers' model. The present method could measure the viscoelasticity of 1-mm long root zones from 2 to 9 mm behind the tip. Among the parameters, E_0 and η_0 were the most significant parameters to represent the whole extensibility of the roots. The parameter η_0 markedly declined in response to the environmental low pH (acid growth), whereas other parameters were not much affected by low pH. Relationship between the change in these physical parameters and the change in cell wall extensibility under low pH was discussed in order to elucidate the rheological processes taking place in the elongating cell walls.

Introduction

Cell-wall extensibility is one of the growth-limiting factors for the elongation growth of roots. In stems and coleoptiles, mechanical properties of the walls has been investigated by the stress-relaxation and/or load-extension analysis to evaluate auxin, gibberellin and acid-induced increase in the wall extensibility (Adams et al., 1975; Carpita and Kanabus, 1988; Hohl and Schopfer, 1992; Matsukura et al., 1998; Nakamura et al., 1975; Tanimoto and Masuda, 1971; Yamamoto et al., 1970, see reviews: Taiz, 1984; Masuda, 1990; Yamamoto, 1996). The creep extension was also carried out to determine the yield threshold load of cell walls (Kutschera and Kende, 1988; Nakahori et al., 1991; Okamoto and Okamoto, 1994; Pritchard et al., 1990; Tomos and Pritchard, 1994). However there are few reports of root cell walls, which analyze the extension process of the creep by the physical terms such as elastic modulus and viscosity coefficient. Since the extension process of plant cell walls are physical processes, the mechanism of wall extension could be described by these physical terms. It gives great advantage for understanding the conformational change of molecules in extending cell walls to analyze and describe the cell-wall extension by the rheological or physical terms, since the viscoelastic model and equation have been used to describe the deformation of protein gels and other biological materials (Katsuta et al., 1990 and its references).

The technical development of a load cell, a position-sensing transducer and a microprocessor-regulated stepping motor enables us to develop the computer-regulated measurement of creep extension process. The rheological analyses of food texture have been performed by the technology in the field of food science(Kamata et al., 1988). An automated creep analyzer equipped with a microprocessor-regulated stepping motor coupled with a load sensor has been employed to measure a creep phenomenon of protein and starch gels under a constant load (Kamata et al., 1988; Katsuta et al., 1990). This type of machine was recently used by ourselves to measure the cell-wall extensibility of pea roots (Tanimoto and Yamamoto, 1997). In the present study, we have applied this method to analyze the mechanical extensibility of root cell walls. We have determined the viscoelastic parameters of root cell walls by using a rheological model.

On the other hand, low pH causes so-called acid growth *in vivo* and *in vitro*. This phenomenon was found in the coleoptile of *Avena sativa* (Harger et al., 1971) and then the elongation process was precisely analyzed in the roots of *Zea mays* (Edwards and Scott, 1974, 1976; Tanimoto et al., 1989) and *Lactuca sativa* (Tanimoto and Watanabe, 1986). In these studies, the cell-wall extensibility was suggested to play a major role in acid growth, since the low pH increases cell-wall extensibility *in vitro* (Cleland et al., 1987; Rayle and Cleland, 1970; Yamagata et al., 1974; Yamamoto et al., 1974) but does not affect osmotic pressure of root cells (Winch and Pritchard, 1999). The acid-induced increase in cell wall extensibility is also conceived as one of the processes of cell elongation induced by auxin (Cleland et al., 1991; Rayle and Cleland, 1992). The effect of low pH on the wall extensibility has been reported to increase plastic extensibility (Virk and Cleland, 1988) and to decrease the yield threshold of the wall (Okamoto and Okamoto, 1994), but there are few reports of root cell walls, in spite of the fact that roots show rapid elongation in response to low pH (Edwards and Scott, 1974, 1976; Tanimoto and Watanabe, 1986). In contrast to over ground organs (Dreyer et al., 1981), roots are highly sensitive to the environmental pH, since the apoplast of roots is naked against the environment without cuticle or waxy layer. The previous studies on creep extension measured reversible and irreversible extension but few report analyzed the process of creep extension by the physical terms.

In this study, we have analyzed the creep-extension process, described the cell-wall extensibility by the physical terms and evaluated the effect of low pH *in vitro* on the viscoelastic properties of the cell-wall in the lateral roots of pea (*Pisum sativum* L.). We used this plant material to obtain fine and uniform roots for the precise localization of wall extensibility in every 1-mm zones of the roots.

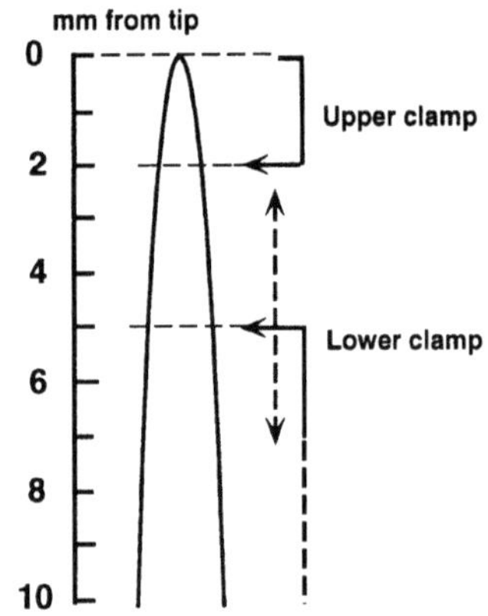

Figure 1. Position of clamping the lateral roots. The apical 2-mm zone was pinched by the upper clamp and placed on the lower clamp of a rheoner creep meter. A few mm of apical part was previously removed for measuring basal one-mm zones. The position of the lower clamp was adjusted by digitized movement of the lower clamp by driving a stepping motor (Figure 2). The distance of extension zone was precisely determined by the digital counter to leave 1-mm or 3-mm distance depending on the experiment.

Materials and methods

Hydroponic culture of pea plants

Germination and selection of seedlings (*Pisum sativum* L. cv. Alaska) were carried out as described by Tanimoto (1988). Selected seedlings with taproots, 2–3 cm long, were transplanted to hydroponic culture with circulation of tap water at 23 °C under 1000 μmol m^{-2} sec^{-1} for 10 days. The lateral roots, 3–6 cm long, were selected for uniformity and the apical 12-mm portion was excised. Excised lateral roots were killed in boiling methanol for 5 min and stored in methanol at 4 °C until use for the creep-extension analysis.

Creep extension analysis

Root segments were rehydrated with 10 m*M* MES (2-(N-morpholino)-ethanesulfate) buffer, pH 6.0, at 0 °C for 15 min. The diameter of the root was measured at the extension zone under a low power microscope to obtain the cross area of the root. The root was secured between two clamps of a Rheoner creep meter (Yamaden RE-33005, Tokyo). For the extension of 1-mm zones of the root, the position of two clamps was changed keeping 1-mm distance (Figure 1) for each root zone. A new root specimen was used for every measurements. The extension zone of the root was kept in a drop of the buffer solution between two clamps.

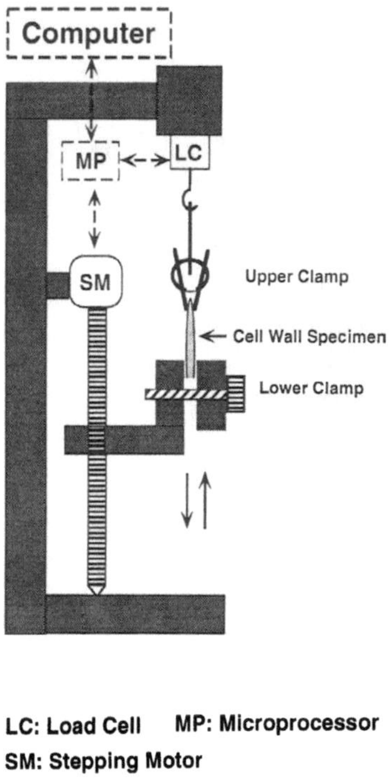

Figure 2. Schematic illustration of the structure of a rheoner creep meter. The microprocessor (MP) controls the up-down movement of the lower clamp which is driven by a stepping motor (SM). The movement is regulated by reading the digitized signal from the load cell (LC). The mode of these feed-back movements and the creep analysis of obtained data are conducted by a computer system.

To test the effect of pH, methanol-killed roots were rehydrated with distilled water at 0 °C for 15 min and transferred to citrate-phosphate buffer (CP buffer) at different pH ranging from 3 to 8 (Edwards and Scott, 1976; Tanimoto et al., 1989). After bathing in CP buffer for at least 10 min at 0 °C, the diameter of root segment was measured and secured between two clamps of a Rheoner creep meter. The apical 2-mm portion was held by the upper clamp and the basal part of the roots was secured by the lower clamp leaving 3-mm for extension (2–5 mm behind the tip) (Figures 1 and 2). The creep extension was carried out at room temperature. A constant load, 25 g mm^{-2}, was applied to the root by driving the lower clamp down in the maximum speed at 0.5 mm sec^{-1}. The amount of load was determined by the previous examination to test the breaking load of this roots and the estimation of turgor pressure *in vivo*. The concentration of osmotica in the apical root of hydroponically cultured Alaska pea was in the range of 0.3–0.4 mol l^{-1} (Tanimoto, 1995) which corresponded to 75–100 g mm^{-2}. However, we found that unidirectional force over 40 g mm^{-2} often caused breakage during extension for 10 min. We also estimated that the net cross area of cell walls is 25–30% of total cross area of roots. By combining these figures, we applied the load of 25 g mm^{-2} to analyze the cell wall extensibility. The amount of this load gave clear and stable extension curves in this plant material. The extension process was recorded by a computer at 0.5-s intervals for 10 min and then the load was released to record the shrinkage of the root for 5 min.

A typical creep curve is shown in Figure 3 (C). It was analyzed by the computer program using Burgers' viscoelastic model to calculate three elastic (E_0, E_1, E_2) and three plastic (η_0, η_1, η_2) parameters involved in the equation of Figure 3 (B). The curve and the equation are simulated by the Kelvin–Voigt–Burgers' viscoelastic model as shown in Figure 3(A). Calculation of these parameters was conducted as described by Kamata et al. (1988). The non-linear extension profiles of pea roots were found to be simulated by a six-element, four-component model (Figure 3 (A)), which consists of one Maxwell model and of two Kelvin-Voigt models, i.e. one Hookean springs (elastic modulus E_0) , two Kelvin-Voigt components (E_1, η_1 and E_2, η_2) and one Newtonian dashpots (viscosity coefficient, η_0).

After 10-min extension under 25 g mm^{-2} load, the lower clamp was returned upward to cancel the load and the shrinkage process was recorded for five min (Figure 3 (C)). After shrinkage for five min, the total extension, the reversible extension (elastic extension) and the irreversible extension (plastic extension) were read from the maximum extension at 10 min and the final length at 15 min, respectively (Figure 3 (C)).

Results

The present method was able to measure the extension of every 1-mm zone, starting from 2 mm behind root tip towards the basal part. Figure 4 shows typical creep-extension profiles of 1-mm zones of lateral roots. Each curve consists of 1800 dots of measurements. The more apical the root zones were, the larger the extensions. The six physical parameters, the total extension and the irreversible (plastic) and reversible (elastic) extensions were indicated for every root zones in Table 1. It is apparent that total extension was biggest at the most apical zone, 2–3 mm behind the tip.

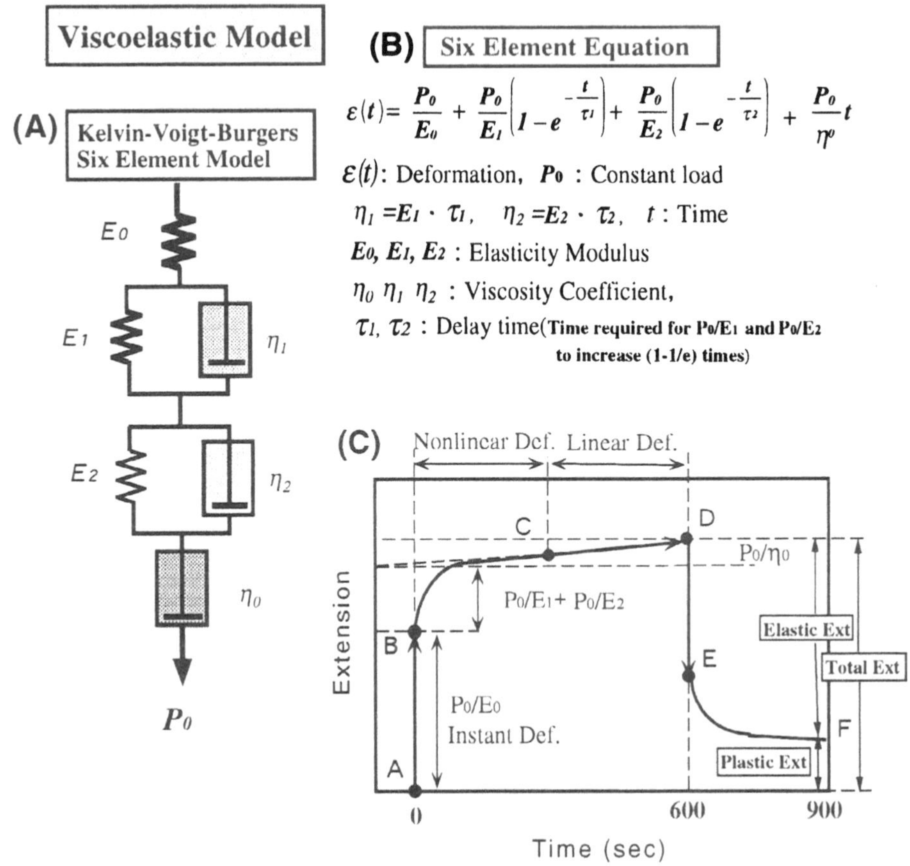

Figure 3. A viscoelastic model and the equation for the analysis of creep extension. (A): The Kelvin–Voigt–Burgers' six-element model composes of one Hookean spring (E_0), one Newtonian dashpot (η_0) and two viscoelastic components (E_1, η_1 and E_2, η_2). (B): The equation for the six-element model. Deformation ($\epsilon(t)$) of such a model under a constant load (P_0) is simulated by the equation of four members. (C): A typical creep extension curve during 10 min extension and 5 min shrinkage. The extension-curve was analyzed by the method of Kamata et al. (1988). Linear instantaneous deformation (A-B), nonlinear deformation (B-C) and final linear deformation (C-D) were simulated to the equation (B) and physical parameters of elastic moduli and viscosity coefficients were calculated. Total extension, plastic extension and elastic extension were also determined by reading the extensions at time 600 and 900 s, respectively.

The extensions rapidly declined towards the basal 5–6 mm zone and stayed almost constant at the 6–9 mm zones. The plastic and elastic extension also decreased in parallel towards the base at 2–5 mm and reached constant at 5–9 mm. All viscoelastic parameters were smallest in the 2–3 mm zone. Since these physical parameters mean elasticity moduli and viscosity coefficients, the smaller these parameters are, the larger the extensibility. The elastic parameters (E_0, E_1, E_2) gradually increased towards the base up to the 5–6 mm zone and were almost constant at 5–9 mm. The η_0 also increased up to the 6–7 mm zone and stayed almost constant in 6–9 mm. The η_1 and η_2 were also smallest in the 2–3 mm zone but stayed nearly constant in 3–9 mm. The absolute value of E_0 was 1/10 of E_1 and E_2, and η_0 was 10 times of η_1, which in turn was 10 times of η_2. Thus E_0 and η_0 contribute the most to the total extensibility. It turned out that the apical four zones, 2–6 mm behind the tip, are highly extensible, whereas the basal 6–9 mm zones are not much extensible. Thus the effect of low pH was tested in the 2–5 mm zone of the root in the next experiments.

The effect of pH on extension profile was shown in Figure 5. The shapes of extension curves were significantly different from each other depending on the pH values. These extension curves were analyzed to

Table 1. Distribution of six viscoelastic parameters and extensibility of cell walls in consecutive one-mm zones of lateral roots. Cell Wall extensibility, elastic moduli and viscosity coefficients were measured as described in caption of Figure 4 and in Material and methods. Average of ten roots with S.E.

Parameters	Zones of lateral roots (distance from tip, mm)						
	2–3	3–4	4–5	5–6	6–7	7–8	8–9
E_0 (N/m^2)	$9.7 \times 10^5 \pm 4.2 \times 10^4$	$1.2 \times 10^6 \pm 1.3 \times 10^5$	$1.5 \times 10^6 \pm 1.2 \times 10^5$	$2.1 \times 10^6 \pm 1.5 \times 10^5$	$2.7 \times 10^6 \pm 3.5 \times 10^5$	$2.6 \times 10^6 \pm$ [illegible] $\times 10^5$	$2.2 \times 10^6 \pm 1.8 \times 10^5$
E_1 (N/m^2)	$6.5 \times 10^6 \pm 6.5 \times 10^5$	$1.3 \times 10^7 \pm 1.7 \times 10^6$	$2.1 \times 10^7 \pm 3.8 \times 10^6$	$2.9 \times 10^7 \pm 4.1 \times 10^6$	$2.7 \times 10^7 \pm 2.5 \times 10^6$	$2.3 \times 10^7 \pm 1.6 \times 10^6$	$2.4 \times 10^7 \pm 1.9 \times 10^6$
E_2 (N/m^2)	$7.2 \times 10^6 \pm 6.6 \times 10^5$	$1.2 \times 10^7 \pm 1.4 \times 10^6$	$1.3 \times 10^7 \pm 1.5 \times 10^6$	$1.6 \times 10^7 \pm 8.1 \times 10^5$	$2.2 \times 10^7 \pm 2.3 \times 10^6$	$2.2 \times 10^7 \pm 2.0 \times 10^6$	$2.4 \times 10^7 \pm 3.9 \times 10^6$
η_0 (pa·sec)	$5.3 \times 10^9 \pm 7.4 \times 10^8$	$1.6 \times 10^{10} \pm 4.3 \times 10^9$	$10 \times 10^{10} \pm 4.5 \times 10^9$	$3.2 \times 10^{10} \pm 4.5 \times 10^9$	$6.7 \times 10^{10} \pm 1.5 \times 10^{10}$	$6.8 \times 10^{10} \pm 1.7 \times 10^{10}$	$8.5 \times 10^{10} \pm 3.7 \times 10^{10}$
η_1 (pa·sec)	$6.4 \times 10^8 \pm 6.9 \times 10^7$	$1.2 \times 10^9 \pm 1.9 \times 10^8$	$1.8 \times 10^9 \pm 2.7 \times 10^8$	$2.5 \times 10^9 \pm 7.5 \times 10^8$	$1.8 \times 10^9 \pm 3.3 \times 10^8$	$1.6 \times 10^9 \pm 6 \times 10^8$	$1.7 \times 10^9 \pm 5.7 \times 10^8$
η_2 (pa·sec)	$8.9 \times 10^7 \pm 1.2 \times 10^7$	$1.3 \times 10^8 \pm 1.2 \times 10^7$	$1.5 \times 10^8 \pm 1.6 \times 10^7$	$1.2 \times 10^8 \pm 9.6 \times 10^6$	$1.7 \times 10^8 \pm 2.6 \times 10^7$	$1.7 \times 10^8 \pm 4.9 \times 10^7$	$1.5 \times 10^8 \pm 2.6 \times 10^7$
Extension (mm)							
Total extension	0.49 ± 0.03	0.34 ± 0.03	0.27 ± 0.03	0.16 ± 0.02	0.14 ± 0.02	0.15 ± 0.02	0.16 ± 0.01
Elastic extension	0.20 ± 0.02	0.13 ± 0.01	0.10 ± 0.01	0.06 ± 0.01	0.06 ± 0.01	0.06 ± 0.01	0.06 ± 0.01
Plastic extension	0.29 ± 0.03	0.20 ± 0.02	0.16 ± 0.02	0.10 ± 0.01	0.08 ± 0.01	0.09 ± 0.02	0.09 ± 0.01

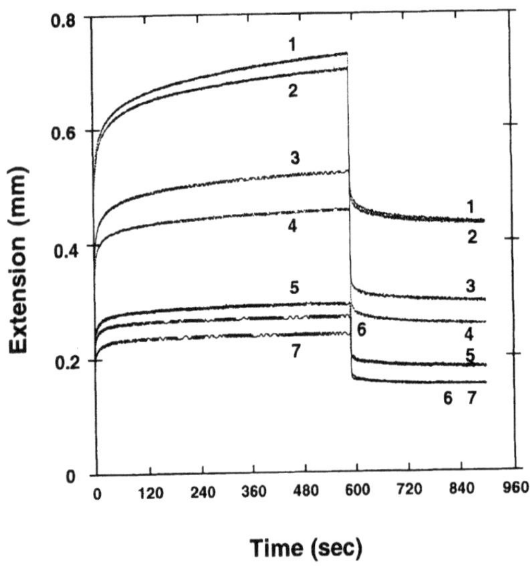

Figure 4. Typical extension curves of one-mm zones of lateral roots. One-mm long zones were secured between two clamps after measuring the diameter and extended by loading 25 g mm^{-2} in MES buffer pH 6.0 for 10 min. The numbers indicate the zones of the root, 1;(2–3), 2;(3–4) 3;(4–5), 4;(5–6), 5;(6–7), 6;(7–8), 7;(8–9) mm behind the tip. Since the extension was measured at 0.5-s intervals, each extension curve consists of 1800 dots of data. Each curve is a typical extension curve of separate root.

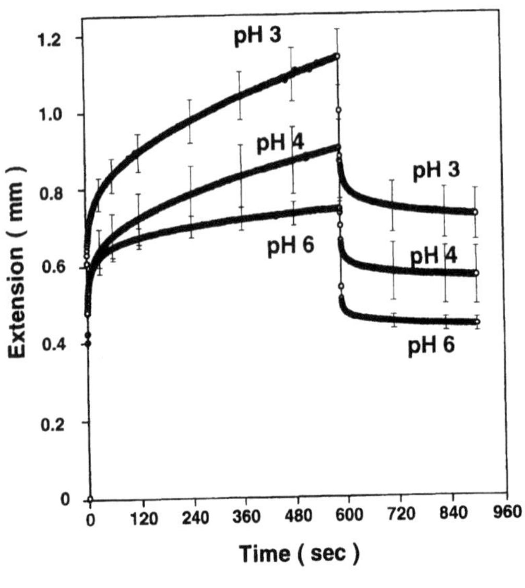

Figure 5. The effect of pH on the extension curves of 2–5-mm zone. The root segments were incubated in citrate-phosphate buffer for at least 10 min and extended by loading 25 g mm^{-2} in the same buffer. Means of 7–9 roots with S.E. (vertical bars).

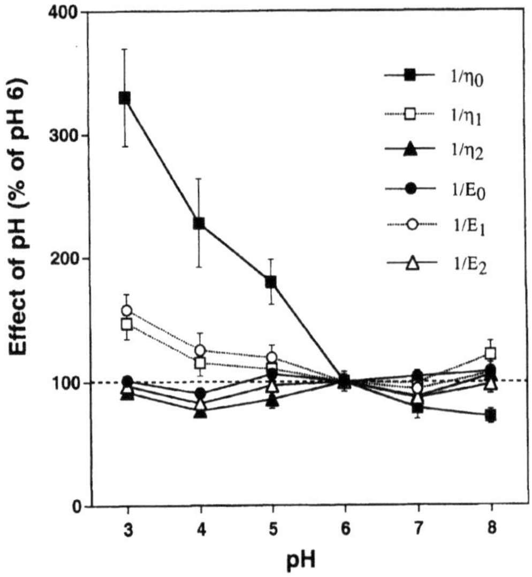

Figure 6. Effect of pH on the six viscoelastic parameters. The extension curve of each root was analyzed by the six-element model of Figure 3. The reciprocal values of six parameters were calculated and the relative values (% of pH 6) were indicated against pH values. Means of 7–9 roots with S.E. (vertical bars).

test the effect of pH on the six physical parameters. The results were summarized and the reciprocal values of six parameters were compared in Figure 6, since the reciprocal value corresponds to the extensibility of the cell walls. In order to compare these values, every datum was indicated as percentages of those at pH 6, since the absolute values of these parameters differed by two orders of magnitude as shown in Table 1. Among the parameters, η_0 exhibited the most remarkable response to the low pH. The $1/\eta_0$ at pH 3 was more than 3 times higher than that at pH 6, whereas other parameters did not change so much. Thus the viscosity coefficient η_0 was most responsible for the acid-induced increase in extensibility of the cell walls.

Discussion

The present method was successful to determine the physical parameters of cell-wall extensibility in the 1-mm zone of the lateral roots. It was found to be applicable to the fine roots of rice and wheat but it was technically difficult for thick roots such as the seminal roots of maize for precise clamping. It is by now successful to measure the seminal roots of maize and main roots of pea in 2-mm zones (Tanimoto et al., 1997).

The extensibility of the root cell walls were closely related to the reciprocal of elastic moduli and viscosity coefficients along the axis of lateral roots (Figure 4 and Table 1). The extensibility of the root cell walls was able to be described by the physical terms of elastic moduli and viscosity coefficients (Table 1). The elastic moduli (E_0, E_1, E_2) of root cell walls were in the order of 10^5–10^7N m^{-2} and the viscosity coefficients (η_0, η_1, η_2) were in the order of 10^7–10^{10}pa·sec, but the real figures of root cell walls may be larger than these values, since we calculated these values by assuming roots as uniform columns. Thus the real physical values must be re-calculated by using real cross area of cell walls. We are trying to measure the cross area of cell walls and the results will be reported elsewhere.

The parameter η_0 markedly declined in response to the environmental low pH (acid growth) (Figures 5 and 6). The relationship between the change in these parameters and the change in cell-wall extensibility clearly indicated that low-pH induced extension of cell walls is due to the decrease in viscosity coefficient of the cell wall. According to the present viscoelastic model, low pH decreased the viscosity of Newtonian dashpot η_0 of Figure 3(A). One of the molecular mechanism of such decrease in viscosity coefficient may be explained by the function of cell wall proteins such as expansins (Cosgrove, 1997) and other proteins (Okamoto and Okamoto, 1995). Another mechanism may involve the removal of calcium from cell walls (Hasenstein and Evans, 1988; Soll and Bottger, 1982; Virk and Cleland, 1988, 1990). Since citric acid is known to be a chelator, the CP buffer may cause calcium solubilization from pectic polysaccharides which is reported to be affected by gibberellin-treatment to enhance the elongation of pea roots (Tanimoto and Huber, 1997). Although the possibility of calcium participation should be confirmed by a further investigation, the acid growth could be detected in phosphate buffer adjusted by HCl solution (Tanimoto and Watanabe, 1986). They also reported that acid growth was clearly observed in excised or decotylized roots which grow slowly, whereas it was hardly detected in intact seedling roots which grow rapidly. These obsevations suggest that cell wall is already acidified in quickly growing intact roots and no more sensitive to low pH, or that strong pH adjustment exists to resist against the artificial acidification of apoplast in the intact roots.

Two major factors are conceived to drive the elongation growth of roots, cell-wall extensibility and speed of water uptake. The latter factor may not contribute to the elongation of roots in hydroponic solution, since there is little resistance of water uptake. In addition, there are some reports of osmotica in root cells which were not influenced by hormonal (Tanimoto, 1994, 1995) and low pH (Winch and Pritchard, 1999) treatments of roots. Thus the low pH-induced rapid growth of excised roots may be regulated by the mechanical property of cell walls such as the decrease in the viscosity parameter η_0 (Figure 6).

This paper described the application of viscoelastic model to the *in vitro* extension process of root cell walls. By such a simulation using the model of Figure 3, we could get an image of physical mechanism for the wall-extension. It may help to understand the molecular mechanism of the elastic and plastic extension in the root cell walls in combination with biochemical investigations.

Acknowledgment

This study was partly supported by Japan Society for the Promotion of Science, The Cooperative Research Program of Arid Land Research Center (Tottori University), and Grant-in-Aid (No. 10640637) from the Ministry of Education, Science, Sports and Culture of Japan to E.T. The authors are very grateful to Drs A Lux, M Lux, J Abe and S Morita for their helpful advice and discussion.

References

Adams P A, Montague M J, Tepfer M, Rayle D L, Ikuma H and Kaufman P B 1975 Effect of gibberellic acid on the plasticity and elasticity of *Avena* stem segments. Plant Physiol. 56, 757–760.

Carpita N C and Kanabus J 1988 Chemical structure of the cell walls of dwarf maize and changes mediated by gibberellin. Plant Physiol. 88, 671–678.

Cleland R E, Cosgrove D and Tepfer M 1987 Long-term acid-induced wall extension, in an *in-vitro* system. Planta 170, 379–385.

Cleland R E, Buckley G, Nowbar S, Lew N M, Stinemetz C, Evans M L and Rayle D L 1991 The pH profile for acid-induced elongation of coleoptile and epicotyl sections is consistent with the acid-growth theory. Planta 186, 70–74.

Cosgrove D J 1997 Relaxation in a high-stress environment: the molecular bases of extensible cell walls and cell enlargement. Plant Cell 9, 1031–1041.

Dreyer S A, Seymour V and Cleland R E 1981 Low proton conductance of plant cuticles and its relevance to the acid-growth theory. Plant Physiol. 68, 664–667.

Edwards K L and Scott T K 1974 Rapid growth responses of corn root segments: Effect of pH on elongation. Planta 119, 27–37.

Edwards K L and Scott T K 1976 Rapid growth responses of corn root segments: Effect of citrate-phosphate buffer on elongation. Planta 129, 229–233.

Hager A, Menzel H and Krauss A 1971 Versuche und Hypothese zur Primarwirkung des Auxins beim Streckungswechstum. Planta 100, 47–75.

Hasenstein K H and Evans M L 1988 The influence of calcium and pH on growth in primary roots of *Zea mays*. Physiol. Plant. 72, 466–470.

Hohl M and Schopfer P 1992 Growth at reduced turgor: irreversible and reversible cell-wall extension of maize coleoptiles and its implications for the theory of cell growth. Planta 187, 209–217.

Kamata Y, Rector D and Kinsella J E 1988 Influence of temperature of measurement on creep phenomena in glycinin gels. J. Food Sci. 53, 589–591.

Katsuta K, Rector D and Kinsella J E 1990 Viscoelastic properties of whey protein gels: Mechanical model and effects of protein concentration on creep. J. Food Sci. 55, 516–521.

Kutschera U and Kende H 1988 The biophysical basis of elongation growth in internode of deepwater rice. Plant Physiol. 88, 361–366.

Masuda Y 1990 Auxin-induced cell elongation and cell wall changes. Bot. Mag. Tokyo 103, 345–370.

Matsukura C, Itoh S, Nemoto K, Tanimoto E and Yamaguchi J 1998 Promotion of leaf sheath growth by gibberellic acid in a dwarf mutant of rice. Planta 205, 145–152.

Nakahori K, Katou K and Okamoto H 1991 Auxin changes both the extensibility and the yield threshold of the cell wall of *Vigna* hypocotyls. Plant Cell Physiol. 32, 121–129.

Nakamura T, Sekine S, Arai K and Takahashi N 1975 Effects of gibberellic acid and indole-3-acetic acid on stress relaxation properties of pea hook cell wall. Plant Cell Physiol. 16, 127–138.

Okamoto H and Okamoto A 1994 The pH-dependent yield threshold of the cell wall in a glycerinated hollow cylinder (*in vitro* system) of cowpea hypocotyl. Plant Cell Environ. 17, 979–983.

Okamoto H and Okamoto A 1995 Two proteins regulate wall extensibility and the yield threshold in glycerinated hollow cylinder of cowpea hypocotyl. Plant Cell Environ. 18, 827–830.

Pritchard J, Wyn Jones R G and Tomos A D 1990 Measurement of yield threshold and cell wall extensibility of intact wheat roots under different ionic, osmotic and temperature treatments. J. Exp. Bot. 41, 669–675.

Rayle D L and Cleland R E 1970 Enhancement of wall loosening and elongation by acid solution. Plant Physiol. 46, 250–253.

Rayle D L and Cleland R E 1992 The acid growth theory of auxin-induced cell elongation is alive and well. Plant Physiol. 99, 1271–1274.

Soll H and Bottger M 1982 The mechanism of proton-induced increase of cell wall extensibility. Plant Sci. Let. 24, 163–171.

Taiz L 1984 Plant cell expansion: Regulation of cell wall mechanical properties. Annu. Rev. Plant Physiol. 35, 585–657.

Tanimoto E 1988 Gibberellin regulation of root growth with change in galactose content of cell walls in *Pisum sativum*. Plant Cell Physiol. 29, 269–280.

Tanimoto E 1994 Interaction of gibberellin A_3 and ancymidol in the growth and cell-wall extensibility of dwarf pea roots. Plant Cell Physiol. 35, 1019–1028.

Tanimoto E 1995 Effect of gibberellin and ancymidol on the growth and cell wall components of pea (Pisum sativum L.) roots. *In* Structure and Function of Roots. Ed. F. Baluška et al., pp 91–98. Kluwer Academic Publishers, Dordrecht.

Tanimoto E and Huber D J 1997 Effect of GA_3 on the molecular mass of polyuronides in the cell walls of Alaska pea roots. Plant Cell Physiol. 38, 25–35.

Tanimoto E and Masuda Y 1971 Role of the epidermis in auxin-induced elongation of light-grown pea stem segments. Plant Cell Physiol. 12, 663–673.

Tanimoto E and Watanabe J 1986 Automated recording of lettuce root elongation as affected by auxin and acid pH in a new rhizometer with minimum mechanical contact to roots. Plant Cell Physiol. 27, 1475–1487.

Tanimoto E and Yamamoto R 1997 Change in cell wall extensibility during gibberellin-regulated growth of pea roots. Zamedelska Technika (Agricultural Engineering) 43, 15–19.

Tanimoto E, Scott T K and Masuda Y 1989 Inhibition of acid-enhanced elongation of *Zea mays* root segments by galactose. Plant Physiol. 90, 440–444.

Tanimoto E, Yamamoto R, Yamaguchi J and Matsukura C 1997 Cell-wall extensibility analyzed by creep method: (2) Comparative studies on visco-elastic parameters of roots, epicotyls, leaf sheath and coleoptiles. Plant Cell Physiol. 38. Supplement, s129.

Tomos D and Pritchard J 1994 Biophysical and biochemical control of cell expansion in roots and leaves. J. Exp. Bot. 45, 1721–1731.

Virk S S and Cleland R E 1988 Calcium and the mechanical properties of soybean hypocotyl cell walls: Possible role of calcium and protons in cell-wall loosening. Planta 176, 60–67.

Virk S S and Cleland R E 1990 The role of wall calcium in the extension of cell walls of soybean hypocotyls. Planta 182, 559–564.

Winch S and Pritchard J 1999 Turgor measurements reveal that acid-induced wall loosening is confined to the accelerating region of the root growing zone. J. Exp. Bot. (*In press*).

Yamagata Y, Yamamoto R and Masuda Y 1974 Auxin and hydrogen ion actions on light-grown pea epicotyl segments II. Effect of hydrogen ions on extension of the isolated epidermis. Plant Cell Physiol. 15, 833–841.

Yamamoto Y 1996 Stress relaxation property of the cell wall and auxin-induced cell elongation. J. Plant Res. 109, 75–84.

Yamamoto R, Shinozaki K and Masuda Y 1970 Stress-relaxation properties of plant cell walls with special reference to auxin action. Plant Cell Physiol. 11, 947–956.

Yamamoto R, Maki K, Yamagata Y and Masuda Y 1974 Auxin and hydrogen ion actions on light-grown pea epicotyl segments I. Tissue specificity of auxin and hydrogen ion actions. Plant Cell Physiol. 15, 823–831.

Section editor: H Lambers

Recent Advances of Plant Root Structure and Function, 31–37
Edited by O. Gašparíková, M. Čiamporová, I. Mistrík and F. Baluška
© 2001 Kluwer Academic Publishers.

Secondary dilatation growth in the root endodermis

Alexander Lux[1] & Mária Luxová[2,†]
[1]*Department of Plant Physiology, Faculty of Natural Sciences, Comenius University, Mlynská dolina B-2, SK 84215 Bratislava, Slovak Republic**. [2]*Institute of Botany, Slovak Academy of Sciences, Dúbravská cesta 14, SK 84223 Bratislava, Slovak Republic*

Key words: Casparian bands, cell division, cell expansion, endodermis density, root anatomy, secondary growth

Abstract

Early secondary growth of roots with persisting cortex is the result of two different growth processes: Cambial growth and dilatation growth. Vascular cambium forms the secondary vascular tissues of the thickening vascular cylinder. At the same time, reactivated, peripherally situated cortex is adequately expanding by the dilatation growth. Dilatation growth of the cortex starts in the endodermis, i.e. the youngest innermost cortical layer. In relation with the species specificity concerning the developmental type of endodermis (species with maturation of endodermis in the state I, II or III), the dilatation growth of endodermis starts in the state I or III. Secondary dilatation activity of the endodermis is manifested by tangential expansion of endodermal cells, by their renewed division, or by the combination of both processes. The additional anticlinal divisions increase the number of circumferential endodermal cells. In some species, the newly-formed radial walls acquire typical endodermal character by the formation of Casparian bands or even by further deposition of suberin lamellae. In other species, additive radial walls remain unmodified. The use of fluorescence staining techniques allows to characterise chemical properties of additional radial endodermal walls, which do not always coincide with the older reports. The density of endodermal network is changed by the dilatation growth: By cell expansion and by the formation of unmodified additional cell walls it is decreased, while the increased number of modified radial cell walls results in the higher density. The described changes of endodermal network, constituting the apoplastic barrier can have a role as a regulating factor.

Introduction

In secondary thickening roots, dilatation growth results in enlargement of peripheral layers appropriate to secondary thickening of vascular cylinder. The secondary growth of roots thus is comprised of *cambial growth* (*cambial thickening*), determined by the function of vascular cambium and cork cambia (phellogens), and the *dilatation (latitudinal) growth*. The latter is present in the primary cortex, in axial phloem parenchyma and in the ray parenchyma in the bark (Clowes, 1961; Esau, 1969; Lev-Yadun and Aloni, 1992). It is caused by the capability of reactivation of these cells. The reactivation is expressed in cell enlargement and/or their renewed division. Contrary to the similar growth processes, occurring in wound tissues (Schultz, 1988), new cellular determination of the newly-formed cells does not occur during dilatation growth (Lev-Yadun and Aloni, 1992).

While explaining the induction of secondary dilatation growth, authors have reported the effect of physical forces and the hormonal regulation of dilatation activity. Mager (1932), Holdheide (1951), Whitmore (1962a,b, 1963), and others postulated the regulation of dilatation growth by the changes of inner pressure, caused by the cambial growth. In *Hordeum vulgare*, a monocotyledon, characterized by the absence of secondary growth, the renewed cell division was induced by application of external pressure. In mechanically impeded roots, the dilatation anticlinal division was present in the proendodermal cells and in the endodermal cells in developmental state I. Some endodermal cells in the state III were divided periclinally and obliquely (Wilson et al., 1977; Wilson

* E-mail: lux@fns.uniba.sk
† deceased

and Robards, 1978). Similarly, additional anticlinal division of endodermis, occurring in the basal region of secondary non-thickening lateral roots of primrose (*Primula acaulis*) during their penetration through peripheral tissues of the main root, can also be considered as the result of pressure (Luxová and Lux, 1998). This example shows, at the same time, the relative independence of cambial and secondary dilatation growth.

The author of the hypothesis of hormonal regulation in dilatation growth is Huber (1961). His idea was based on the appearance of microscopic ruptures of cell walls during tangential tension. The factors inducing dilatation growth supposed by him were wound hormones induced due to the tension. Lev-Yadun and Aloni (1992) also studied hormonal regulation of dilatation. Their results show that ethylene can be the major promoting factor in the induction of dilatation growth.

The dilatation growth is not realized equally within the cortex. Uniseriate endodermis with its specific wall modifications and an important role of apoplastic barrier has already drawn attention of earlier authors in connection with the dilatation growth. In the present study, the review of older data is supplemented with the new results obtained by fluorescence microscopy techniques with a higher sensitivity for evaluation of endodermal wall development. They showed interspecific differences of dilatation growth related to the developmental type of endodermis and accompanied by the changes of chemical composition of new cell walls developing as a result of dilatation activity. The possibility of functional regulation of the density of endodermal network results from the changes of this density induced by dilatation growth.

Development of endodermis

The endodermis is characterized by the development of specific wall modifications, important from the physiological aspect. In the state I of development (primary endodermis), the Casparian bands are formed by the deposition of lipidic and phenolic substances into the radial and transverse cell walls. In the species with primary endodermis, the development of this sheath is completed at this stage. In other species, the development proceeds by the state II of development, i.e. the deposition of suberin lamellae on the inner surface of endodermal cells (secondary endodermis). Distinction between the non-lamellar suberin of Casparian bands and lamellar suberin is enabled by fluorescence staining techniques introduced by Brundrett et al. (1988, 1991). The development of endodermal cells can be followed by the state III, during which secondary cellulose wall is produced internal to the suberin lamella (tertiary endodermis). Later, this wall may become suberized or lignified. A detailed recent review of the structure of these wall modifications and their potential effect on ion uptake in the roots is available (Peterson and Cholewa, 1998).

Ontogenesis of the endodermis is complicated by asynchronous appearance of the developmental states that may occur simultaneously at a certain distance from the root tip. This asynchrony is influenced by the relationship between the endodermis and vascular cylinder development. The vascular cylinder is divided into phloem and xylem radial sectors with specific individual programs. Developmental preference of endodermal cells, located opposite phloem sectors, is related to the higher level of nutrition (sucrose or substances the distribution of which is regulated by sucrose) closely to the phloem (Barlow and Adam, 1989; Heimsch, 1960). Contrary to that, endodermal cells, adjacent to the xylem sectors, are developmentally delayed, and in some species they remain permanently in the state I as passage cells. In tertiary endodermis of the complete type, even these cells complete their development up to state III.

The development of endodermis need not be finished by the state I, II or III. In the secondary thickening roots with persisting cortex, the endodermis takes part in the secondary dilatation growth. By this process, the primary cortex adjusts to the secondary thickening of vascular cylinder. Cambial growth and dilatation growth proceed usually isochronally. Although a centripetal radial developmental gradient is characteristic in the majority of species for the inner part of the cortex (Von Guttenberg, 1943; Williams, 1947), reactivation of the cortical cells connected with the secondary dilatation growth starts in endodermis, the youngest innermost cortical layer.

Participation of endodermis in secondary dilatation growth

The dilatation growth of endodermal cells is realized by their lateral expansion in tangential direction, by renewed anticlinal division or by the combination of both processes.

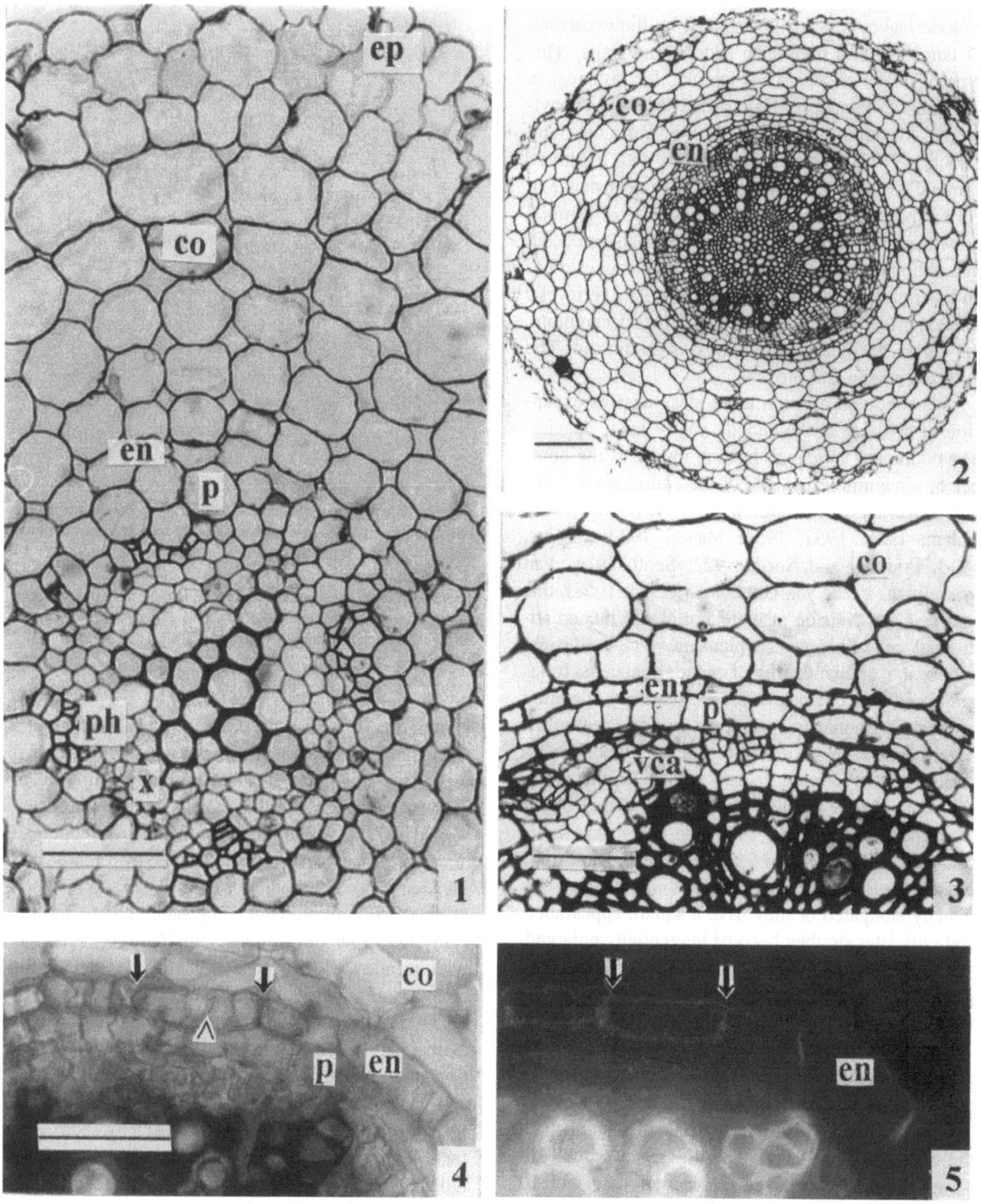

Figure 1–5. Cross sections of *Plantago lanceolata* L. roots in primary and various stages of secondary growth. *Figure 1.* Primary structure, tetrarch root of 520 μm diameter. *Figures 2 and 3.* Root of ca 1000 μm diameter. Advanced secondary thickening, persisting cortex in spite of the prominent cambial activity. Endodermal cells with additional radial walls. The number of cells in endodermal ring is increased approximately four times, comparing with the primary root pattern. *Figures 4 and 5.* Freehand section stained with the berberine-aniline blue procedure. Intensively tangentially expanded endodermal cells with newly-formed radial walls without Casparian bands viewed with white light (Figure 4) and with ultraviolet light (Figure 5). Figures 4 and 5 are courtesy of A. Štefanovičová. ep – epidermis; co – cortex; en – endodermis; ph – phloem; vca – vascular cambium; x – xylem; p – pericycle; arrows – original radial walls; arrowhead – newly-formed radial wall. Scale bar = 50 μm (Figures 1 and 3–5) and 100 μm (Figure 2).

In the first case, the increase in circumference of endodermal cylinder is determined by the expansion of tangential and transverse endodermal walls. This type of dilatation activity is present, for instance, in the family Solanaceae (Hayward, 1951), Caryophyllaceae, Leguminosae and Linaceae (Kutschera and Sobotik, 1992). It occurs in species with a limited cambial thickening, connected with the persistence of cortex for some time. Due to the increase of cambial growth, periclinal division of pericycle is also induced, in addition to the induction of tangential expansion of endodermis and of other cortical cells. The pericyclic phellogen is formed. By activity of this cork cambium, the cortex, including endodermis, is eliminated. The cortex is broken and shed.

Renewed cell division takes place besides expansion of cell size in the endodermis of many species. The additional anticlinal division increases the number of circumferential cells of the endodermal cylinder. In earlier works (Bond, 1930; Kroemer, 1903; Luhan, 1951, 1952, 1954; Mager, 1932; Mylius, 1913; Priestley and North, 1922; Scott, 1928; Van Wisselingh, 1926; Von Guttenberg, 1943, 1968), this renewed cell division of endodermis was named additional cell division, supplementary cell division, subdivision or chambering of endodermal cells. It has rather common characteristics of secondary dilatation growth. Its frequency is a species specific feature. It ranges from a single division of an individual endodermal cell (*Primula*), to cell packets with several cells e.g. 16 new radial walls found in *Viola tricolor* (Kutschera and Sobotik, 1992), up to 24 new radial cell walls in *Gentiana asclepiadea.* The divisions may proceed fast one after another. In this case, the cells are radially elongated or they have a square appearance, and only later on they become tangentially enlarged. Within the endodermal ring, individual cells or small groups of cells can remain without the additional division. In some species, along with the anticlinal divisions, the original endodermal cells can be divided also transversally (*Gentiana nana*, Luhan, 1954), or oblique or periclinal divisions can be also present.

It was already found earlier that in relation with the endodermal cell ontogeny, the renewed cell division starts during its developmental state I or III (Bond, 1930; Mylius, 1913). The observations were completed by the conclusion that the difference was expressed also in a different development of newly-formed radial walls.

After the renewed divisions of endodermal cells during the state I, even the new radial walls get typical endodermal character in majority of species. Along with the Casparian bands persisting in the original walls, the Casparian bands are rapidly also formed in the newly-formed radial walls. The hypothesis assumed by Bond (1930) that the function of endodermis as an apoplastic barrier is not impaired by renewed cell division was confirmed by Weerdenburg and Peterson (1984) using fluorescent apoplastic tracer. In this study, *Impatiens balsamina* and *Helianthus annuus* with primary endodermis, and *Vicia faba* with the endodermis maturing in the state II were used.

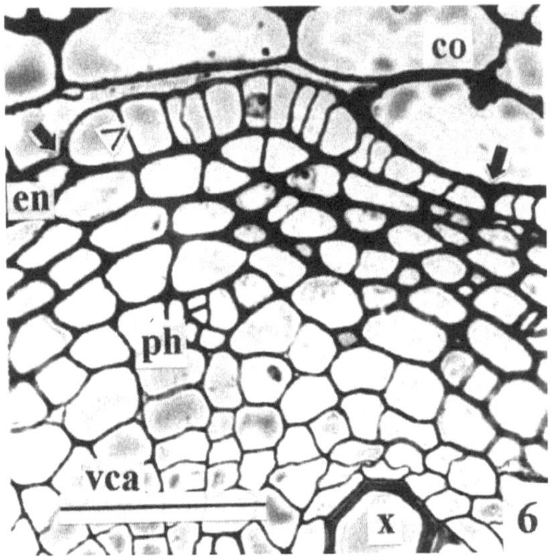

Figure 6. Cross-section of *Gentiana asclepiadea* L. root during the early secondary growth. Suberised endodermal cells are intensively tangentially expanded and additionally divided by several radial walls (without Casparian bands). The cell in the center has 14 newly-formed radial walls. Semi-thin section, toluidine blue-basic fuchsin staining. co – cortex; en – endodermis; ph – phloem; vca – vascular cambium; x – xylem; arrows – original radial walls; arrowhead – newly-formed radial walls; Scale bar = 50 μm.

During the renewed divisions of endodermal cells in the state III, the new radial walls remain without Casparian bands (Bond, 1930; Mager, 1932; Mylius, 1913 and others). These cells have a character of cellulose prop walls (partition walls, Zwischenwände), protecting strongly expanded endodermal walls against collapse. The examples are mentioned in the following species: *Fuchsia*, *Oenothera*, *Lythrum*, *Viola*, many species of Rosaceae (Mylius, 1913); *Digitalis*, *Anthriscus*, *Papaver*, *Plantago*, *Hedera*, *Campanula*, *Phlox*, *Achillea* and other Compositae (Mager, 1932); *Androsace*, *Primula*, *Gentiana* (Luhan, 1951, 1954). It has been confirmed (using fluorescence methods)

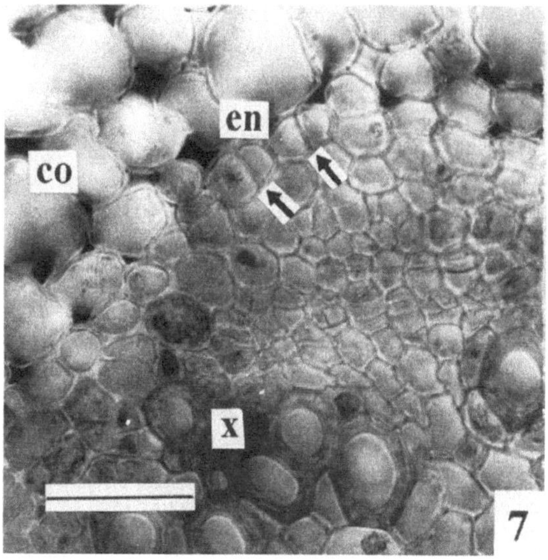

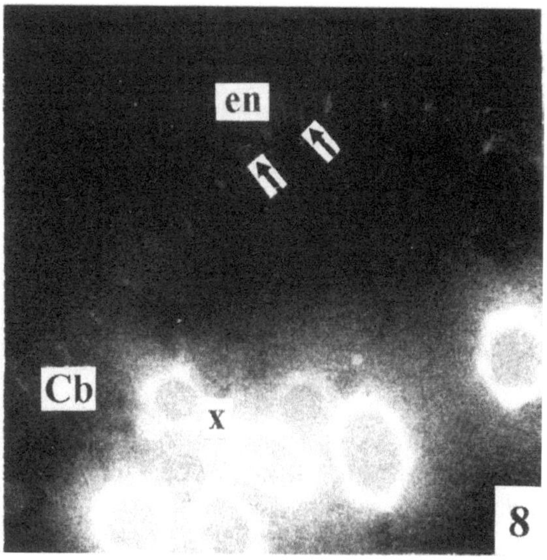

Figure 7-8. Freehand cross-section of *Primula acaulis* L. root in early secondary growth stained with the berberine-aniline blue procedure. After tangential expansion of endodermal cells, dilatation growth continues by their additional anticlinal division. *Figure 7.* White light. *Figure 8.* The same section as in Figure 7 viewed with the ultraviolet light. Casparian bands are formed in new radial walls. co – cortex; en – endodermis; x – xylem; Cb – Casparian bands; arrows – new radial walls with Casparian bands. Scale bar = 50 μm.

that these characteristics of cellulosic additional radial walls in the endodermis were correct (Figures 1–6) in some of the above mentioned species (*Plantago*, *Gentiana*). However, in some other species, it was possible to show the presence of Casparian bands and suberin lamellae also in the newly-formed radial walls. One of these species is primrose (*Primula acaulis* Figures 7 and 8), with a persisting cortex, having a character of reserve tissue. A single anticlinal additional division of individual endodermal cells characterizes homorhize roots with limited amount of secondary cambial growth in this species (Luxová and Lux, 1998).

Dilatation growth and density of endodermal Casparian network

Endodermal character of new radial walls formed during the dilatation growth of endodermis indicates the possibility of regulating the endodermal network density. Endodermal network is formed by the Casparian bands, which correspond to each other in neighbouring cells. Casparian bands of individual endodermal cells represent meshes of this cylindrical network in the root. The mesh density is determined by:

(a) Width of Casparian bands.

(b) The distance between the radial walls of an individual endodermal cell giving the length of Casparian band on each transverse wall.

(c) The distance between the transverse walls of an endodermal cell, determining the length of Casparian band on each radial wall.

Little attention has been paid to these parameters, with the exception of the width of Casparian bands. The role of the width of Casparian bands is related with the increased adherence of plasma membrane to the impregnated region of cell wall. The function of the network formed by Casparian bands as an apoplastic barrier is more effective when the protoplasts are attached to larger areas of the walls of endodermal cells (Fahn, 1964). The appearance of additional anticlinal divisions during the secondary dilatation growth of endodermis confirms the importance of the distances between the endodermal radial walls. If the dilatation growth of endodermis is represented only by cell expansion, the number of cells in the endodermal circle remains constant, but the distances between the radial walls are increased. The density of endodermal network is reduced. A similar result is obtained by the additional division with newly formed radial walls without Casparian bands (prop walls). In such case, the Casparian bands are present only in the original walls of "chambered" cells. Contrary to that, the endodermal character of new additional walls with Casparian bands results in a denser network. The combination of endodermal cell expansion and formation of new radial walls with Casparian bands cause regulation of the density of endodermal network during the renewed anticlinal division. It is dependent on the ratio between the tangential expansion of cells and the frequency of additional anticlinal divisions with modified new walls.

Conclusions

The existence of various developmental types of root endodermis (species with endodermis maturing by the state I, II or III) is related with the successive modifications of endodermal cell walls and their variable permeability. The development of endodermis need not be finished by the primary growth. In secondary thickening roots with persisting cortex, the endodermis has also to expand in perimeter to accommodate the increasing vascular cylinder inside. This is done by tangential expansion of endodermal cells and/or by their renewed anticlinal division during the secondary dilatation growth. Tangential expansion and development of new radial walls, that can be modified or nonmodified in individual species, result in the changed density of endodermal network during secondary growth of the root.

Root tissue density (i.e. dry matter increment per tissue volume) is an important anatomical trait. Significance of tissue density for ecological behavior of plants is the reason of recent interest in this feature. However, up till now little attention has been paid to the density of endodermal network from the viewpoint of its ontogenetic, interspecific and phenotypic structural and functional variation. Relating the modifications of endodermal walls and their function, the evaluation of this tissue density requires specific criteria.

Acknowledgements

We thank Anna Štefanovičová for her help with hand sections. The authors express their gratitude to the Department of Molecular Biology, Faculty of Natural Sciences, Comenius University, Bratislava and Department of Agricultural and Environmental Biology, Graduate School of Agricultural and Life Sciences, University of Tokyo, for the possibility to use their fluorescence microscopes. The research was partially financially supported by Slovak Grant Agency VEGA, grant No.1/4102/97 and grant No.3009.

References

Barlow P W and Adam J S 1989 Experimental control of cellular patterns in the cortex of tomato roots. *In* Structural and Functional Aspects of Transport in Roots. Eds. BC Loughman, O Gašparíková and J Kolek. pp 21–24. Kluwer Academic Publishers, Dordrecht, The Netherlands.

Bond G 1930 The occurrence of cell division in the endodermis. Proc. R. Soc. Edinb. 50, 38–50.

Brundrett M C, Enstone D E and Peterson C A 1988 A berberin – aniline blue fluorescent staining procedure for suberin, lignin and callose in plant tissue. Protoplasma 146, 133–142.

Brundrett M C, Kendrick B and Peterson C A 1991 Efficient lipid staining in plant material with Sudan red 7B or Fluorol yellow 088 in polyethylene glycol – glycerol. Biotech. Histochem. 66, 111–116.

Clowes F A L 1961 Apical meristems. Blackwell Scientific Publications Oxford 217 p.

Esau K 1969 Pflanzenanatomie. Gustav Fisher Verlag, Stuttgart. 594 p.

Fahn A 1964 Some anatomical adaptations of desert plants. Phytomorphology 14, 93–102.

Hayward H E 1951 The Structure of Economic Plants. The Macmillan Company, Third Printing, New York. 674 p.

Heimsch C H 1960 A new aspect of cortical development in roots. Am. J. Bot. 47, 195–201.

Holdheide W 1951 Anatomie mitteleuropäischer Gehölzrinden. *In* Hanbuch der Mikroskopie in der Technik Band V Teil I. Ed. H Freund pp 193/235. Umschau Verlag, Frankfurt am Main.

Huber B 1961 Grundzüge der Pflanzenanatomie. Springer Verlag, Berlin. 319 p.

Kroemer K 1903 Wurzelhaut, Hypodermis und Endodermis der Angiospermenwurzel. Biblioth. Bot. 12, 1–159.

Kutschera L and Sobotik M 1992 Wurzelatlas mitteleuropäischer Grünlandpflanzen. Bd. 2, Teil 2 Anatomie. Gustav Fischer Verlag. Stuttgart-Jena-New York. 261 p.

Lev-Yadun S and Aloni R 1992 Experimental induction of dilatation meristems in *Melia azedarach* L. Ann. Bot. 70, 379–386.

Luhan M 1951 Zur Wurzelanatomie unserer Alpenpflanzen. I. Primulaceae. Sitz. – Ber. d. Wiss. Wien, math.-nat. Kl, Abt.I 160, 481–507.

Luhan M 1952 Zur Wurzelanatomie unserer Alpenpflanzen. II. Saxifragaceae und Rosaceae. Sitz. – Ber. d. Wiss. Wien, math.-nat. Kl, Abt.I 161, 199–237.

Luhan M 1954 Zur Wurzelanatomie unserer Alpenpflanzen. III. Gentianaceae. Sitz. – Ber. d. Wiss. Wien, math.-nat. Kl, Abt.I 163, 89–107.

Luxová M and Lux A 1998 Growth and differentiation of endodermis in the roots of *Primula acaulis*. In 8th Days of Plant Physiology, Book of Abstracts. Olomouc. 144 p.

Mager H 1932 Beiträge zur Kenntnis der primären Wurzelrinde. Planta 16, 666–708.

Mylius G 1913 Das Polyderm. Biblioth. Bot. 18, 1–119.

Peterson C A and Cholewa E 1998 Structural modifications of the apoplast and their potential impact on ion uptake. Z. Pflanzenernähr. Bodenk. 161, 521–531.

Priestley J H and North E E 1922 The structure of the endodermis in relation to its function. New Phytol. 21, 113–139.

Schulz A 1988 Vascular differentiation in the root cortex of peas: Premitotic stages of cytoplasmic reactivation. Protoplasma 143, 176–187.

Scott L I 1928 The root as an absorbing organ. II. The delimitation of the absorbing zone. New Phytol. 27, 141–174.

Van Wisselingh C 1926 Beitrag zur Kenntnis der inneren Endodermis. Planta 2, 27–43.

Von Guttenberg H 1943 Die physiologischen Scheiden. *In* Handbuch der Pflanzenanatomie. Eds. K Linsbauer, G Tischler and A Pascher. 217 p. I.Abt. 2.Teil, Bd. V. Berlin.

Von Guttenberg H 1968 Der primäre Bau der Angiospermenwurzel. *In* Handbuch der Pflanzenanatomie Vol. 8. Ed. K Linsbauer. 472 p. Gebrüder Borntraeger, Berlin.

Weerdenburg C A and Peterson C A 1984 Effect of secondary growth on the conformation and permeability of the endodermis of broad bean (*Vicia faba*) sunflower *(Helianthus annuus)* and garden balsam *(Impatiens balsamina).* Can. J. Bot. 62, 907–910.

Whitmore T C 1962a Studies in systematic bark morphology I. Bark morphology in Dipterocarpaceae. New Phytol. 61, 191–207.

Whitmore T C 1962b Studies in systematic bark morphology II. General features of bark construction in Dipterocarpaceae. New Phytol. 61, 208–220.

Whitmore T C 1963 Studies in systematic bark morphology IV. The bark of beech, oak and sweet chestnut. New Phytol. 62, 161–169.

Williams B C 1947 The structure of the meristematic root tip and origin of the primary tissues in the roots of vascular plants. Am. J. Bot. 34, 455–462.

Wilson A J, Robards A W and Goss M J 1977 Effects of mechanical impedance on root growth in barley *(Hordeum vulgare* L.). II. Effects on cell development in seminal roots. J. Exp. Bot. 28, 1216–1225.

Wilson A J and Robards A W 1978 The ultrastructural development of mechanically impeded barley roots. Effects on the endodermis and pericycle. Protoplasma 95, 255–265.

Recent Advances of Plant Root Structure and Function, 39–47
Edited by O. Gašparíková, M. Čiamporová, I. Mistrík and F. Baluška
© 2001 *Kluwer Academic Publishers.*

Changes of root structure in connection with the development of lateral root primordia in wheat and pumpkins

K. N. Demchenko & N. P. Demchenko
*Komarov Botanical Institute, Russian Academy of Sciences, Prof. Popov St. 2, 197376, St.-Petersburg, Russia**

Key words: Cucurbita pepo, DNA synthesis, growth, lateral root, proliferation, *Triticum aestivum*

Abstract

The root structure and its changes in connection with development of lateral root primordia were studied in *Triticum aestivum* and *Cucurbita pepo* seedlings. Seedlings of these two species differ in respect of the localisation of parent root site where primordia development occurs. In *T. aestivum* roots, primordia initiation takes place at a considerable distance from root tip opposite phloem poles, while in *C. pepo* roots, it occurs in meristem opposite xylem poles. The structure of parent root site where primordia initiation occurs determines the tissues and the cells which take part in this process. In *T. aestivum* roots, primordium initiation is a result of the resumption of cell proliferation in stelar parenchyma, pericycle and endodermis. The resumption of cell divisions is preceded by the resumption of DNA synthesis in some cells of these tissues. In *C. pepo* roots, the local reduction of the mitotic cycle duration occurs in cells of pericycle, endodermis and 4–5 inner cortical layers participating in primordium initiation. The plane of the first cell division is anticlinal in both investigated species. We suggest that the increase of number of anticlinal cell walls in any site of the pericycle (the founder root tissue between central cylinder and cortex) creates conditions for the origin of the growth vector during initiation of lateral root primordia. The sequence of involving cells in different tissues in the primordium initiation, as well as their contribution to formation of primordium and lateral root, were investigated in both species.

Introduction

The change of any plant organ structure results from local changes in rates and directions of cell growth and proliferation. The impulses that induce changes of organ structure and the causes of the competence of some cells still remain unknown. The solution to these problems attracts attention of researchers to study the initiation and development of lateral roots (Blakely et al., 1982; Charlton, 1991; Casero et al., 1993, 1995, 1996; Laskowski et al., 1995; Malamy and Benfey, 1997; Sussex et al., 1995). In some plant species, the formation of lateral root primordia occurs in meristem, in others at a considerable distance from the apex of the parent root (outside the meristem and elongation zone). In the latter species, the earliest stage of lateral root (LR) initiation is connected with anticlinal divisions of the pericycle cells. Our studies on *Triticum aestivum* roots show (Demchenko and Demchenko, 1996) that the resumption of proliferation in connection with lateral root primordium (LRP) initiation begins with the resumption of DNA synthesis in nuclei of stelar parenchyma, pericycle and endodermis cells. The resumption of DNA synthesis occurs much closer to the root tip, than first dividing cells appear in these tissues. A new vector of growth is established after the several anticlinal cell divisions are completed (Demchenko and Demchenko, 1996; Laskowski et al., 1995). During LRP initiation, the change of the developmental program occurs in some cells of the parent root. In *Arabidopsis thaliana* roots, some genes have been proved to be responsible for the initiation and development of lateral roots (Celenza et al., 1995; Sussex et al., 1995). The different stages of the primordium development and their genetic markers are determined (Malamy and Benfey, 1997). However, there are only a few works in which the search of the earliest stages of the LR initiation has been made (Casero et al., 1993, 1995, 1996; De-

* E-mail: dkn@knd.stud.pu.ru

mchenko and Demchenko, 1996). The structure of the parent root sites, where the initiation and development of LRP occur, has not been adequately explored. In the present paper, the general features of and differences in root structure, growth, proliferation and differentiation of cells during initiation and LR development have been studied in two plant species, *Triticum aestivum* and *Cucurbita pepo*, differing in the localisation of the parent root site where LRPs appear.

Materials and methods

Caryopses of winter wheat *Triticum aestivum* L. (*Poaceae*) var. *lutescens* cv. 'Besostaya-1' were germinated in the darkness at 25 ± 1 °C on moistened filter paper. After 6 days of germination, the seedlings with roots 10–12 cm long were placed in ^{3}H-thymidine solution (concentration 80 MBq/l, specific activity 629 TBq/mol) for 30 min and then after the washing the roots were fixed in Navashin's fixative (chromic acid, formalin, glacial acetic acid, 10:4:1 v/v/v). The root tips were sectioned into four segments 5 mm long each. The remains of the roots were sectioned at 3 mm long segments, in which primordia or LRs occurred in different developmental stages. The roots were stained with the Feulgen's reaction for 4 h, hydrolysis being accomplished for 40 min in 5 N HCl solution at 22 °C. Cell walls were stained for 10 h with 1.0% solution of Alcian blue 8GS in 3% acetic acid. The root segments were embedded in paraffin wax and cut into serial longitudinal or transverse sections, 7 μm thick. Cell walls were stained once more for 40 min with 1.0% solution of Alcian blue. Part of the sections was dehydrated, and was mounted under a coverslip in Canada Balsam. Fine liquid emulsion (Moscow's factory of technical photographic) was applied to other sections. The exposure time was 20 days at 4 °sC before development. The emulsion was developed using the amidol developer for 3 min and was fixed in 40% sodium thiosulfate for 6 min. The changes of the index labelled (^{3}H-thymidine) cells and the mitotic index in stelar parenchyma, pericycle, endodermis and cortex were determined in longitudinal sections of root segments located at 0–5, 5–10, 10–15 and 15–20 mm from the root tip. Apical segments (0–5 mm) were analysed in 5 roots, the following ones in 20 roots.

Cucurbita pepo L. (*Cucurbitaceae*) seeds cv. 'Gribovsky-37' were germinated in the darkness at 30±1 °C on moistened filter paper. After 2 days of germination, the seedlings were placed in aerated glass beaker with 1/10 Knopp's solution on plate with holes 3 mm in diameter. Then after 2 days of germination, the seedlings with roots 4–5 cm long were selected and fixed in Navashin's fixative (see above). Root tips 6 mm long were sectioned into two 3 mm long segments. The remains of the roots were sectioned at segments 3 mm long, in which primordia or LRs occurred in different developmental stages. Root segments were embedded in paraffin wax and cut into serial longitudinal or transverse sections, 8 μm thick. Sections were stained with iron hematoxylin (Sass, 1958) and 1.0% solution of Alcian blue 8GS (as described above). Twenty roots of both species were investigated at all primordium developmental stages.

Results

Triticum aestivum *root structure and its changes related to the initiation and development of lateral roots*

The initial cells in wheat root tip have a closed type of organisation (Figure 1A). The root cap has its own initial cells. The average length of the columella file is about 250 μm. Rhizodermis and cortex originate from the same initial cell. A separate group of initial cells gives rise to the tissues of the central cylinder. The vascular system in the root tip has a polyarch type of organisation (Figure 1B). It is composed of 6–8 phloem and xylem groups. The cells of different tissues complete their division and DNA synthesis at different distances from the root tip. There are no more dividing cells at the distance of 1.5–2.0 mm from the root tip. The most proximal ^{3}H-thymidine labelled (DNA synthesising) cell was found in endodermis at a distance about 3 mm from the root tip. The analysis of change of average cell length of three outer cortex layers along the longitudinal axis of a root has shown, that the elongation zone extends at the distance of 1.2–4.5 mm from the root tip (Figure 2). During elongation, the average cell length of any tissue increases 10-fold. The average cell length of three outer cortex layers before onset of elongation was 28±1 μm, and after its completion 288±14 μm. In the following root sites, the cell length practically does not change. Only protophloem cells are able to start and to complete the differentiation in the meristem. The nucleus and cytoplasm degrade in those cells (sieve elements, SE) already in the middle of the meristem. In some rhizodermis cells (trichoblasts) located at the end of

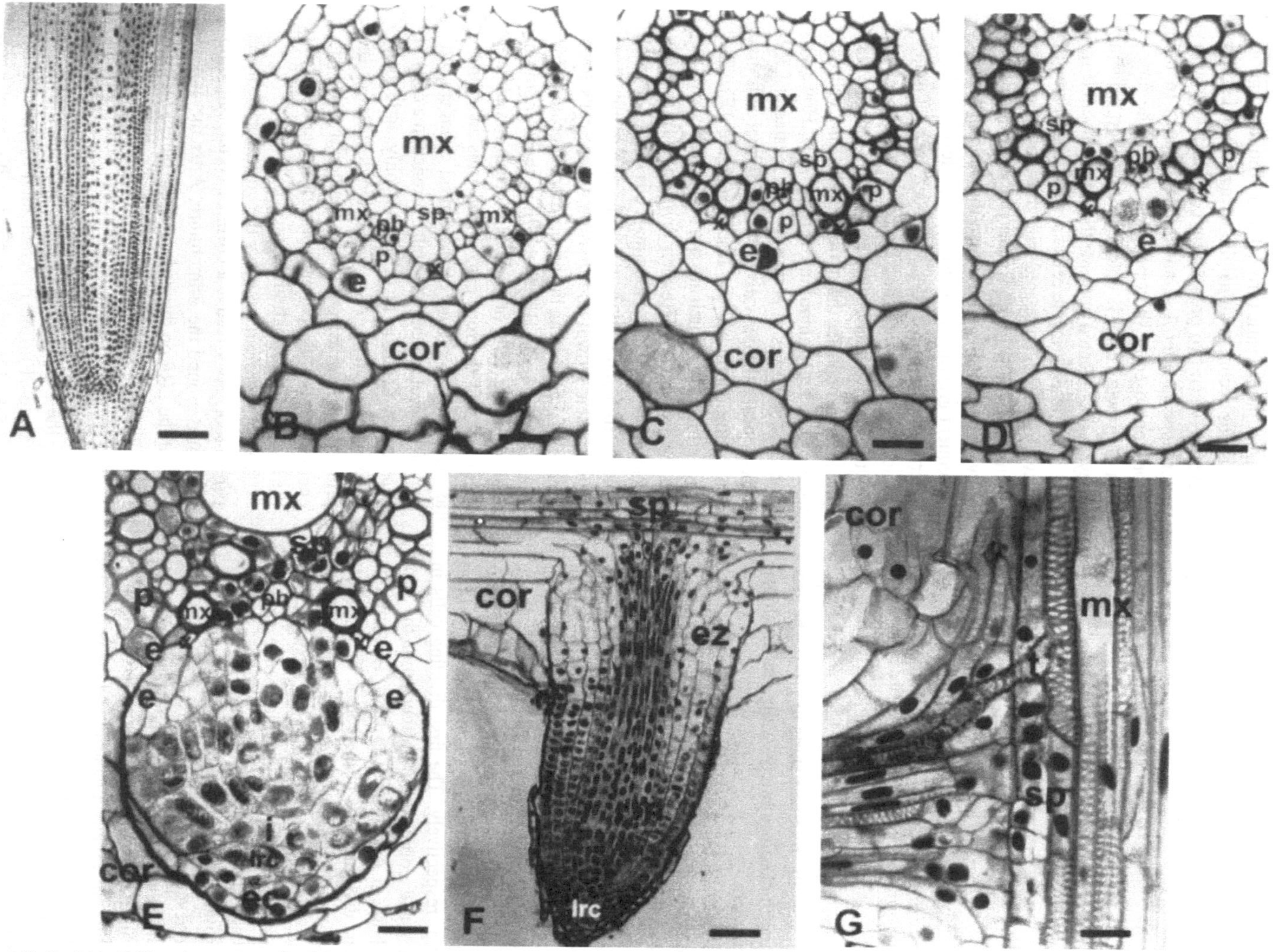

Figure 1. Longitudinal (A, F, G) and transverse (B, C, D, E) sections of *Triticum aestivum* roots at various distances from the root tip. (A) root apex, (B) root tissues at 2 mm, (C) pericycle cells with thin walls in the sector of lateral root initiation at 24 mm, (D) the appearance of transverse growth vector in pericycle cells at 26 mm, (E) formation of endodermal cap and initial cells of some lateral root tissues at 60 mm, (F) onset of elongation in lateral root base and the formation of connection zone at 80 mm, (G) differentiation of stelar parenchyma cells into vascular elements at 120 mm. Cor – cortex; e – endodermis; ec – endodermal cap; ez – elongation zone; i – common initial cortex and rhizodermis cell in lateral root; lrc – lateral root cap; mx – metaxylem; p – pericycle; ph – phloem group; sp – stelar parenchyma; t – connective tracheal elements; x – protoxylem. Scale bars: 150 μm (A), 25 μm (B,C,D,E,G), 50 μm (F).

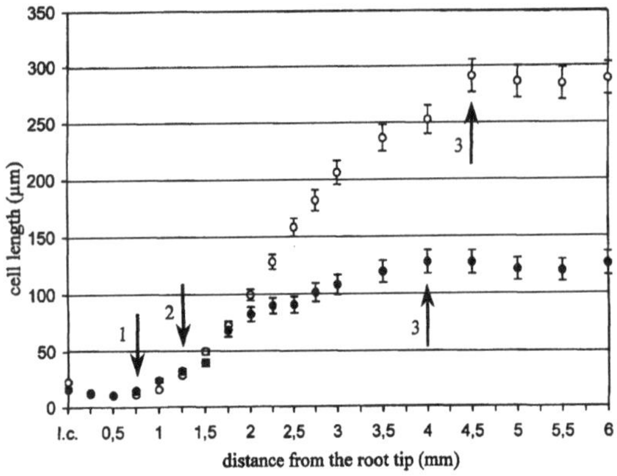

Figure 2. Changes of cell length of three outer cortex layers along longitudinal root axis in *Triticum aestivum* (○) and *Cucurbita pepo* (●) seedlings. I.c., initial cell. Arrows show: 1 – the cessation of cell division, 2 – the start point of cell elongation, 3 – the completion of cell elongation.

elongation zone, the root hair starts to grow. The secondary spiral thickenings of cell walls are developed in the protoxylem.

The analysis of the distribution of ^{3}H-thymidine labelled in the root sites located above the elongation zone showed that the resumption of DNA synthesis in stelar parenchyma cells starts at the distance of 6–7 mm from the root tip, in pericycle cells at 8–9 mm, in endodermis at 12–13 mm, in cortex at 16–17 mm. The pictures of autoradiographs are presented in our previous work (Demchenko and Demchenko, 1996). The first dividing cells appear in stelar parenchyma and pericycle at a distance 12–13 and 16–17 mm from the root tip, respectively. The first dividing cells in endodermis and cortex occur farther from the root tip than those in pericycle. The first ^{3}H-thymidine labelled and dividing cells in the cortex were exclusively opposite dividing pericycle cells, which initiate primordium development. No structural changes have been found in the root site in which some cells of stelar parenchyma and pericycle had resumed DNA synthesis (Figure 1B). During the resumption of divisions in pericycle cells, their walls remain thin in the sector of the future primordium, whereas they become thickened in the other sectors (Figure 1C). The protoxylem cells are already anucleate and the peripheral metaxylem cells have thickened cell walls.

The division plane in the first dividing cells is perpendicular to the longitudinal axis of a root (i.e. anticlinal). The first division of cells is both symmetrical and asymmetrical. In the pericycle file, the resumption of divisions takes place mostly in one or two neighbouring cells. The divisions of two neighbouring cells in a file not infrequently are asymmetrical. The number of pericycle files, which participate in primordium initiation, varies from 4 to 6. The resumption of divisions occurs in all neighbouring pericycle cells located between two protoxylem cell files. A growth vector appears after 1–2 anticlinal divisions in the sector, composed of short pericycle cells. The formation of growth vector is determined by short cells in pericycle files located opposite the protophloem (Figure 1 D). These pericycle cells have the largest area of transverse section compared with those of other files (Figure 1B, C). During growth, the short cells start to divide periclinally earlier than the other ones. The subsequent growth and divisions of pericycle and endodermis cells form LRP surrounded by a 'Tasche' of endodermal origin. The 'Tasche' is formed by the growth and divisions of the cells in 3–4 endodermal files located opposite the pericycle files, the cells of which form the primordium (Figure 1D). However, soon the derivatives of pericycle cells located at LRP flanks squeeze the derivatives of endodermal cells disturbing the continuity of 'Tasche' (Figure 1E). The endodermal cells, which are situated at the base of the 'Tasche', soon complete their division and the vacuolisation begins. The endodermal cells of the 'Tasche' which surround the upper part of the primordium continue in their growth and division. These cells form endodermal cap before the primordium appears on the root surface (Figure 1E). At this time, the primordium formed by the cells of pericyclic origin has already initial cells of the root cap, rhizodermis and cortex, as well as some tissues of the central cylinder. During LRP development, the stelar parenchyma cells of parent root continue to divide not only anticlinally but also periclinally. As a result, the diameter of the primordium base as well as the number of stelar parenchyma cells of the parent root increase noticeably at the level of LR formation (Figure 1E).

The primordium squeezes the cortex cells of the parent root. As a result, a cover of cortex cell walls is formed around the primordium. After the appearance of the LRP on the surface of the parent root, the longitudinal cell files of different tissues are formed by growth and division of the initial cells and their derivatives. When the LR approaches 500–600 μm in length, the elongation zone starts to form in its basal part (Figure 1F). Cell differentiation of the vascular system of the LR and contact zone with vascular system of the parent root begins after LR appears on the parent root surface. Nucleus degradation of pycnotic type occurs in stelar parenchyma cells contacting the phloem group. Then the processes of nucleus degradation extend towards the LR base and progress in acropetal direction. The cells of stelar parenchyma contacting the metaxylem go into xylem developmental pathways (Figure 1G). In these cells, the secondary thickenings of cell walls are formed. The process of xylem differentiation expands into the LR (Figure 1G). During LR development, the number of m[illegible]re xylem and phloem elements increases in the s[illegible] of LR formation of the parent root.

Cucurbita pepo *root structure and development of lateral root primordia*

The root tip of *C. pepo* has an open type of organisation of initial cells (Figure 3A). Peripheral layers of cap cells have common initials with the rhizodermis and cortex, whereas the cap columella has common initials with the central cylinder. The root meristem is entirely covered with the cap. The vascular system of the root has polyarch organisation. The protophloem cells complete their divisions and differentiation closer to the root tip than the cells of other tissues (Figure 3C). Analysis of average cell length changes of the three outer cortex layers along the longitudinal axis of the root has shown that the elongation zone extends 1.2–4.0 mm from the group of initial cells (Figure 2). The average cell length prior to the onset of elongation was 24±1 μm, while it was 124±4 μm after completion of the elongation. During elongation, the cell length increased 5-fold.

LRPs could be found already in root meristem (Figure 3A). Their initiation obviously starts as a result of anticlinal cell divisions. A group of 3–6 pairs of short pericycle cells appears at a distance of 200–250 μm from the initial cortex cells. Their length is noticeably less than that of neighbouring cells (Figure 3B). The endodermal cells are involved in the initiation almost simultaneously with the pericycle cells. Then the cells of the other cortex layers are sequentially and very quickly involved in the primordium development. Only the cells of the three outermost cortex layers do not participate in the LRP formation (Figure 3C). These cells stop their divisions at the distance of 600 μm from the initial cells. The appearance of a new growth vector (in the radial direction) takes place after completing the first anticlinal divisions of pericycle cells. The new growth vector causes the appearance of periclinal and oblique divisions in pericycle and endodermis cells during LRP initiation (Figure 3C, D).

LRPs are formed opposite xylem pole between two files of protophloem cells (Figure 3C). The earliest stages of LRP development take place after the protophloem cells of the parent root completed their divisions and differentiation (Figure 3C). In the basal part of the meristem, the secondary thickenings of cell walls appear in protoxylem cells (Figure 3C, D). Over the whole length of the elongation zone, the volume of LRP significantly increases (Figure 3D, E). The diameter of the primordium base at the basal boundary of parent-root elongation zone increases 2.5-fold in comparison with the diameter at the distal boundary. However, the average size of LRP cells was the same. A primordium, increasing in its volume, moves aside three cortex layers and rhizodermis, which did not participate in its formation (Figure 3E). More basipetally on the parent root, a primordium starts to extend fast through the outer cortex layers and rhizodermis (Figure 3F). The cap originated from cortex cells is formed at the primordium top. During the growth of the LRP through the cortex, the cortical cells are lysigenously destroyed. As a result, cavities in cortex are formed near the primordium (Figure 3F). Anticlinal cell divisions take place in the stelar parenchyma (Figure 3F). These divisions form the contact zone of vascular systems of the parent and lateral roots. The processes of differentiation of phloem and xylem during subsequent development of LRP spread quickly in acropetal direction. The LRs appear on the parent root surface at a distance of 12–15 mm from the root tip. The longitudinal files of all the tissues are already distinguished in the LRs about 1 mm long (Figure 3G). The formation of the LR meristem is completed and the formation of elongation zone starts. The LR cap is composed mainly of the cells of cortex origin. The rhizodermis and 2–3 cortex layers have the same origin. However, the common initial cells of the columella and the tissues of the central cylinder have already arisen and they are of pericyclic origin. During the following LR development, all the initial cells of cortex origin are successively replaced by the initials of pericyclic origin.

Discussion

The results of the present investigation show essential differences in changes of parent root structure depending on the localisation of the root site, in which the

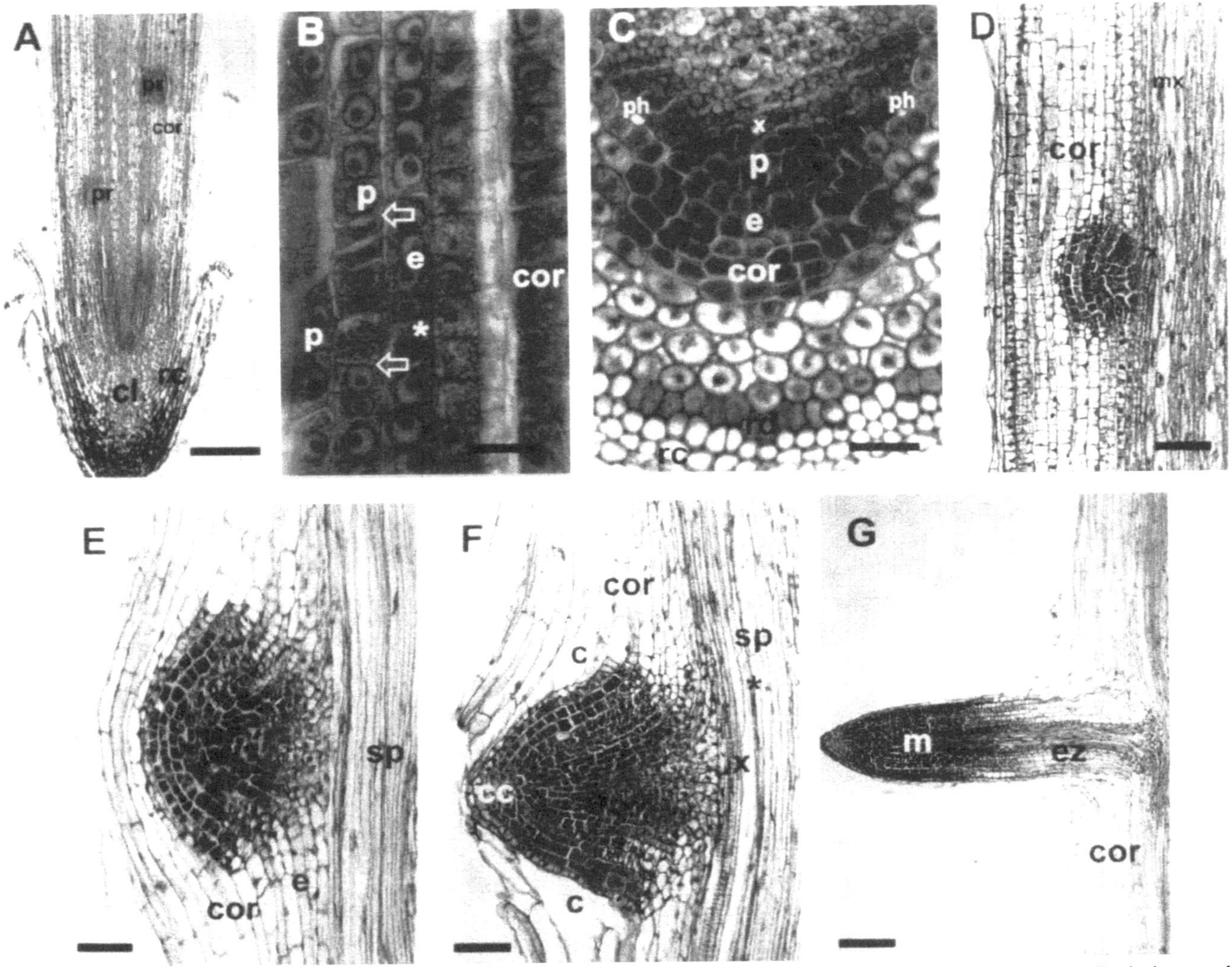

Figure 3. Longitudinal (A, B, D, E, F, G) and transverse (C) sections of *Cucurbita pepo* roots at various distances from the root tip. (A) root apex, (B) short cell pairs in external pericycle file (between arrows) at 200–250 μm from the initial cortex cells, mitosis in endodermis cell (asterisk), (C, D) root tissues at about 1 mm from the initial cortex cells, (E) lateral root primordium at 5 mm from the root tip, (F) formation of cavities in cortex located around the primordium at 10 mm; cortical cap is formed on the primordium; mitosis in stelar parenchyma cell (asterisk), (G) emerged lateral root with vascular connection between primary and lateral roots at 20 mm from the root tip; the formation of elongation zone begins at the lateral root base. C – cavity; cc – cortex cap; cl – columella; pr – primordium; rc – root cap; rd – rhizodermis, other designations as in Figure 1. Scale bars: 150 μm (A,G), 10 μm (B), 25 μm (C), 50 μm (D,E,F).

initiation and development of LRPs occur. These differences are conditioned mainly by the fact that in *C. pepo* roots, the LRPs initiation occurs in the meristem. In *T. aestivum* roots, the LRPs initiation takes place above the elongation zone and cell proliferation has to resume for the initiation of LRPs. Most probably, for this reason, the cells of the greater number of cortex layers participate in the initiation and development of primordia in *C. pepo* roots than in *T. aestivum* roots. Therefore, cortex cells in the former case make greater contribution to LRP formation. The LRP formation with the participation of several cortex layers is characteristic of *Cucurbitaceae* (Dubrovsky, 1986a,b; Mallory et al., 1970; Whiting, 1938). The LRPs of many *Magnoliophyta* are formed at considerable distance from the parent root tip, above the elongation zone. Most commonly, pericycle and/or endodermis cells in these plants take part in the initiation and formation of primordia (Guttenberg, 1968; Janczewski, 1874; McCully, 1975; Van Tieghem and Duliot, 1888).

The earliest stage of LRP initiation is marked by the appearance of the first anticlinal divisions of pericycle cells in roots of those plants in which the LR initiation occurs at a considerable distance from the apex of the parent root (Bell and McCully, 1970; Blakely et al., 1982; Casero et al., 1995; Lloret et al., 1989; Seago and Marsh, 1990). The earliest stage of LRP initiation is difficult to recognise if it occurs in meristem. In *Ipomoea purpurea* roots, an early indication of LRP initiation includes the enlargement of the nucleus, increased protoplast stainability, and slight radial expansion of 4–8 cells (Seago, 1973). Similar changes in stainability and nucleus enlargement were observed also in a few immature endodermis cells adjacent to these pericycle cells. The earliest divisions in this group of cells in pericycle and endodermis were anticlinal. In the present investigation, the group of short pericycle and endodermis cells have been also revealed (Figure 3B). These cells could appear as a result of anticlinal divisions having significantly shortened mitotic cycle as compared to the adjacent cells. The decrease of the mitotic cell cycle duration during LRP development has been found also in *Zea mays*, *Vicia faba*, *Phaseolus vulgaris, Pisum sativum* (MacLeod and Thompson, 1979; Thompson and MacLeod, 1981), *Cucumis sativum* (Dubrovsky, 1986b) and *Raphanus sativus* (Laskowski et al., 1995).

An important question arises during the investigation of LRP initiation outside the elongation zone. Did the pericycle cells participating in LRP initiation leave or not at the mitotic cycle at the end of the meristem? Blakely et al. (1982) suggested that cells could be displaced by elongation processes towards the root base while they are still engaged in the mitotic cycle. Analysis of the data on the distance of the first anticlinal divisions of pericycle cells initiating the LRP development from the root tip in *Raphanus sativus*, *Helianthus annuus*, *Zea mays* and *Daucus carota* confirmed the hypothesis that pericycle cells continue dividing when they leave the apical meristem (Casero et al., 1993, 1995; Lloret et al., 1989). Blakely et al. (1982) suggest that in *Raphanus sativus* roots, only a small part of pericycle cells continue to divide after leaving the meristem, and then participate in LRP initiation. Most of the pericycle cells leave the cell cycle in G2-phase when they leave the apical meristem (Blakely and Evans, 1979). These cells can be stimulated by exogenous auxin to resume the mitotic activity and to participate in LRP formation (Blakely et al., 1982; John et al., 1993). The results of our investigation show that in *T. aestivum* roots, the LRP initiation occurs on the background of the resumption of cell proliferation of certain tissues. It starts with the resumption of DNA synthesis in nuclei of some cells. The latter occurs much closer to the root tip, than the first dividing cells appear. We suggest that the cells having completed the mitotic cycle in G1-phase in the meristem are capable only to resume proliferation and subsequently to initiate the LRP development. The resumption of DNA synthesis in nuclei and the cell divisions occur first in stelar parenchyma, then in pericycle, and later in endodermis. Such sequence of the resumption of cell proliferation in tissues most likely reflects the direction of the movement of regulators that carry out the removal of a block on the cell transition to S-phase. According to the opinion of Casero et al. (1995), the first asymmetric divisions of pericycle cells play a key role in LRP initiation. The detailed pattern of cell proliferation in pericycle during LRP initiation in *Allium cepa* roots is described by Casero et al. (1996).

Owing to rigid walls, the cells compose a part of supporting subsystem of an organ (Barlow, 1994). The increase of number of anticlinal cell walls in any site of the pericycle (a boundary tissue between central cylinder and cortex) causes the change in this subsystem and creates the conditions for the appearance of new growth vector required for LRP development. In the species with LRP initiation occurring in meristem, the increase of the number of anticlinal cell walls is a result of the decrease of the mitotic cycle duration in a cell group. In other species, in which LRPs initiation occurs at a considerable distance from the root tip (above elongation zone), this occurs as a result of the resumption of division in some cells.

It has been found that in roots of all studied *Magnoliophyta*, the central cylinder of lateral root is pericyclic by origin, while the stelar parenchyma cells form the contact zone of vascular systems of parent and lateral roots (McCully, 1975). The contribution of endodermis and cortex cells to the formation of LRP and LR tissues depends on the plant species. The larger is the number of cortex layers involved in LRP formation, the more significant is their contribution to the formation of a LR. The present observations that the continuity of the endodermal 'Tasche' breaks during the LRP development confirm the data of other researches who studied the roots of many dicotyledons (Guttenberg, 1968; McCully, 1975), and also *Hordeum vulgare* (Danilova et al., 1992). Destruction of the continuity of the endodermal 'Tasche' occurs before the primordium appears on the parent root surface, as a result of both completion of cell divisions in the endodermal 'Tasche' and subsequent rapid growth of the LRP. The disintegration of middle lamella takes place between the neighbouring cells at the base of primordium but the cells at the base of LRP remain intact (Danilova et al., 1992). However, in *Zea mays* roots, the initial cells of the permanent root cap originate from the cells of the endodermal 'Tasche' (McCully, 1975).

Acknowledgements

This research was supported by a grant from Russian Foundation for Basic Research (96-04-48719).

References

Barlow P W 1994 Cell divisions in meristems and their contribution to organogenesis and plant form. Ontogenez 25, 5–28 (in Russian).

Bell J K and McCully M E 1970 A histological study of lateral root initiation in *Zea mays*. Protoplasma 70, 179–205.

Blakely L M and Evans T A 1979 Cell dynamics studies on the pericycle of radish seedling root. Plant Sci. Let. 14, 79–83.

Blakely L M, Durham M, Evans T A and Blakely R M 1982 Experimental studies on lateral root formation in radish seedling roots. 1. General methods, developmental stages and spontaneous formation of laterals. Bot. Gaz. 143, 341–352.

Casero P J, Casimiro I and Lloret P G 1995 Lateral root initiation by asymmetrical transverse divisions of pericycle cells in four plant species – *Raphanus sativus, Helianthus annuus, Zea mays*, and *Daucus carota*. Protoplasma 188, 49–58.

Casero P J, Casimiro I and Lloret P G 1996 Pericycle proliferation pattern during the lateral root initiation in adventitious roots of *Allium cepa*. Protoplasma 191, 136–147.

Casero P J, Rodriguez-Gallardo L, Martin-Partigo G and Lloret P G 1993 Lateral root initiation by means of asymmetrical transversal divisions of the pericycle cells in adventitious roots of *Allium cepa*. Protoplasma 176, 138–144.

Celenza J L Jr, Grisafi P L and Fink G R 1995 A pathway for lateral root formation in *Arabidopsis thaliana*. Genes and Development 9, 2131–2142.

Charlton W A 1991 Lateral root development. *In* Plant Roots: The Hidden Half. Eds. Y Waisel, A Eshel and U Kafkafi. pp 103–128. Marcel Dekker Inc., New York, Basel, Hong Kong.

Danilova M F, Serdjuk H M and Kharitonova T M 1992 The lateral root initiation and transport of nutrients. *In* Root Ecology and its Practical Application. Proc. 3 ISRR Symp. pp 29–32. Vienna.

Demchenko K N and Demchenko N P 1996 Early stages of lateral root development in *Triticum aestivum* L. Acta Phytogeogr. Suec. 81, 71–75.

Dubrovsky I G 1986a The origin of tissues of the embryo lateral root in the cucumber. Interactions between tissues and positional control in its development. Ontogenez 17, 176–189 (in Russian).

Dubrovsky I G 1986b Dynamics of cell reproduction and cell complexes (cell packets) in the embryo lateral root primordium of the cucumber. Ontogenez 17, 525–533 (in Russian).

Guttenberg H 1968 Der primare Bau der Angiospermenwurzel. Gebruder Borntraeger, Berlin.

Janczewski E 1874 Recherches sur l'accroissement des racines dans les Phanerogames. Ann. Sci. Nat. Bot. 5, 208–233.

John P C L, Zhang K, Dong C, Diederich L and Wightman F 1993 p34 cdc2 related proteins in control of cell cycle progression, the switch between division and differentiation in tissue development, and stimulation of division by auxin and cytokinin. Aust. J. Plant Physiol. 20, 503–526.

Laskowski M J, Williams M E, Nusbaum H C and Sussex I M 1995 Formation of lateral root meristems is a two-stage process. Development 121, 3303–3310.

Lloret P G, Casero P J, Pulgarin A and Navascues J 1989 The behaviour of two cell populations in the pericycle of *Allium cepa, Pisum sativum* and *Daucus carota* during early lateral root development. Ann. Bot. 63, 465–475.

MacLeod R D and Thompson A 1979 Development of lateral root primordia in *Vicia faba, Pisum sativum, Zea mays* and *Phaseolus vulgaris*: Rates of primordium formation and cell doubling times. Ann. Bot. 44, 435–449.

Malamy J E and Benfey P N 1997 Organisation and cell differentiation in lateral roots of *Arabidopsis thaliana*. Development 124, 33–44.

Mallory T E, Chiang S -H, Cutter E G and Gifford E M 1970 Sequence and pattern of root formation in five selected species. Am. J. Bot. 57, 800–809.

McCully M E 1975 The development of lateral roots. *In* The Development and Function of Roots, Third Cabot Symposium. Eds. JG Torrey and DT Clarkson. pp 105–124. Acad. Press, London, New York, San Francisco.

Sass J E 1958 Botanical microtechnique. The Iowa State College Press, Iowa, USA. 3rd edn. 228 p.

Seago J L 1973 Developmental anatomy in roots of *Ipomoea purpurea*. 2. Initiation and development of secondary roots. Am. J. Bot. 60, 607–618.

Seago J L and Marsh L C 1990 Origin and development of lateral root in *Typha glauca*. Am. J. Bot. 77, 713–721.

Sussex I M, Godoy J A, Kerk N M, Laskowski M J, Nusbaum H C, Welsch J A and Williams M E 1995 Cellular and molecular events in a newly organizing lateral root meristem. Philosophical Transaction: Bilogical Sciences 350 (1331), 39–43.

Thompson A and MacLeod R D 1981 Increase in size and cell number of lateral root primordia on the primary roots on intact plants and in excised roots of *Pisum sativum* and *Vicia faba*. Am. J. Bot. 68, 955–964.

Van Tieghem P and Douliot H 1888 Recherches comparatives sur l'origine des membres endogenes dans les plantes vasculaires. Ann. Sci. Nat. Bot. 7 Ser. 8, 1–660.

Whiting A 1938 Development and anatomy of structures in the seedlings of *Cucurbita maxima*. Bot. Gaz. 99, 497–528.

Recent Advances of Plant Root Structure and Function, 49–54
Edited by O. Gašparíková, M. Čiamporová, I. Mistrík and F. Baluška
© 2001 Kluwer Academic Publishers.

Some aspects of endodermis and cortex structure in *Dryas* roots

K. M. Barmicheva & M. F. Danilova
Komarov Botanical Institute of Russian Academy of Sci., Prof. Popov Str., 2, 197376 St.- Petersburg, Russia

Key words: Casparian strip, cortical cells, development, endodermis, *phi*-thickening, suberin

Abstract

Some specific features of the endodermal development and cortical anatomy in *Dryas* mycorrhizal roots were studied. These features of the endodermis were: 1) the absence of typical suberin lamellae and the presence of dendrites in the secondary cell wall; 2) the direct contact of the Casparian strip with tertiary (polysaccharide) cell wall. The presence of *phi*-thickenings on the anticlinal walls in the cells of the inner cortical layer (adjacent to the endodermis) was noted. The possible functional roles of these features in *Dryas* roots are discussed in the paper.

Introduction

The development of the endodermis in apple and lime roots was described by MacKenzie (1979, 1983). This author showed that the Casparian strips of the endodermis at the second stage remain uncovered by suberin lamellae. In addition, there were 'passage' cells (i.e. cells totally devoid of suberin lamellae and tertiary walls) among suberized cells of the endodermal layer. The 'passage' cells were usually located over xylem poles. MacKenzie also studied deposition of special *phi*-thickenings on the anticlinal cell walls of the innermost cortical layer in roots of these plants. *Phi*-thickenings are the modification of the middle sections of radial and transverse walls of some root cortical cells. Those cells often occur as an uniseriate layer. The formation of *phi*-thickenings involves the deposition of large amounts of cellulose and, later, lignin (Weerdenburg and Peterson,1983). The structure and function of *phi*-thickenings in cortical and exodermal cells also have been discussed (Haas et al., 1976; MacKenzie, 1979, 1983; Weerdenburg and Peterson, 1983). The aim of the present communication was to trace the development of the endodermis and cortical cells with *phi*-thickenings during the growth of *Dryas* mycorrhizal roots using electron microscopy.

Materials and methods

The plants of *Dryas octopetala* L. (Rosaceae) were collected in the Hibins (Kola Peninsula, July 1990 and 1994), Austrian Alps (June 1994); the plants of *Dryas punctata Juz.* were collected in the Chukchi Peninsula (August 1990, 1991). Five or seven plant roots of each fixation have been taken for the investigation. The plant roots were fixed in the glutar-formaldehyde solution (2.5% glutaraldehyde and 2% paraformaldehyde) and postfixed with 2% osmium tetroxide (pH 6.8–6.9 for all solutions) at room temperature. Samples were embedded in Epon-Araldite mixture. Ultrathin sections were cut using LKB ultramicrotome, stained with uranylacetate and lead citrate and viewed with a Hitachi H-600 electron microscope.

Results and discussion

Dryas typically form ectomycorrhizae on the roots of the second and third order. The growing mycorrhizal roots are short (10–25 mm). The meristematic zone of such root is reduced and the differentiation of all tissues proceeds close to the root tip, which favours the study of developmental processes. Similar structure of the endodermis and the cortex was observed in all studied plant roots of *Dryas*.

Endodermis

Three stages of endodermal development have been followed in *Dryas* roots. Depending on the root length, the presence of endodermis in secondary and tertiary stages can be noted at the distance of 0.3–0.9 mm from root tip. Figure 1 shows relative position of the endodermis and cortical cells with *phi*-thickenings

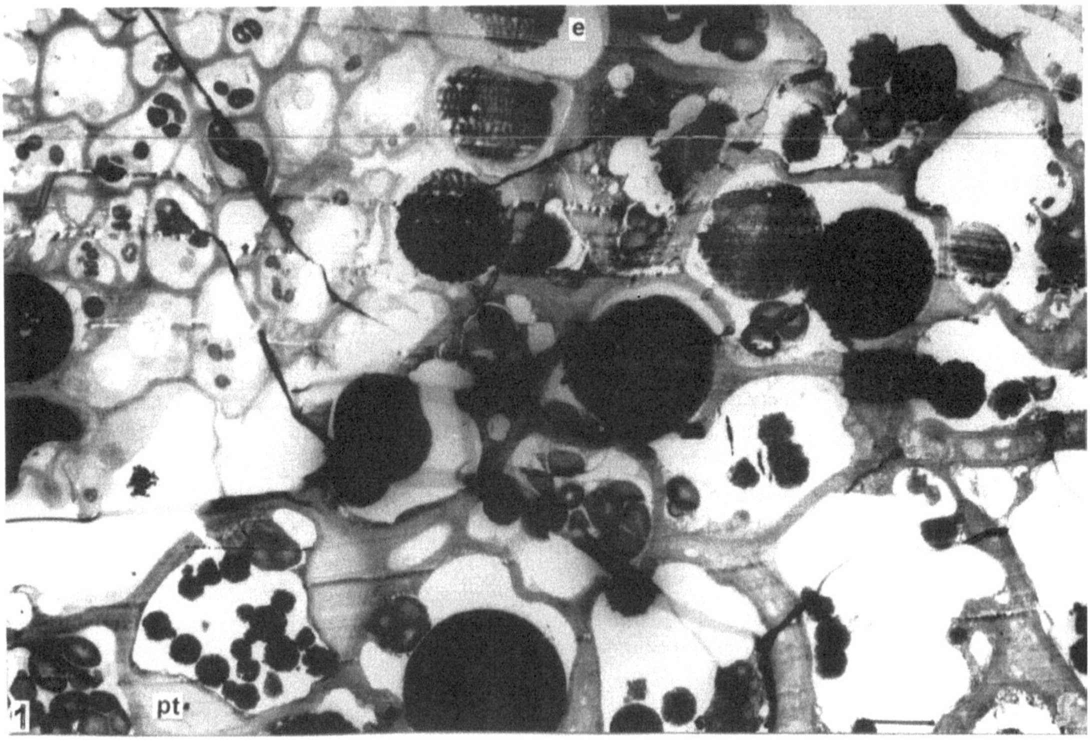

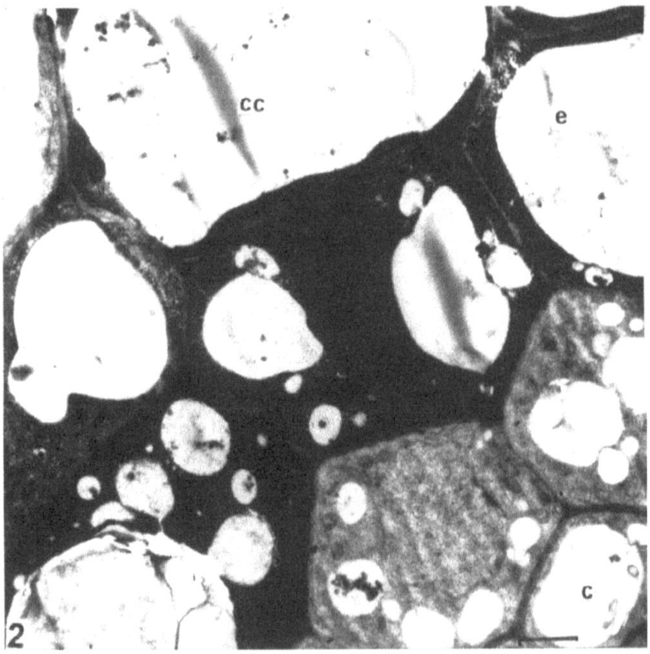

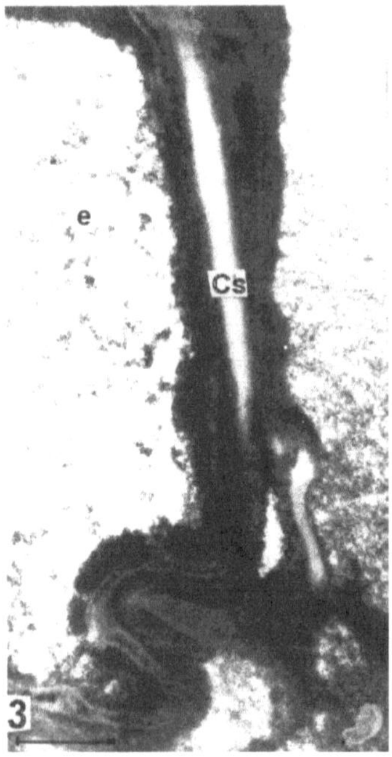

Figures 1–11. Transverse sections of *Dryas* roots.
Figure 1 The arrangement of internal root tissues, 0.4 mm from root tip: central cylinder (c), endodermis (e), cortical cells (cc) with *phi*-thickenings (pt). Scale bar = 5 μm.
Figure 2 Meristematic endodermal cells with electron-opaque cytoplasm (0.2 mm from root tip). Scale bar = 2 μm.
Figure 3 Casparian strip (Cs) on the radial endodermal cell walls. Scale bar = 1 μm.

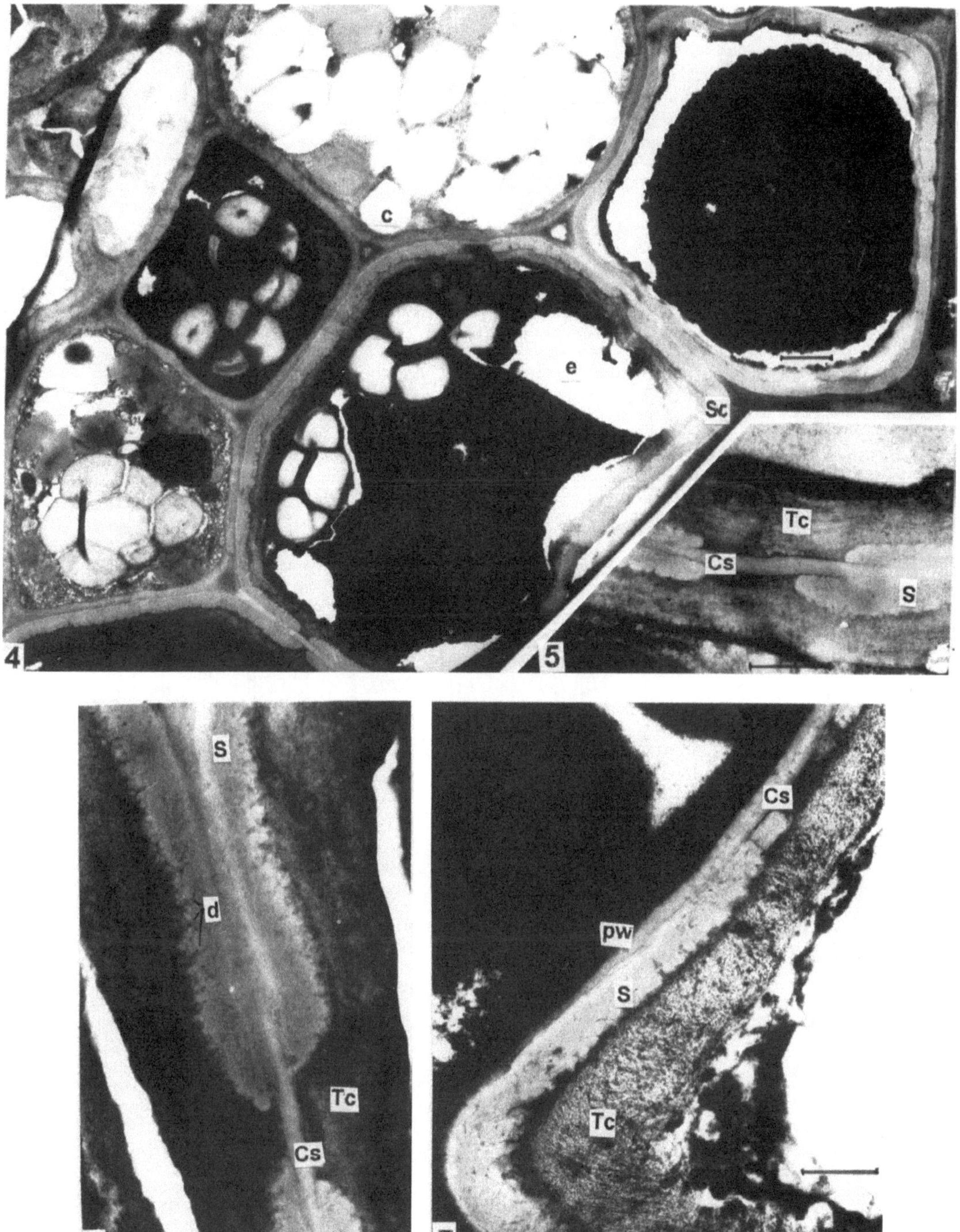

Figures 4–7.

Figure 4 Secondary and tertiary stages of endodermal cell development. Sc – secondary cell wall; Tc – tertiary cell wall. Scale bar = 2 μm.

Figure 5 Casparian strip between suberin-like secondary cell walls. S – suberin. Scale bar = 0,5 μm.

Figure 6 Secondary suberin-like cell wall with dendrites (d) on the interface between secondary and tertiary cell walls of the endodermis. Scale bar = 0.5 μm.

Figure 7 Casparian strip between suberized endodermal cell wall and 'passage' cell wall (pw). Scale bar = 0.5 μm.

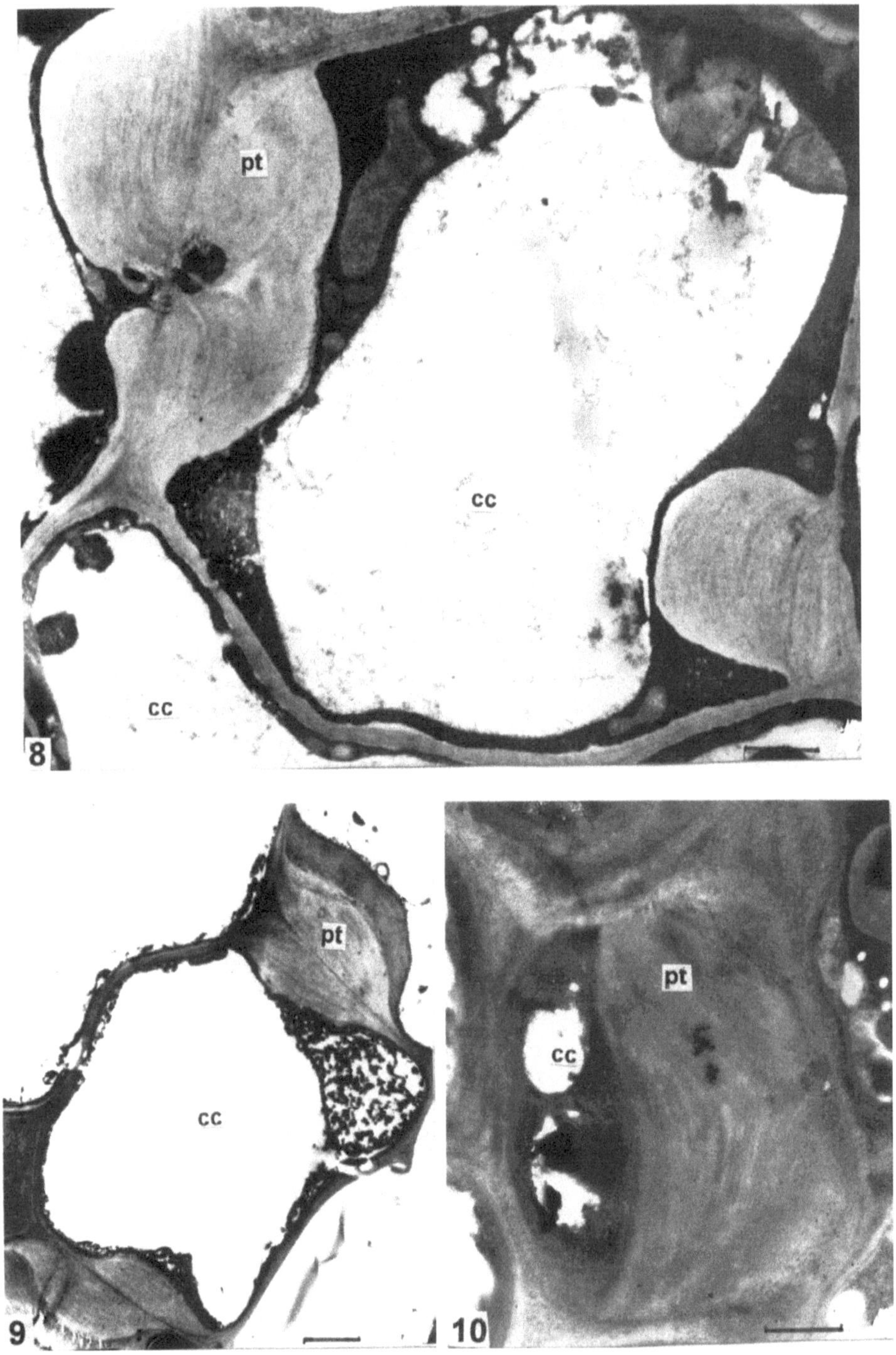

Figures 8–10.
Figure 8 Phi-thickenings at the onset of their development in cortical cells. Scale bar = 1 μm.
Figure 9 The *phi*-thickenings on the radial cell walls of cortical cells. Scale bar = 1 μm.
Figure 10 The final stage of *phi*-thickening development in cortical cells. Scale bar = 1 μm.

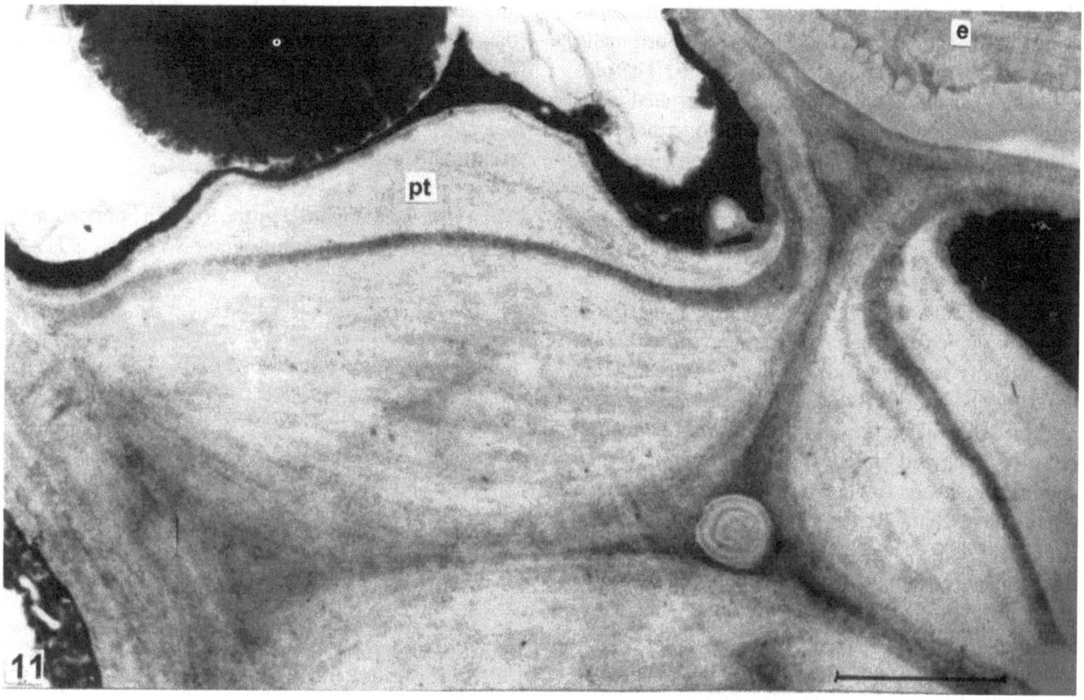

Figure 11. Structure of the *phi*-thickenings. Scale bar = 1 μm.

in *Dryas* root transverse section. Meristematic endodermal cells are characterized by the accumulation of tannins (Figure 2). At the first stage, homogeneous Casparian strips are usually developed in the inner third of radial endodermal cell walls (Figure 3). The Casparian strips are not stained by osmium tetroxide and appear electron-transparent. At the second stage, lipid, suberin-like layer is laid down on the inner side of the primary wall. We called this layer suberin-like layer because it is electron-translucent, however, lamellate structure characteristic of suberin (Kolattukudy, 1984) is not revealed. It appears more similar to the inner layer of cuticle in the epidermis of the aboveground organs (Miroslavov, 1980). This similarity is confirmed by the presence of the network of dendrites in the interface between the secondary and tertiary endodermal cell walls. However, the Casparian strips remain uncovered. At the third stage, tertiary cellulose layer (with lignin) was laid down on the inner side of the suberin-like layer. Tertiary walls are more electron-dense than the secondary cell walls. The second and third stages occured nearly simultaneously, and the tertiary cellulose cell wall came into contact with the Casparian strip itself (Figures 4–7). The presence of 'passage' cells without secondary and tertiary cell walls in the endodermis was noted. Earlier, MacKenzie (1979, 1983) described a similar structure of the endodermis in apple and lime roots. This author mentioned the presence of the second amorphous lobed layer on several endodermal cell walls (MacKenzie, 1979). The specific features of *Dryas* endodermis are: (1) the absence of typical suberin lamellae and the presence of dendrites in the secondary cell walls; (2) the direct contact of the Casparian strip with tertiary (polysaccharide) cell wall. The Casparian strip could be impregnated by the lipid (suberin-like) substance which later was laid down in the secondary cell walls. Usually at the secondary stage, the plasmalemma of the endodermal radial cell walls are separated from Casparian strip. According to our observation, this process in *Dryas* roots occured in tertiary stage. We suggest that all specific features of *Dryas* endodermis did not affect the barrier role of endodermis of apoplast transport.

Cortex

The uniseriate layer with *phi*-thickenings develops in cortex adjacent to the endodermis of *Dryas* roots

(Figure 1). These thickenings were formed at the junctions of radial and transverse cell walls and originally were more pronounced on the radial walls. The *phi*-thickenings encircle the cells in a manner analogous to that of the Casparian strips in the endodermal cells. The development of these thickenings coincides with the development of the secondary depositions in the endodermal cell walls. Incipient thickenings appeared as small swellings of the radial cortical cell walls (Figure 8); then these depositions became similar to the Greek character Φ (Figure 9). Finally, these thickenings occupied the major part of the cortical cells, leaving only a little space for the cytoplasm (Figure 10). Only the wall adjacent to the endodermis remained thin. Transverse sections of *phi*-thickenings revealed a layered structure with alternating lighter and darker layers (Figure 11), suggesting a helicoidal pattern characteristic of collenchyma (Vian and Roland, 1987). The chemical composition of *phi*-thickenings remains to be determined. Weerdenburg and Peterson (1983) discussed the functional role of similar *phi*-thickenings in cortical cells of apple roots and suggested that they serve as a major strengthening agent. We support their idea and also suppose that this cortical layer may prevent the penetration of mycorrhizal fungi into the central part of the root. *Phi*-thickenings differ from both the Casparian strips and the secondary and tertiary walls of the endodermis, and according to Weerdenburg and Peterson (1983) they cannot serve an additional physiological barrier for the transport of the ions.

Acknowledgements

This work was supported by the Russian Foundation of Fundamental Investigations, grant No 97-04-49712.

References

Haas D L, Z B Carothers and R R Robins 1976 Observations on the phi-thickenings and Casparian strips in *Pelargonium* roots. Am. J. Bot. 63, 863–867.

Kolattukudy P E 1984 Biochemistry and function of cutin and suberin. Can. J. Bot. 62, 2918–2933.

MacKenzie K A 1979 The development of the endodermis and phi layer of apple roots. Protoplasma 100, 21–32.

MacKenzie K A 1983 Some aspects of the development of the endodermis and cortex of *Tilia cordata* and *Picea sitchenensis*. Plant soil 71, 147–153.

Miroslavov E A 1980 Epidermis. *In* Atlas of Plant Tissue Ultrastructure. Eds. MF Danilova and GM Kozubov. pp 41–78. Karelia, Petrozavodsk (in Russian).

Vian Br and Roland J-C The helicoidal cell wall as a time register. New Phytol. 105, 345-357.

Weerdenburg C A and C A Peterson 1983 Structural changes in phi thickenings during primary and secondary growth in roots. 1. Apple (*Pyrus malus*) Rosaceae. Can. J. Bot. 61, 2570–2576.

Recent Advances of Plant Root Structure and Function, 55–61
Edited by O. Gašparíková, M. Čiamporová, I. Mistrík and F. Baluška
© 2001 *Kluwer Academic Publishers.*

IAA and ZR content in leek (*Allium porrum* L.), as influenced by P nutrition and arbuscular mycorrhizae, in relation to plant development

A. Torelli[1], A. Trotta[2*], L. Acerbi[2], G. Arcidiacono[1], G. Berta[3] and C. Branca[1]
[1]*Dipartimento di Biologia Evolutiva e Funzionale, viale delle Scienze, I-43100 Parma, Italy,* [2]*Dipartimento di Biologia Vegetale dell'Università di Torino, viale Mattioli 25, I-10125, Torino, Italy and* [3]*Dipartimento di Scienze e Tecnologie Avanzate dell'Università del Piemonte Orientale 'Amedeo Avogadro', corso Borsalino 54, I-15100 Alessandria, Italy.* **Corresponding author*

Key words: Arbuscular mycorrhizae, hormone balance, leek, P nutrition, root morphogenesis

Abstract

Leek plants (*Allium porrum* L.), infected or not with the arbuscular mycorrhizal (AM) fungus, *Glomus mosseae*, were grown in a sand-hydroponic system, fed with a nutrient solution containing 3.2 or 96 μM P and analyzed for root IAA and ZR content, to assess the role played by the fungus and P nutrition on host hormonal balance. IAA was analyzed by HPLC-fluorimetry, ZR by HPLC-UV coupled with a bioassay based on the expression of a phytohormone-regulated GUS reporter gene. Shoot and root weights and shoot FW-DW ratio enhancements, as well as root-to-shoot DW ratio decrement in mycorrhizal plants, were related to P nutrition. Shoot P concentration was increased by mycorrhizae at both P levels, but was comparable in AM plants grown at 3.2 μM P and non mycorrhizal (NM) plants at 96 μM P. Mycorrhizae and P increased IAA at substantially similar values, while P increased ZR much more than mycorrhizae did. These results are discussed in relation to root architecture modifications induced by the AM fungus.

Abbreviations: AM – arbuscular mycorrhizal, DW – dry weight, FW – fresh weight, IAA – indole-acetic acid, IPA – indole-propionic acid, MU – 4-methyl-umbelliferyl-glucuronide, NAA – naphthalene-acetic acid, NM – non mycorrhizal, P – phosphorus, ZR – zeatin riboside

Introduction

Plant development is influenced by several internal and environmental factors (Marschner, 1995). The latter include mycorrhizae, symbiotic associations between soil fungi and the roots of most land plants. This has long been known for ectomycorrhizae due to the obvious modifications they induce in the short roots of their hosts (Harley and Smith, 1983). Recent work has also shown that endomycorrhizae, mainly AM, influence plant development (Berta et al., 1993a). The best known effect of AM is the enhanced growth of AM plants in relatively infertile soils, due to more efficient absorption of low mobile minerals, mainly P (Smith and Gianinazzi-Pearson, 1988). Moreover their modifications of host root development are related to root size and architecture (Berta et al., 1993a; 1995, and references). Our previous results on leek plants infected with the AM fungus *Glomus* sp. strain E3 showed remarkable differences between NM and AM plants in root architecture when grown at low P nutrition. The latter had more numerous, shorter and more branched adventitious roots (Berta et al., 1990). Subsequent work on root apical meristem activity related these results to a higher percentage of inactive and necrotic apices, lower mitotic activity and lengthening of the mitotic cycle (Berta et al., 1991). Higher P nutrition resulted in NM plants with a root system substantially similar to that of mycorrhizal AM plants grown at low P nutrition (Trotta et al., 1991b). These effects may be due to improved P nutrition by the fungus and/or its more direct influence on root development. In both cases, changes in the level of plant hormones may be expected, since they mediate in-

* Tel: +39 11 670 7446/7/9. Fax No: + 39 11 670 7459. E-mail: atrotta@bioveg.unito.it

ternal and environmental signals in plant development (Davies, 1995). This work was aimed to ascertain the role played by the AM fungus and mineral nutrition on the host root hormone balance, starting with IAA and ZR, and to relate hormonal changes to the morphogenetic modifications to the host root induced by the endophyte (Berta et al., 1993a).

Materials and methods

Experimental cultures

Seeds of leek cv Mostruoso di Carentan were sown on quartz sand in 1.5 l plastic pots. Some pots were inoculated immediately before sowing with 3 g of roots infected with *G. mosseae* (Nicol and Gerd.) Gerdemann and Trappe (BEG 12). Root fragments were mixed with the top 0.2 l of sand. The cultures were raised in a growth chamber at 16/8 hours light/dark photoperiod, 24–20 ° C light-dark thermoperiod, 150 $\mu mol.m^{-2}.s^{-1}$ light irradiance at pot height (alternate cool white and day light 36 W Sylvania fluorescent lamps). On alternate days, plants were fed with a nutrient solution composed of: 1 *mM* K_2SO_4, 1 *mM* $NaNO_3$, 2 *mM* $CaCl_2$, 0.75 *mM* $MgSO_4$ and micronutrients, containing 3.2 μM P (low P level) or 96 μM P (high P level).

Seeds of transgenic *Nicotiana tabacum* cv Petit Avana Pg5-GUS, kindly supplied by Dr.Boerjan (University of Gent, Belgium), were sterilized and sown on MS medium. Plantlets were selected on Kanamycin 100mg/l and used for protoplast preparation.

Plant harvest

Leek plants were harvested two months after sowing and subjected to weight analysis. Roots were collected, weighed, immediately frozen in liquid nitrogen, and lyophilized before extraction for hormone analysis. Each sample (6–24 plants) was weighed again after lyophilization and dry tissue amounts corresponding to 5 g of fresh weight were used for each extraction. Shoots were oven dried to constant weight at 80 °C for 24–48 hours.

P analysis

Dry shoot material was wet-digested (Kuttner and Lichtenstein, 1932) and analyzed for total P content (Fogg and Wilkinson, 1958). Pulverized samples of 30–40 mg were digested on a microburner in 100 ml Kjeldahl flasks with 1 ml of 10 *M* H_2SO_4, to which 10–20 drops of 30% H_2O_2 was added. After neutralization and addition of an ammonium molybdate-sulfuric acid solution, samples were boiled for 1 min with 100 mg of ascorbic acid. After colour development, they were diluted to 50 ml, and the optical density determined with a Varian DMS 100 spectrophotometer at 660 nm. P content values were obtained from a 10–350 μg P calibration curve.

IAA and ZR analyses

General conditions

IAA analysis was carried out by HPLC coupled with fluorescence detection. ZR analysis employed HPLC in association with a bioassay based on expression of the *Agrobacterium tumefaciens* chimeric Pg-5 gene in protoplasts of transgenic *N. tabacum* cv Petit Avana (Boerjan et al., 1992). This chimeric gene, obtained by coupling the promoter of a phytohormone regulated gene to the coding sequence of the GUS reporter gene, is slightly inducible by cytokinin or auxin alone, but strongly activated when both are included in the medium.

Extraction

Extraction of IAA and ZR from leek roots was performed according to Maldiney et al. (1986) with slight modifications. Lyophilized roots were ground in liquid nitrogen and extracted with 80% methanol, 10 ml/g of fresh weight, containing butylhydroxytoluol 200 mg/l and a fixed amount of IPA as internal standard to reach a final concentration of 5×10^{-6} *M* in the resuspended organic phase. Only free IAA was measured and the internal stardard (IPA) was used to calculate for losses during extraction. Extraction was carried out by stirring overnight at 4 °C in the dark. Extracts were filtered on Whatman 3 MM, methanol concentration was adjusted to 50% and the pH was increased to 8 with 1 *M* K_2HPO_4. The samples were loaded on Sep-Pak C18 cartridge (Waters, USA) to remove part of the pigments. The eluate was evaporated to the aqueous phase under vacuum at 30 °C in a VR 1 vacuum concentrator (Heto Lab Equipment). The aqueous phase was adjusted to pH 2.8 with acetic acid and extracted with an equal volume of ethyl acetate. The aqueous and organic phases were separated and evaporated to dryness in a vacuum concentrator in the dark. The residues were resuspended in CH_3CN: H_2O 40: 60 pH 2.8 and analyzed by HPLC. Unless stated otherwise all the extraction steps were carried out in the dark or under dim red light to avoid photo-oxidation of IAA.

HPLC

The extracts were injected on to the HPLC column C18 (Kontron 322 HPLC system with a spherisorb ODS2 column from Kontron Instruments) and eluted with a discontinuous gradient of CH_3CN: H_2O adjusted to pH 2.8 with acetic acid (CH_3CN: H_2O 15:85 for the first 8 minutes, CH_3CN: H_2O 40:60 for the remaining 15 minutes). Column effluent was monitored with a Kontron 430 UV detector (λ 268 nm) to detect ZR and a Kontron SFM125 spectrofluorimetric detector (excitation 280 nm, emission 354 nm) to detect IAA (Crozier et al., 1980; Sandberg et al., 1985). Retention times for the different growth substances were: ZR 11.32 min, IAA 15.09 min, IPA 17.20 min, respectively. Peaks corresponding to the ZR retention time were collected from each phase of the initial extract, dried, redissolved in the bioassay culture medium, filter-sterilized and used in dilutions in the bioassays.

Protoplast preparation

N. tabacum seedlings were apex propagated on MS medium and used for protoplast preparation after 21–28 days. Protoplasts were prepared as described by Branca et al.(1993), except that they were cultured at a density of 2.5×10^5/ml.

Bioassay

Protoplasts were incubated in 2 ml culture medium containing the extracts, using 2.2 cm diameter, 12-well clusters (Costar). Aliquots of 2×10^{-6} *M* NAA were added to test for cytokinin activity. Since the protoplast batches might respond differently, a standard curve to interpolate activity values and quantitate cytokinin content in the biologically active fractions was constructed for each experiment on a ZR concentration range (8–9 doses between 1×10^{-11} *M* and 1×10^{-9} *M*) determined in preliminary experiments. Only experiments in which the standard curve showed a linearity coefficient > 0.98 were considered. After 72 h of culture at 26 °C in the dark, protoplasts were collected, transferred into 2 ml Eppendorf tubes and centrifuged at 1000 rpm (centrifuge ALC 21, rotor 6728 C) for 5 min. The supernatant was removed by aspiration with a vacuum pump. The protoplasts were resuspended in 300 μl of GUS extraction buffer (Jefferson, 1987), frozen in liquid nitrogen and sonicated with a Labsonic 1510 sonicator at 100 Watts. Each sample was supplemented with dithiothreitol 1×10^{-3} *M* and phenylmethyl-sulphonyl fluoride 1×10^{-4} *M*. After centrifugation at 12000xg, the protein concentration of the supernatant was measured using the Bio-Rad protein assay (Bio-Rad Laboratories GmbH, München, Germany). Fluorimetric GUS assay, using 12.5 μg of total proteins, was performed according to Jefferson (1987) using 4-methyl-umbelliferyl-glucuronide (Sigma) as substrate.

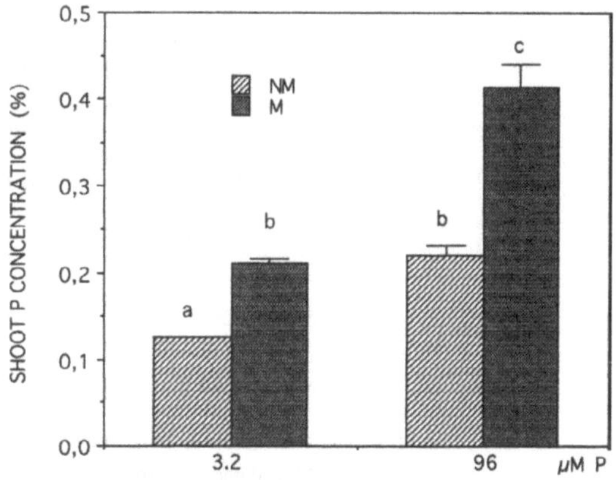

Figure 1. Shoot P concentration in mycorrhizal (M) and non-mycorrhizal (NM) plants grown at 3.2 *μM* or 96 *μM* P. Values followed by different letters are significantly different ($p \leq 0.05$).

Statistical analysis

Three replicates, each made of 6-24 plantlets, per measurement were analyzed. Data were compared by analysis of variance, with significance assumed $p = 0.05$. Standard errors were calculated for all data.

Results

Growth and P content analyses

Shoot FW increased with P applied and was greater in AM plants than in NM plants at both P levels (Table 1a). Shoot DW as well as root FW and DW increased in NM plants with P applied while the responses of AM plants at both P levels were substantially similar to those of NM plants at high P. Shoot FW-DW ratio increased with P applied and the response was higher in AM plants compared to the NM plants at each P level. Root FW-DW ratio was less increased by P applied. The response was similar in both treatments at low P and higher in NM plants than in AM plants at high P (Table 1b). Root-to-shoot DW ratio of NM plants at low P was higher compared to AM plants and decreased with P applied, while AM plants had similar root to shoot DW ratio at both P levels. Shoot P concentration was higher in AM plants than NM plants at each P level. Shoot P concentration

Table 1. *a.* Shoot and root fresh and dry weights of mycorrhizal (M) and non mycorrhizal (NM) plants grown at 3.2 μM or 96 μM P. Values followed by different letters are significantly different ($p \leq 0.05$)

		Shoot FW (mg)	Root FW (mg)	Shoot DW (mg)	Root DW (mg)
3.2 μM P	NM	0.175 ^{a}	0.300 ^{a}	34 ^{a}	19 ^{a}
	M	1.797 ^{b}	0.878 ^{b}	209 ^{b}	59 ^{b}
96 μM P	NM	1.269 ^{c}	0.914 ^{b}	170 ^{b}	44 ^{c}
	M	2.374 ^{d}	1.006 ^{b}	228 ^{b}	55 ^{bc}

b. Shoot and root fresh weight-dry weight ratio and root-to-shoot dry weight ratio in mycorrhizal (M) and non-mycorrhizal (NM) plants grown at 3.2 μM or 96 μM P. Values followed by different letters are significantly different ($p \leq 0.05$)

		Shoot FW-DW ratio	Root FW-DW ratio	Root-shoot DW ratio
3.2 μM P	NM	5.11 ^{a}	15.58 ^{a}	0.56 ^{a}
	M	8.69 ^{b}	14.98 ^{a}	0.28 ^{b}
96 μM P	NM	7.60 ^{b}	20.49 ^{b}	0.27 ^{b}
	M	10.43 ^{c}	18.23 ^{c}	0.24 ^{b}

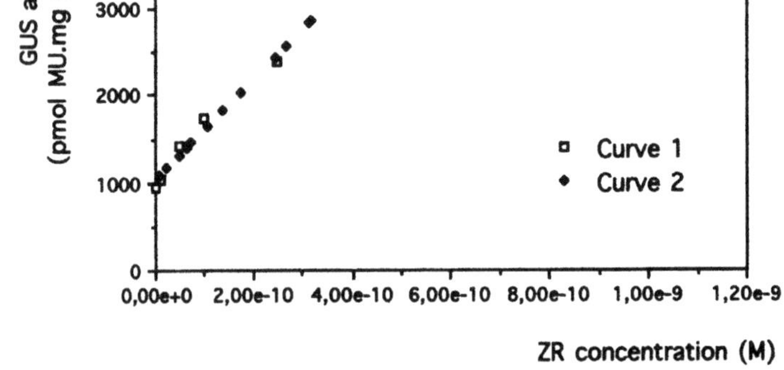

Figure 2. Dose response curve from a typical enzymatic assay. The induction curve (curve 1) was obtained culturing protoplasts with increasing doses of ZR in the presence of a fixed dose of auxin (NAA 2 × 10^{-6} *M*). Only the experimental points (curve 2) included within the linear part of the curve (linearity coefficient = 0.996) where interpolated.

in AM plants at low P was similar to that in NM plants at high P (Figure 1).

IAA and ZR analysis

HPLC analysis for ZR, monitored with an UV detector, showed a detection limit of about 1×10^{-5} *M*. The fluorimeter detected IAA up to 5×10^{-8} *M*. The bioassay improved sensitivity for ZR bv about 1×10^{6} times and the limit of sensitivity (ɒed to 1×10^{-11} *M*. The optimal ZR concentratioı .t can be accurately interpolated on a linear regression by the bioassay ranged between 5×10^{-11} and 5×10^{-10} *M* (Figure 2). At higher doses, the dose-response curve became flat. Sensitivity for IAA was sufficiently high with HPLC-fluorimetry and not improved by the bioassay. HPLC was therefore considered the final step for IAA and used only as a purification step for ZR.

Root IAA content increased with P applied in NM plants, while it did not vary appreciably in AM plants, which had values similar to those of NM plants at high P (Figure 3). Root ZR content increased with P applied in both treatments, but much more in NM plants than in AM plants: in particular AM plants had higher ZR content than NM plants at low P, but much lower ZR content than NM plants at high P (Figure 4).

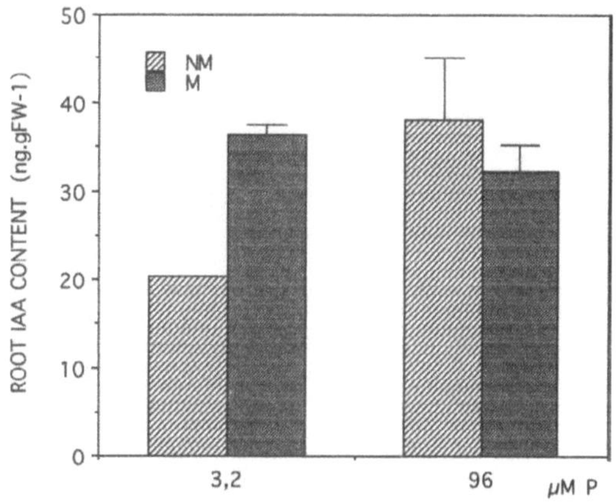

Figure 3. Root IAA content in mycorrhizal (M) and non-mycorrhizal (NM) plants grown at 3.2 μM or 96 μM P.

Discussion and conclusions

The main goal of this work was to relate hormonal changes to the root morphogenetic modifications previously described in our mycorrhizal system (Berta et al., 1993a). Therefore we chose to harvest leek plants two months after inoculation, the mean time required to observe significant differences in root development between NM and AM plants (Berta et al., 1990; Trotta et al., 1991b).

Growth analysis of leek plants showed that mycorrhizae increased root and shoot FW and DW, as well as the shoot FW-DW ratio when grown at low P. These differences disappeared at high P, when the growth of the controls approached that of AM plants. Root FW-DW ratios of AM and NM plants were substantially similar at both P levels. NM plants had a higher root-to-shoot DW ratio at low P, and approached that of AM plants at high P. The effect of mycorrhizae on growth can thus be related to P nutrition. This is particularly true for AM plants grown at low P and NM plants grown at high P, which are comparable in shoot P concentration.

Mycorrhizae and P increased root IAA levels at comparable values. The simplest explanation is that improved P nutrition creates a sink for photosynthate in the root and thus increases auxin delivery from the shoot with the photosynthate (Marschner, 1995). This interpretation is consistent with our previous results on root development (Berta et al., 1993a). Increased auxin delivery may: (i) lengthen the mitotic cycle in the root apical meristem, leading to the decrease of individual adventitious root length, since auxin is involved in the control of cell division cycle (Dudits et al., 1993); (ii) increase the adventitious root number, as new root production may be related to the inactivation or elimination of existing apices (Torrey, 1986); and (iii) enhance lateral root production, a well-known auxin effect (Charlton, 1996). Moreover, more recent data on the length of the mitotic cycle at high P nutrition show that lengthening of the cycle by P is comparable to that measured in AM plants (Berta et al., 1993b). However, other kinds of AM fungal influences on root development cannot be ruled out, since first and second-order lateral roots are shorter in AM plants, even at high P nutrition (Berta et al., 1993a; Trotta et al., 1991a). These results are not in agreement with those of Danneberg et al. (1992), who did not find alterations of IAA levels induced by the AM fungus during development in maize, from juvenility to reproductive stage. Similarly Ludwig-Müller et al. (1997) did not find changes of IAA levels in AM maize, though indole-butyric acid increased in the early stages of infection. However, plants of different species or groups display different, even contrasting,

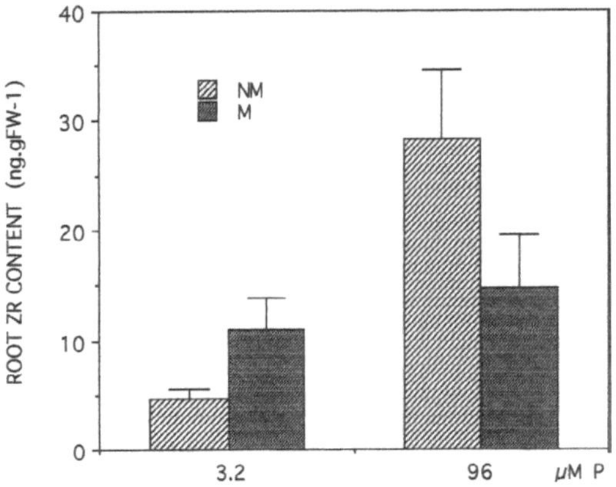

Figure 4. Root ZR content in mycorrhizal (M) and non-mycorrhizal (NM) plants grown at 3.2 μM or 96 μM P.

morphogenetic responses to AM infection (Berta et al., 1993a). This is especially true when we compare plants with coarse root systems like those of leek and plants with fine, highly branched root systems like those of maize. Therefore, it is not surprising to find different responses in hormonal balance to AM infection in these two plants.

ZR was increased in AM plants grown at low P, in agreement with several studied of cytokinin content in AM plants (Allen et al., 1980; Baas and Kuiper, 1989; Danneberg et al., 1992; Dixon et al., 1988; Drüge and Schönbeck, 1992; Edriss et al., 1984). Higher P nutrition enhanced root ZR content more than the fungus, pointing to a mycorrhizal effect, not mediated by P nutrition, on the host hormone balance. ZR is considered a transport form for cytokinins (Dixon et al., 1988) and hence may be involved in the root-to-shoot communication of environmental signals (Kuiper et al., 1988; 1989). According to the results of this group, root apices are very sensitive to the soil mineral nutrition status, especially nitrogen and phosphorus. Roots respond by modulating cytokinin production and/or export to the shoot and adjusting their development to the prevailing soil conditions. Our results on root cytokinin content are in agreement with those of Baas and Kuiper (1989), who found that P increased cytokinins more than mycorrhizae in *Plantago*. They suggested that the lower levels of cytokinins in AM plants at low P, compared with NM plants at high P, could be related to a lower number and/or activity of root apices, or to a direct stimulation of cytokinin synthesis in root tips, proportional to the P concentration in the soil solution. However, in this study AM plants had lower ZR levels than high P NM plants, even when grown at the same P level. Besides this, mycorrhizal leek plants had a higher number of root apices at low P, compared to NM plants, but a comparable number at high P (Fusconi et al., personal communication). Therefore, the results on ZR content cannot be explained merely on the basis of the number of production sites. Our previous results on leek apical meristems (Fusconi et al., 1994) however, showed a lower metabolic activity in root apices of AM plants. Consequently cytokinin levels could be the result of an interaction between number and activity of root tips, though the contribution of a direct stimulation of production by P cannot be dismissed. Obviously ZR cannot account for the whole root cytokinin pool, which comprises free bases and conjugated forms (Baas and Kuiper, 1989). Moreover, other plant hormones, such as abscisic acid and gibberellin, may regulate mycorrhizal symbioses (Allen et al., 1982; Danneberg et al., 1992; Esch et al., 1994). Plant hormone levels can be regulated by synthesis, degradation/inactivation, transport and compartmentation (Davies, 1995). Mycorrhizae may influence the host hormone balance by intervening in each of these processes and/or by sending hormone messages. There are some indications that AM fungi produce hormones (Barea and Azcon-Aguilar, 1982; Esch et al., 1994). Whether these substances are transferred to the plant, to regulate the symbiosis and/or some plant function, is unknown.

References

Allen M F, Moore T S Jr and Christensen M 1980 Phytormone changes in *Bouteloua gracilis* infected by vesicular-arbuscular mycorrhizae. I. Cytokinin increases in the host plant. Can. J. Bot. 58, 371–374.

Allen M F, Moore T S and Christensen M 1982 Phytohormone changes in *Bouteloua gracilis* infected by vesicular-arbuscular mycorrhizae. II. Altered levels of gibberellin-like substances and abscisic acid in the host plant. Can. J. Bot. 60, 468–471.

Baas R and Kuiper D 1989 Effects of vesicular-arbuscular mycorrhizal infection and phosphate on *Plantago major* ssp. *pleiosperma* in relation to internal cytokinin concentrations. Physiol. Plant. 76, 211–215.

Barea J M and Azcon-Aguilar C 1982 Production of plant growth regulating substances by the vesicular-arbuscular mycorrhizal fungus *Glomus mosseae*. Appl. Environ. Microbiol. 43, 810–813.

Berta G, Fusconi A, Trotta A and Scannerini S 1990 Morphogenetic modifications induced by the mycorrhizal fungus *Glomus* strain E3 in the root system of *Allium porrum* L. New Phytol. 114, 207–215.

Berta G, Tagliasacchi A M, Fusconi A, Gerlero D, Trotta A and Scannerini S 1991 The mitotic cycle in root apical meristems of

Allium porrum L. is controlled by the endomycorrhizal fungus *Glomus* sp. strain E3. Protoplasma 161, 12–16.

Berta G, Fusconi A and Trotta A 1993a VA mycorrhizal infection and the morphology and function of root systems. Environ. Exp. Bot. 33, 159–173.

Berta G, Fusconi A, Trotta A, Brazzaventre S, Scannerini S, Tagliasacchi A M 1993b Duration of the mitotic cycle in an endomycorrhizal system as related to phosphorus nutrition. Abstracts of the IV International Symposium 'Structure and Function of Roots'. p 18. Starà Lesnà, Slovakia, June 20–26.

Berta G, Trotta A, Fusconi A, Hooker J E, Munro M, Atkinson D, Giovannetti M, Marini S, Fortuna P, Tisserant B, Gianinazzi-Pearson V and Gianinazzi S 1995 Arbuscular mycorrhizal induced changes to plant growth and root system morphology in *Prunus cerasifera* L. Tree Physiol. 15, 281–293.

Boerjan W, Genetello C, Van Montagu M and Inzè D 1992 A new bioassay for auxins and cytokinins. Plant Physiol. 99, 1090–1098.

Branca C, Ricci A, Torelli A, Amorosi S, Gaetani E, Laureri C F, Vitto M, Bolchi A, Brunelli M and Ottonello S 1993 Non-radioactive detection of glucuronidase and chloramphenicol acetyltransferase activities in co-transformed protoplasts by HPLC. Plant Cell Reports 12, 361–365.

Charlton W A 1996 Lateral root initiation. *In* Plant Roots the Hidden Half. Eds. Y Waise, A Eshel and U Kafkafi. pp 149–173. Marcel Dekker, New York, USA.

Crozier A, Lofersky K, Zaerr J B and Morris R O 1980 Analysis of picogram quantities of indole-3-acetic acid by high performance liquid chromatography-fluorescence procedures. Planta 150, 366–370.

Danneberg G, Latus C, Zimmer W, Hundeshagen B, Schneider-Poetsch Hj and Bothe H 1992 Influence of vesicular-arbuscular mycorrhiza on phytohormone balances in maize (*Zea mays* L.). J. Plant Physiol. 141, 33–39.

Davies P J 1995 Plant Hormones – Physiology, Biochemistry and Molecular Biology. Kluwer Academic Publishers, Dordrecht, NE. 833 p.

Dixon R K, Garrett H E and Cox G S 1988 Cytokinins in the root pressure exudate of *Citrus jambhiri* Lush. colonized by vesicular-arbuscular mycorrhiza. Tree Physiol. 4, 9–18.

Druge U and Schonbeck F 1992 Effect of vesicular-arbuscular mycorrhizal infection on transpiration, photosynthesis and growth of flax (*Linum usitatissimum* L.) in relation to cytokinin levels. J. Plant Physiol. 141, 40–48.

Dudits D, Bögre L, Bakò L, Dedeaglu D, Magyar Z, Kapros T, Felföldi F and Györgyey J 1993 Key components of cell cycle control during auxin-induced cell division. *In* Molecular and Cell Biology of the Plant Cell Cycle. Eds. J C Ormrod and D Francis. pp 111–131. Kluwer Academic Publishers, Dordrecht, NE.

Edriss M H, Davis R M and Burger D W 1984 Influence of mycorrhizal fungi on cytokinin production in sour orange. J. Am. Soc. Hortic. Sci. 109, 587–590.

Esch H, Hundeshagen B, Schneider-Poetsch Hj, Bothe H 1994 Demonstration of abscisic acid in spores and hyphae of the arbuscular-mycorrhizal fungus *Glomus* and in the N_2-fixing cyanobacterium *Anabaena variabilis*. Plant Science 99, 9–16.

Fusconi A, Berta G, Tagliasacchi A M, Scannerini S, Trotta A, Gnavi E and De Padova S 1994 Root apical meristems of arbuscular mycorrhizae of *Allium porrum* L. Environ. Exp. Bot. 34, 181–193.

Fogg D N and Wilkinson N T 1958 The colorimetric determination of phosphorus. Analyst 83, 406–414.

Harley J L and Smith S E 1983 Mycorrhizal Symbiosis. Academic Press, London, UK. 483 p.

Jefferson R A 1987 Assaying chimeric genes in plants: the GUS gene fusion system. Plant Mol. Biol. 5, 387–405.

Kuiper D, Schuit J and Kuiper P J C 1988 Effects of internal and external cytokinin concentrations on root growth and shoot to root ratio on *Plantago major* ssp. *pleiosperma* at different nutrient conditions. Plant Soil 111, 231–236.

Kuiper D, Kuiper P J C, Lambers H, Schuit J and Staal M 1989 Cytokinin concentration in relation to mineral nutrition and benzyladenine treatment in *Plantago major* ssp. *pleiosperma*. Physiol. Plant. 75, 511–517.

Kuttner T and Lichtenstein L 1932 Microcolorimetric studies. III. Estimation of organically bound phosphorus. A system of analysis of phosphorus compounds in blood. J. Biol. Chem. 95, 661–670.

Ludwig-Müller J, Kaldorf M, Sutter E G, Epstein E 1997 Indole-3-butyric acid (IBA) is enhanced in young maize (*Zea mays* L.) roots colonized with the arbuscular mycorrhizal fungus *Glomus intraradices*. Plant Science 125, 153–162.

Maldiney R, Leroux B, Sabbagh I, Sotta B, Sossountzov L and Miginiac E 1986 A biotin-avidin-based enzyme immunoassay to quantify three phytohormones: auxin, abscisic acid and zeatin riboside. J. Immunol. Methods 90, 151–158.

Marschner H 1995 Mineral Nutrition of Higher Plants. Academic Press, London, UK. 889 p.

Sandberg G, Ljung K and Alm P 1985 Precision and accuracy of radio immunoassays in the analysis of endogenous 3-indole acetic acid from needles of scots pine. Phytochemistry 24, 1439–1442.

Smith S E and Gianinazzi-Pearson V 1988 Physiological interactions between symbionts in vesicular-arbuscular mycorrhizal plants. Annu. Rev. Plant Physiol. Plant Mol. Biol. 39, 221–244.

Torrey J G 1986 Endogenous and exogenous influences on the regulation of lateral root formation. *In* New Root Formation in Plants and Cuttings. Ed. M B Jackson. pp 31–66. Martinus Nijhoff, Dordrecht, NE.

Trotta A, Berta G, Fusconi A and Scannerini S 1991a Root development in a VA mycorrhiza as related to phosphorus nutrition. Abstracts of III ISRR Symposium 'Root Ecology and its Practical Application'. p 134. Vienna, 2–6 September.

Trotta A, Carminati C, Schellenbaum L, Scannerini S, Fusconi A and Berta g 1991b Correlation between root morphogenesis, VA mycorrhizal infection and phosphorus nutrition. *In* Plant Roots and their Environment. Eds. B L McMichael and H Persson. pp 333–339. Elsevier Science Publishers, Amsterdam, NE.

Section editor: J H Graham

Recent Advances of Plant Root Structure and Function, 63–70
Edited by O. Gašparíková, M. Čiamporová, I. Mistrík and F. Baluška
© 2001 Kluwer Academic Publishers.

Polyploidy in tomato roots as affected by arbuscular mycorrhizal colonization

G. Berta[1]*, A. Fusconi[2], S. Sampò[1], G. Lingua[1], S. Perticone[1] and O. Repetto[1]
[1]*Dipartimento di Scienze e Tecnologie Avanzate, Università del Piemonte Orientale 'A Avogadro', Corso Borsalino 54, I-15100 Alessandria, Italy;* [2]*Dipartimento di Biologia Vegetale dell'Università di Torino, Viale Mattioli 25, I-10125 Torino, Italy*

Key words: flow and static cytometry, *Glomus mosseae*, *Lycopersicon esculentum*, polyploidization, respiratory activity

Abstract

Nuclear changes in roots of tomato (*Lycopersicon esculentum*), a plant with a small genome, during the establishment of arbuscular mycorrhizal (AM) colonization were studied using light and electron microscopy, as well as flow and static cytometry. Nuclei of mycorrhizal root cortex cells were larger and had more decondensed chromatin than those of controls. Significant ploidy distribution differences were observed between nuclei of AM colonized and control roots, and a strong correlation between nuclear polyploidization and AM colonization was found. Polyploidization and decondensation are usually associated with high metabolic activity. The metabolic activity of mycorrhizal root cells, evaluated in this work as respiratory activity by using a cytochemical assay for succinate dehydrogenase combined with image analysis, increased in comparison to controls. The meaning of polyploidization is discussed in relation to the structural and metabolic modifications induced by mycorrhization.

Abbreviations: AM – arbuscular mycorrhizae; DAPI – 4′6-diamidino-2-phenylindole; MT – microtubule; NBT – nitrobluetetrazolium; PI – propidium iodide; SDH – succinate dehydrogenase

Introduction

Mycorrhizal symbiosis is a close association between the roots of most land plants and many soil fungi, from which both partners benefit; arbuscular mycorrhizae (AM) are the most common and a universal model for analysis of the molecular mechanisms controlling a symbiotic state and fungal-root interactions. Many data are available on the cellular and physiological compatibility between the partners (see Gianinazzi-Pearson, 1996; Koide and Schreiner, 1992 and references therein) and the modifications that occur during the establishment of the symbiosis (Bonfante-Fasolo and Perotto 1995). Arbuscule formation causes invagination and proliferation of the host cell membrane, and cells with developing arbuscules are very active in the synthesis of new components. Modifications induced in arbuscule-containing cells include deposition of wall material around the hyphae during their intracellular growth, decrease in vacuolar size as the relative cytosol volume increases and proliferation of cell organelles (mitochondria, plastids, dictyosomes and endoplasmic reticulum).

The structural and quantitative changes of the host cell nuclei, however, and their involvement in host gene expression (Berta and Fusconi, 1998) have received little attention. In *Allium porrum* most cortical cell nuclei are round and centrally placed in arbuscule colonized root cells, whilst they are lenticular and lateral in uninfected roots (Balestrini et al., 1992). The presence of the fungus results in significant hypertrophy in host cell nuclei in various AM systems (Berta and Fusconi, 1998). In *Lotus pedunculatus*, *L. corniculatus* and *Trifolium pratense* + *G. versiforme* (Blair et al., 1988), *A. porrum* + *Glomus* E3 (Berta et al., 1990) and *Pisum sativum* + *G. mosseae* (Sgorbati et al., 1993), this effect cannot be explained by polyploidization, as small if any quantitative variations are

* FAX No: +39 0131 254410. E-mail:berta@al.unipmn.it

found. In *A. porrum* and *P. sativum* it has been related to chromatin decondensation (Berta et al., 1990; Sgorbati et al, 1993).

Tomato (*Lycopersicon esculentum*) is a plant with a small genome (1.5–2 pg of DNA) (Bennet and Leich, 1995). Preliminary results indicate that it is a multiploid species with 2C, 4C and 8C ploidy levels in its roots; both differentiation and colonization by symbiotic or pathogen fungi induce variations in the percentages of nucleus populations in its root cells (Fusconi et al., 1997). Nuclear ploidy is linked to cell size (De Rocher et al., 1990; Valente et al., 1998) and metabolic activity, as the increased availability of nuclear templates enhances the amount and the rate of RNA synthesis (D'Amato, 1998). It is thus probably related to the respiratory activity.

In this paper we investigate the chromatin structure and quantitative changes in tomato root cell nuclei following AM colonization, by light and transmission electron microscopy, and flow and static cytometry. Rapid evaluation of the DNA of many isolated nuclei by flow cytometry increased the reliability of sampling, and static cytometry on sections was used to quantify the DNA content of single and localized nuclei. We also quantified the activity of the mitochondrial enzyme succinate dehydrogenase (SDH) by a combination of cytochemical assay and image analysis, as an increase in metabolic activity is related to an increase in respiratory activity.

Materials and methods

Plant growth and AM infection

Seeds of *Lycopersicon esculentum* L. cv. Early mech were germinated in Petri dishes on moist filter paper for 3–4 days. The seedlings were then transferred to quartz sand (0.5 mm coarse grade) in 1.5 L polyethylene pots.

Mycorrhizal treatments were inoculated on transplanting with chopped roots of leek (*Allium porrum* L.) colonized by the AM fungus *Glomus mosseae* (Nicol and Gerd) Gerdmann and Trappe (BEG 12).

The pots were placed in growth chambers, 16/8 hours light/dark photoperiod, 24/22 °C light/dark thermoperiod, 12Klux (130 mmol m^{-2} s^{-1}) light irradiance at bench height and watered on alternate days, with 100 mL of half-strength Long Ashton nutrient solution, phosphorus 32 μM according to Trotta et al. (1996).

In mycorrhizal plants, the percentage of root colonization (M%) was calculated according to Trouvelot et al. (1986) using 50 root fragments (about 1 cm) per plant. Only plants with M% values higher than 40% were used for further analyses.

Light and transmission electron microscopy

Segments of adventitious roots 6 cm from the apex excised from 5 plants per treatment were fixed in 3% glutaraldehyde in a 0.1 *M* phosphate buffer (PH 7) for 3 h at room temperature, rinsed postfixed in 1% O_sO_4 in the same buffer for 2 h at room temperature, dehydrated and embedded in Spurr medium. Semithin sections were prepared on a Reichert-Jung Ultramicrotome and stained with 1% toluidine blue in 1% sodium tetraborate to determine the mean nucleus diameter and locate nuclei for further electron microscopy. Nucleus sizes were measured with an ocular micrometer scale on about 100 nuclei per plants.

Thin sections were stained with uranyl acetate and lead citrate and examined in a Philips CM 10 electron microscope. Nuclear morphology was examined for both treatments.

Metabolic activity

Five complete root systems per treatment were immersed in a reaction mixture containing 4 mg/ml nitrobluetetrazolium (NBT), 1.25 *M* sodium succinate and 0.2 *M* Tris HCl to determine the SDH activity (and therefore the respiratory activity) according to Tisserant et al. (1993). In the presence of the reducing agent $FADH_2$, NBT is reduced to blue formazan crystals visible by light microscopy. The degree of viability of zones 6 cm away from the apices of adventitious roots deprived of the central cylinder was evaluated by image analysis: at least 20 images in each treatment were digitized with a Scion PCI system (Scion Corp: Frederick, Maryland) using a Liege MTI camera. Digitized images were analyzed by NIH Image software and the percentages of coloured areas were determined.

Flow cytometry

Nuclei of three 60-day plants per treatment were extracted from complete root systems without apices.

The roots were chopped with a razor blade in a few drops of extraction buffer (0.1 *M* citrate, 0.5 % Tween 20). Nuclei were extracted after 30 min incubation in this buffer by filtering through a 20 μm nylon mesh and fixed adding two volumes of ice cold ethyl alcohol – acedic acid (3:1) to one volume of nucleus suspension. Two milliliters of each fixed sample were centrifuged on a 1 *M* sucrose cushion, in order to eliminate debris, the small fungal nuclei and the fixative. The pellet, containing the plant nuclei, was resuspended in PBS (phosphate buffered saline, 0.1 *M*, pH 7.2). The samples were microscopically checked and the nucleus concentration was adjusted with PBS to 10^5 mL^{-1}. They were then stained with the DNA specific dye 4′, 6-diamidino-2-phenylindole (DAPI, Sigma) at a saturating dye concentration (5.6 μM) established from saturating curves. Polyscience beads (4.2 μm) were used as a standard. A mercury arc lamp was used in conjunction with BG1, UG1 filters and a TK420 dichroic mirror to select the exciting wavelength (UV) of a Partec PAS III instrument. DAPI fluorescence was detected with a GG415 barrier filter (all filters by Partec GmbH, Münster, Germany). About 3000 nuclei were measured per plant.

Static cytometry

Segments of adventitious roots 6 cm from the apex of three 60-day plants per treatment were fixed in ethanol:acetic acid (3:1) and gradually infiltrated with 1.5 *M* saccharose. Thick sections (15 μm) obtained with a cryomicrotome (Microm, Heidelberg), were stained with DAPI and dissolved in phosphate buffer (0.1 *M*, pH 7), to a final saturating concentration (5.6 μM).

The fluorescence intensity of 100 cortex nuclei stained per plant was measured with a Leitz-Wetzlar Dialux 22/22B microscope equipped with a MPV compact photometer and a Leitz Ploemopak incident light illuminator, employing a 100W mercury lamp, excitation filter BP 340–380, dichroic mirror RKP 4000 and barrier filter LP 430. The photomultiplier was operated at constant voltage (500 V); measurement time was 0.25 s, the field diaphragm was 0.2 mm in diameter, and a FL 100/132 objective was used. The output was recorded on a scale of 0–1000 mV, with all accuracy of 0.1 mV. The background output of the immediately adjacent field was subtracted from that of the nucleus.

An internal biological standard was used to check instrumental and staining variability. Nuclei of chicken erythrocytes prepared according to Sgorbati et al. (1993) were squashed on each slide near the sections before staining.

Statistical analyses

All data were compared by analysis of variance (ANOVA) with $P<0.05$ and $P<0.001$ as statistically significant and highly significant.

Results

Cortex cell nuclei were generally morphologically active, i.e. their structure resembled that of meristematic cells. Only a small percentage of the nuclei was degenerated by chromatolysis (gradual loss of stainable content) or pyknosis (formation of very dense masses of chromatin) (Figure 1). Active nuclei displayed differences in structure in the two treatments. In uninfected cells they were lenticular or round with coarse and numerous patches of peripheral chromatin, the chromocentres, near the nuclear envelope (Figure 1a). Whereas, in arbuscule colonized cells they were irregularly lobed, with very decondensed and homogeneous chromatin and rare chromocentres (Figure 1d). Active nuclei were significantly larger in mycorrhizae whereas chromatolytic and pyknotic nuclei did not differ significantly in size (Table 1). Chromatolytic nuclei were round and swollen, and significantly larger than the other types of nuclei, with chromatin fibrils dispersed in the nucleoplasm (Figure 1b). Pyknotic nuclei displayed very dense masses of chromatin (Figure 1c) and small diameters.

The metabolic activity of cortex cells was evaluated cytochemically by quantification of SDH activity, The zones coloured by blue formazan crystals corresponded closely to the cytoplasmic areas. In controls, where most or the cell volume was occupied by a large central vacuole, the crystals were mainly located around the nucleus (Figure 2a). Whereas in the arbuscule host cells, they were spread all over because the cytoplasm volume increased and the central vacuole fragmented following colonization (Figure 2b). Since the SDH activity of the cortex cells could not be distinguished from that of the fungus, only contiguous cortical zones not directly colonized were considered for quantification. The metabolic activity of cortex cells was greater in mycorrhizae: the cell area coloured by NBT was 5.897±0.267% and 1.689±0.171% in the controls.

Flow cytometry of the extracted nuclei gave characteristic fluorescence peaks closely corresponding to base two arithmetic progression. The lowest peak was recognized as corresponding to 2C nuclei, relying on

Table 1. Mean diameter (μ) of nuclei in control (C) and mycorrhizal (M) roots

Nucleus type	C	M	
Active	7.33±0.11	9.42±0.29	p<0.0001
Chromatolytic	10.08±0.46	10.40±0.33	p=0.5167
Pyknotic	6.05±0.29	4.93±0.18	p=0.0015

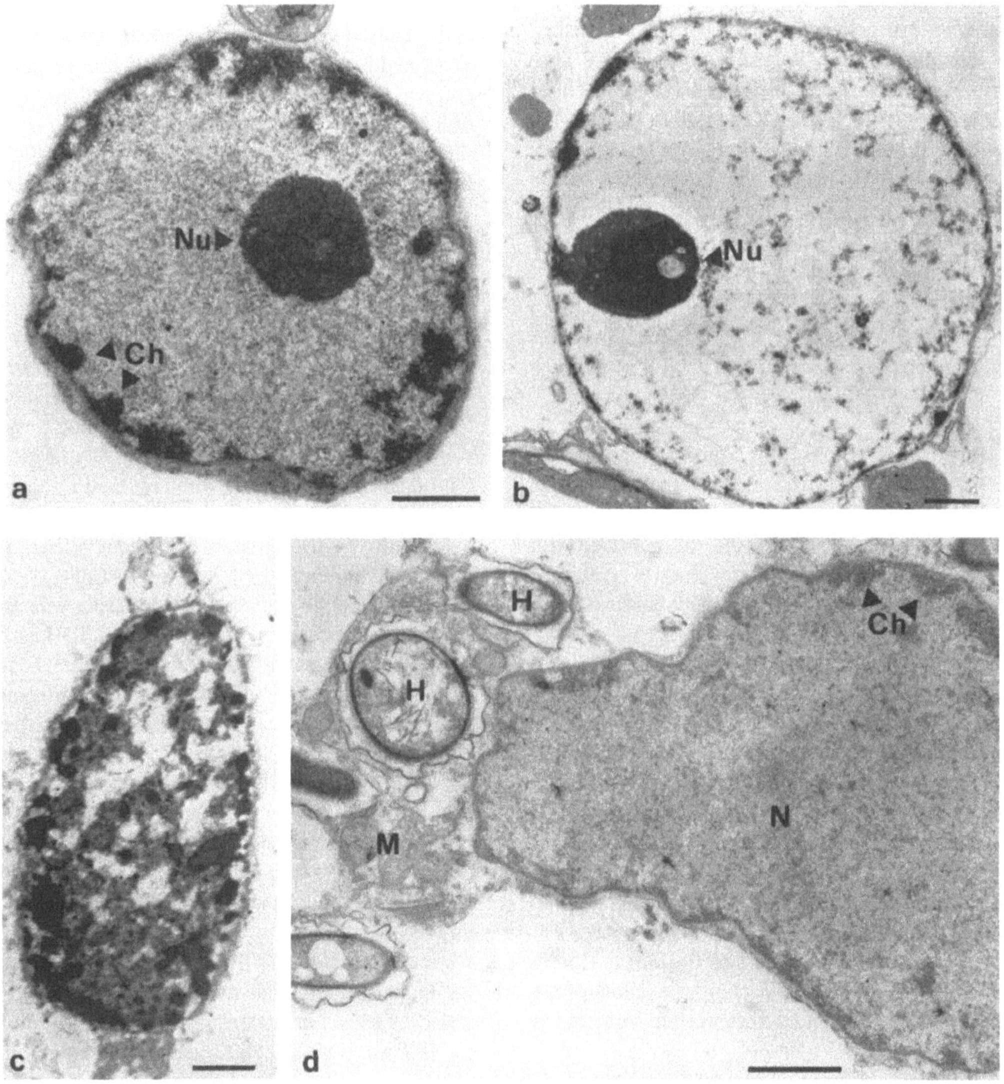

Figure 1. Transmission electron micrographs of cortex cell nuclei of tomato roots. (a) Active control nucleus showing large patches of peripheral chromatin (chromocentres); (b) chromatolytic nucleus with swollen and fragmented chromatin; (c) pyknotic nucleus with dark and clustered chromatin; (c) nucleus of an arbuscule-hosting cell, the chromatin is decondensed and shows few chromocentres. A mitochondrion is visible. Ch, chromocentres; H, arbuscular hyphae; M, mitochondria; N, nucleus; Nu, nucleolus. Bars = 1 μm.

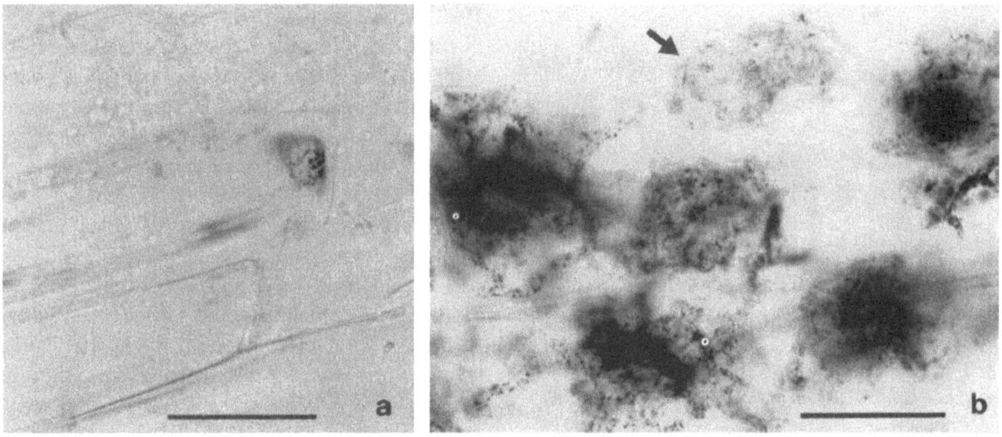

Figure 2. Light micrographs of root cortex cells of tomato plants, after the cytochemical reaction for SDH activity. (a) Controls; the formazan crystals are mainly located around the nucleus. (b) Mycorrhizae; the crystals are spread all over the arbuscule hosting cells. Arrow indicates a cortical cell not directly colonised. Bars= 50 μ.

previous data (Lingua et al., 1996), where human lymphocytes were used as internal biological standard. The others were taken to represent 4C and 8C nuclei. Highly significant differences were found in 4C nucleus populations ($P<0.001$) and significant differences in 2C and 8C nucleus populations ($P < 0.05$) between controls an and mycorrhizae. The 4C populations were higher in controls, whilst the 2C and the 8C populations were higher in mycorrhizae (Figure 3b).

Static cytometry was used to measure the DNA content of cortex cells only. In control cells, only 2C and 4C populations were found, with prevalence of 4C; in presence of the fungus, a high percentage of 8C nuclei and a reduction of 4C and 2C were found (Figures 3d, e, f). All the percentage differences were significant (P<0.05). The increase of 8C nuclei in mycorrhizae was higher in the arbuscule colonized cells, than in the indirectly colonized cells whereas the 2C population was almost absent (Figure 3e, f).

Discussion

These results show that tomato nuclei are highly modified by AM colonization. Active nuclei of arbuscule colonized cells were hypertrophic and their chromatin was more decondensed with less chromocentres than in controls, they were neither chromatolytic or pyknotic. These findings were consistent with a previous flow cytometric study of nuclei stained with propidium iodide (P1), showing that the mean fluorescence intensity of the prevalent nucleus population, 4C, is lower in controls (Lingua et al., 1996). Since PI is an intercalating dye, highly decondensed chromatin binds more PI molecules and is more fluorescent than condensed chromatin. Nuclear hypertrophy is common in AM root cortex cells and has been investigated in detail *Allium porrum* (Berta et al., 1990) a plant with a high nuclear DNA content (about 45 pg) (Berta, unpublished results). This hypertrophy mainly depends on decondensation of the chromatin, as demonstrated by electron microscopy and by accessibility to DAPI used at less than saturating concentration. Due to their lower degree of chromatin condensation, allowing a greater availability of the DNA binding regions (i.e. the A-T zones) to the dye, AM cortex nuclei were more fluorescent than control nuclei (Berta et al., 1990). Chromatin decondensation can be attributed to increased transcription and metabolic activities in AM colonised cells (Berta and Fusconi, 1998).

However in tomato, a multiploid species (Fusconi et al., 1997), like other plants with a small genome, such as *Arabidopsis thaliana* (Galbraith et al, 1991) and succulent plants, such as *Mesembryanthemum crystallinum* (De Rocher et al., 1990), nuclear hypertrophy could also be explained by polyploidization.

In mycorrhizal roots the proportion of nucleus populations with different ploidy was modified in comparison to the control with an increase of 2C and 8C nuclei. The former increase could be related to variations in root morphogenesis, e.g. in mycorrhizal plants larger roots and consequently larger cribro-vascular cambium tissues have been detected. These meristems are generally formed by cycling diploid cells (D'Amato,

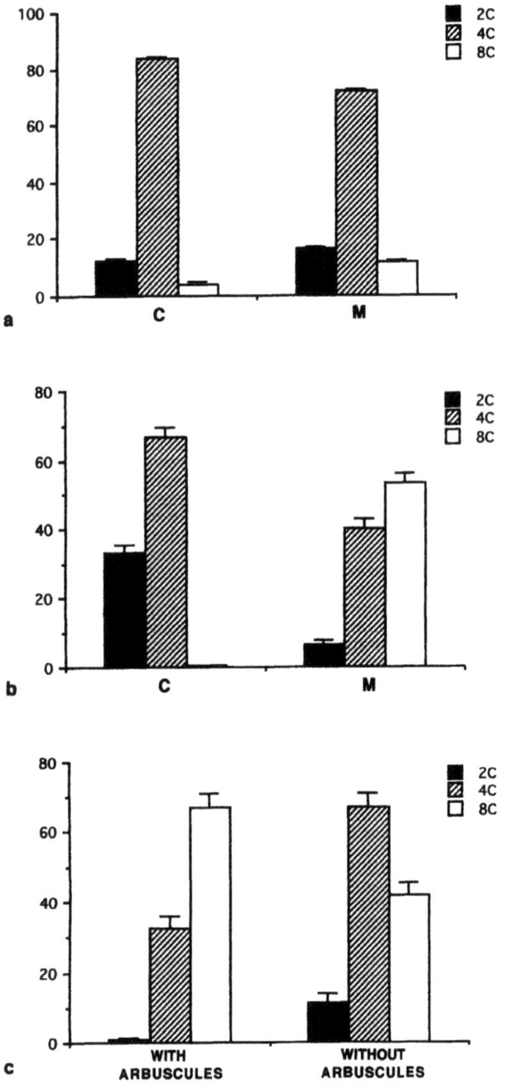

Figure 3. Histograms showing the ploidy distribution of nuclei in control (C) and mycorrhizal (M) tomato roots. (a) Flow cytometry; each graph shows three populations of nuclei corresponding to 2C, 4C and 8C content; mycorrhization caused an increase of the 8C population. (b) Static cytometry of cortex cell nuclei; the population of 8C nuclei is high in mycorrhizae (M) whilst it is absent in controls (C). (c) Ploidy distribution in root cortex cells of mycorrhizal roots assessed by static cytometry; cells hosting arbuscules show a higher percentage of 8C nuclei than non-hosting cells.

1998). This situation is different from that observed in pathogen infected roots where the number of 2C nuclei also increases, but a stop in endoreduplication cycle has been suggested. The increase of 8C nuclei can be explained with a more direct influence of the AM fungus as the arbuscule-containing cells have a high pecentage of 8C nuclei.

Endopolyploidy is a widespread phenomenon in plants, particularly the angiosperms (D'Amato, 1998). It is developmentally regulated according to the age of tissues and their position within the plant (De Rocher et al., 1990; Galbraith et al., 1991). Some investigations suggest that increases in ploidy level are related to increases in cell size (Galbraith et al., 1991; Melaragno et al., 1993) and to high metabolic activity: plant cells with small genomes may increase the number of their 'useful' genes more simply by redoubling the genome itself rather than selective gene amplification (Galbraith et al., 1991). It has been suggested that a given mass of DNA is required both to maintain specific regulatory and functional states and for correct interactions of nuclear and organellar genomes (Galbraith et al., 1991). During plant cell development, the number of mitochondria and plastids per cell increases, particularly in arbuscule colonized cells (Gianinazzi-Pearson et al., 1996). This agrees with our observation of increased respiratory activity of mycorrhizal roots, as has also been shown by the enhanced transcription of another mitochondrial enzyme, isocitrate dehydrogenase (Tahiri-Alaoui, personal communication).

A second effect or polyploidization is regulation of gene expression. Replication disrupts both active and repressed chromatin structure and the entire nucleus can then be remodelled after each replication event (Baluska et al., 1995; Wolfe, 1991).

We conclude that in the root cortex cells of tomato decondensed structure of the chromatin and increased ploidy may be complementary means by which the cells increase and modify their metabolic activity following mycorrhization. The mechanisms that trigger polyploidization remain to be established.

Poyploidization of arbuscule colonized cells could be favoured by microtubule (MT) reorganization: the MT cyoskeleton connects postmitotic plant nuclei with the cell walls and its disorganization has been reported to trigger enhancement of the G1/S transition and the subsequent initiation of DNA synthesis (Baluska et al., 1995). Drastic MT reorganization, linking arbuscular branches to each other, to the host cell cortex and to the nucleus, has been observed in arbuscule-containing cells of *Nicotiana tabacum* (Genre and Bonfante, 1998). Our observation of the nuclear hypertrophy and the lobed shape of nuclei of arbuscule colonized cells provides further evidence of this reorganization.

Our results may be related to variations in nutrition factors (Nagl, 1995) and the hormonal balance that follow AM colonization (Torelli et al., in press, and references therein). Endogenous auxin and cytokinin levels influence the synthesis and activation of the protein kinase p34^{odc2}, which is probably involved in the control of endoreduplication (Nagl, 1995). In single cells of *Nicotiana tabacum*, an auxin-only signal induces increase in the cell volume and DNA endoreduplication and a subsequent cytokinine+auxin signal induces cell division and DNA deduplication (Valente et al., 1998).

Lastly, polyploidization could be related to enhanced transcription in AM tomato roots of a recently isolated gene that codes for a protein belonging to the new family of the cullins and is probably involved in control of the cell cycle (Tahiri-Alaoui and Antoniw, 1997). Comparison with other genes of this family suggests that it could be involved in promoting the synthetic phase of the cell cycle (Lisztwan et al., 1998).

Acknowledgements

This work was supported by an E.U. grant (ERBIC18CT970180) and, partly, by Italian MURST

References

Balestrini R, Berta G and Bonfante-Fasolo P 1992 The plant nucleus in mycorrhizal roots: positional and structural modifications. Biol. Cell. 75, 235–243.

Baluska F, Bacigálová K, Oud J L, Hauskrecht M and Kubica S 1995 Rapid reorganization of microtubular cytoskeleton accompanies early changes in nuclear ploidy and chromatin structure in posmitotic cells of barley leaves infected with powdery mildew. Protoplasma 185, 140–151.

Bennet M D and Leich I J 1995 Nuclear DNA amounts in angiosperms. Ann. Bot. 76, 113–176.

Berta G and Fusconi A 1998 Effects of arbuscular mycorrhizal and ericoid fungi on the structure and activity of host cell nuclei. *In* Plant cell proliferation and its regulation in growth and development. Eds J A Bryant and D Chiatante. pp 213–217. J Wiley & Sons. Chichester.

Berta G, Sgorbati S, Soler V, Fusconi A, Trotta A, Citterio A, Bottone M G, Sparvoli B and Scannerini S 1990 Variations in chromatin structure in host nuclei of a vesicular arbuscular mycorrhiza. New Phytol. 114, 199–205.

Blair D A, Peterson R L and Bowley S R 1988 Nuclear DNA content in root cells of *Lotus* and *Trifolium* colonized by the VAM fungus *Glomus versiforme* New Phytol. 109, 167–170.

Bonfante-Fasolo P and Perotto S 1995 Strategies of arbuscular mycorrhizal fungi when infecting host plants. New Phytol. 130, 3–21.

D'Amato F 1998 Chromosome endoreduplication in plant tissue development and function. *In* Plant cell proliferation and its regulation in growth and development, Eds Bryant J A and Chiatante D, pp 153–166. J Wiley & Sons, Chichester.

De Rocher B J, Harkins K R, Galbraith D W and Bohnen H J 1990 Developmentally regulated systemic endopolyploidy in succulent plants with small genomes. Science 250, 99–101.

Fusconi A, Sampó S, Lingua G, Gnavi B and Berta G 1997 Ploidy in tomato roots and its variations induced by *Glomus mosseae* and *Phytophthora nicotianae*. Atti dell'Accademia dei Fisiocritci Tomo XXVI Serie XV: pp 55–58.

Galbraith D W, Harkins K R and Knapp S 1991 Systemic endopolyploidy in *Arabidopsis thaliana*. Plant Physiol. 96, 985–989.

Genre A and Bonfante P 1998 Actin versus tubulin configuration in arbuscule-containing cells from mycorrhizal tobacco roots. New Phytol. 140, 745–752.

Gianinazzi-Pearson V 1996 Plant cell responses to arbuscular mycorrhizal fungi: getting to the roots of symbiosis, Plant Cell 8, 1871–1883.

Gianinazzi-Pearson V, Gollotte A, Cordier C and Gianinazzi S 1996 Root defence responses in relation to cell and tissue invasion by symbiotic microorganisms: cytological investigations. *In* Histology, ultrastructure and molecular cytology of plant- microorganism interactions. Eds M Nicole and V Gianinazzi-Pearson. pp 177–191. Kluwer Academic Publishers, Dordrecht.

Koide R T and Schreiner R P 1992 Regulation of the vescicular-arbuscular mycorrhizal symbiosis. Annu. Rev. Plant Physiol 43. 557–581.

Lingua G, Varese O C, Trotta A, Fusconi A, Gnavi E and Berta G 1996 Flow cytometry as a tool to investigate nuclear senescence in symbiotic and pathogenic systems. Agronomie 16, 663–670.

Lisztwan J, Marty A, Sutterlüty H, Gstaiger M, Wirbelauer C and Krek W 1998. Association of human CUL-1 and ubiquitin-conjugating enzyme CDC 34 with the F-box protein p45SKP2: evidence for evolutional conservation in the subunit composition of the CUC 34-SCF pathway. EMBO J. 17, 368–383.

Melaragno E J, Mehrotra B and Coleman A W 1993 Relationship between endopolyploidy and cell size in epidermal tissue of *Arabidopsis*. Plant Cell 5, 1661–1668.

Nagl W 1995 Cdc2-kinases, cyclins, and the switch from proliferation to polyploidization. Protoplasma 188, 143–150.

Sgorbati S, Berta G, Trotta A, Schellenbaum L, Citterio S, Dela Pierre M, Gianinazzi-Pearson V and Scannerini S 1993 Chromatin structure variation in successful and unsuccessful arbuscular mycorrhizas of pea. Protoplasma 175, 1–8.

Tahiri-Alaoui A and Antoniw J F 1997 Cloning of genes associated with the colonization of tomato roots by the arbuscular mycorrhizal fungus *Glomus mosseae*. Agronomie 16, 699–707.

Tisserant B, Gianinazzi-Pearson V. Gianinazzi S and Gollotte A 1993 *In Planta* histochemical staining of fungal alkaline phosphatase activity for analysis of efficient arbuscular mycorrhizal infections. Mycol. Res. 97, 245–250.

Torelli A, Trotta A, Acerbi L, Arcidiacono G, Berta G and Branca C 1999 IAA and ZR content in leek (*Allium porrum* L.) as influenced by P nutrition and arbuscular mycorrhizae in relation to plant development. Plant Soil (*In press*).

Trotta A, Varese G C, Gnavi E, Fusconi A, Sampò S and Berta G 1996 Interactions between the soilborne root pathogen *Phytoph-*

thora nicotianae var. *parasitica*. and the arbuscular mycorrhizal fungus *Glomus mosseae* in tomato plants. Plant Soil 185, 199–209.

Trouvelot A, Kough J L and Gianinazzi-Pearson V 1986 Mesure du taux de mycorhization VA d'un sistème radiculaire. Recherche de méthodes d'estimation ayant une signification fonctionelle. *In* Mycorrhizae: physiology and genetics. Eds V Gianinazzi-Pearson and S Gianinazzi. pp 217–221. INRA, Paris.

Valente P, Tao W and Verbelen J-P 1998 Auxin and cytochinins control DNA endoreduplication and deduplication in single cells of tobacco. Plant Sci. 134, 207–215.

Wolfe A P 1991 Implications of DNA replication for eukarotic gene expression. J. Cell Science 99, 201–206.

Section editor: J H Graham

Recent Advances of Plant Root Structure and Function, 71–82
Edited by O. Gašparíková, M. Čiamporová, I. Mistrík and F. Baluška
© 2001 *Kluwer Academic Publishers.*

Water uptake by plant roots: an integration of views

Ernst Steudle
*Lehrstuhl Pflanzenökologie, Universität Bayreuth, D-95440 Bayreuth, Germany**

Key words: apoplast, composite transport model, endodermis, exodermis, hydraulic conductivity, root, water channels, water flow

Abstract

A COMPOSITE TRANSPORT MODEL is presented which explains the variability in the ability of roots to take up water and responses of water uptake to different factors. The model is based on detailed measurements of 'root hydraulics' both at the level of excised roots (root hydraulic conductivity, Lp_r) and root cells (membrane level; cell Lp) using pressure probes and other techniques. The composite transport model integrates apoplastic and cellular components of radial water flow across the root cylinder. It explains why the hydraulic conductivity of roots changes in response to the nature (osmotic vs. hydraulic) and intensity of water flow. The model provides an explanation of the adaptation of plants to conditions of drought and other stresses by allowing for a 'coarse regulation of water uptake' according to the demands from the shoot which is favorable to the plant. Coarse regulation is physical in nature, but strongly depends on root anatomy, e.g. on the existence of apoplastic barriers in the exo- and endodermis. Composite transport is based on the composite structure of roots. A 'fine regulation' results from the activity of water channels (aquaporins) in root cell membranes which is assumed to be under metabolic and other control.

Variability of root hydraulics

Water supplied to the plant by the root contributes to the overall water balance of the shoot. Despite this important function of roots, relatively little is known about the processes that govern or even regulate root water uptake. There is much evidence that the force driving water across roots is usually provided by the tension (negative pressure) created by transpiration from the shoot and extending to root xylem (Steudle, 1995; Tyree, 1997). Hence, the force driving water across the root cylinder is usually a gradient in hydrostatic pressure. However, there are conditions under which the root acts as an osmometer rather than just a hydraulic resistor. They occur, when, in the absence of transpiration, the uptake of nutrient ions into the xylem causes an osmotic water flow and a build up of root pressure. Instead of being dragged up to the shoot by negative pressure, the positive root pressure pushes xylem sap into the shoot under these conditions. Usually the resistance for axial flow along the xylem is much smaller than that for radial flow across the root cylinder (Steudle and Peterson, 1998). For a given plant, both hydraulic resistances are variable for different reasons. Reasons of the variability of radial hydraulic resistance of roots are complex. They are closely related to the complex structure of roots. The phenomenon has been known for a long time (Brewig, 1937; Brouwer, 1954; Fiscus, 1975; Kramer and Boyer, 1995; Passioura, 1988; Steudle, 1989, 1994; Steudle and Frensch, 1996; Steudle and Peterson, 1998). However, no satisfactory model(s) have been presented so far to explain the finding which is crucial for our understanding of plant water relations.

There are changes in root hydraulic conductivity which are due to root development and aging. Moreover, for a given root, the hydraulic conductance (inverse of hydraulic resistance) may vary in response to external (e.g. drought or salinity) or internal factors such as the nutritional state and water status of the plant or the demand for water from shoot caused by transpiration. Mechanisms are poorly understood. For the plasticity of root hydraulics, different reasons have been discussed, i.e.:

* FAX No.: 921 55 2564. E-mail: ernst.steudle@uni-bayreuth.de

- changes in root morphology, structure and anatomy,
- changes in the transport pattern or mechanism of water flow,
- interactions between water and solute (nutrient) flow, and
- changes in the activity of water channels.

In the following review these factors are briefly discussed. A model is put forward which, at least qualitatively, explains the variability in root hydraulics. The model is quite flexible and allows to integrate different views which have been dsicussed in the past. Some recent findings are summarized which support the model.

Root anatomy

With respect to water permeability, changes in the suberization of roots are most relevant. Suberization of roots increases with age and during stress (drought, high salinity, nutrient deprivation, anoxia, etc.). Young roots suberize when they pass through different stages of development of the endo- and exodermis. During state I, Casparian bands are forming in radial walls of the endodermis. During state II, suberin lamellae are laid down in both radial and tangential walls. Eventually, cell walls are thickened during state III which results in the well-known u-shaped cross section of endodermal cells. It has been shown that the exodermis also develops Casparian bands and that this structure then passes through similar states as the endodermis which includes the formation of passage cells (Peterson and Enstone, 1996). In young maize roots, the formation of a Casparian band in the exodermis did not affect hydraulic conductivity during state I (Peterson et al., 1993; Steudle et al., 1993). However, this may change during later states. For the exodermis, experiments indicated a strong effect on the radial hydraulic conductivity (root Lp_r; Zimmermann and Steudle, 1998). In recent years, considerable information has been collected of how changes in root structure caused by drought, high salinity, anoxia, and nutrient deprivation are reflected into changes of root hydraulic properties (Azaizeh et al., 1992; Birner and Steudle, 1993; Carvajal et al., 1996; Cruz et al., 1992; North and Nobel, 1991; Peyrano et al., 1997; Stavosky and Peterson, 1993). Fairly little is known about the contribution of older thickened roots to overall water uptake. These roots are covered with several layers of suberized cells, e.g. in woody species. Usually, it is thought that, because of suberization, these arrays do not contribute much to overall water uptake. However, this has been questioned by Kramer some time ago (see discussion and literature in Kramer and Boyer, 1995, pp. 184). For technical reasons, it is still difficult to quantify the uptake of different parts of roots or even root zones. Different from the situation in the shoot, much less is known about the 'hydraulic architecture of roots'. This is due to the lack of techniques which would allow to resolve the hydraulic conductivity of root zones in an extensive root system sitting in soil with sufficient accuracy.

On top of these uncertainties, which are largely related to transport properties of the root apoplast, there are open questions with respect water channels (aquaporins) and how they would contribute to the overall water permeability of roots. Water channel activity considerably affects water transport at the cellular (membrane) level (see below; Azaizeh et al., 1992; Frensch et al., 1996; Jones et al., 1983; Steudle and Jeschke, 1983; Steudle et al., 1987; Zhu and Steudle, 1991). For technical reasons, there are, to date, only a few experimental approaches to measure the hydraulic conductivity at the root cell level and to relate this to the overall hydraulic conductivity of the entire organ during different stages of development (Steudle and Jeschke, 1983; Jones et al., 1983; Steudle et al., 1987; Zhu and Steudle, 1991; Azaizeh et al., 1992; Frensch et al., 1996). However, this information is badly needed to work out appropriate models for the water uptake of roots and how this would be controlled or regulated under different conditions. Detailed information on water transport has to be related to root anatomy which can be quite variable. There are differences in root anatomy and structure between species, habitats and growth conditions. When general conclusions are drawn about transport mechanisms and models are worked out, the strong relation between structure and function of roots has to be carefully considered (Steudle and Peterson, 1998).

Transport pattern and mechanisms

Besides other factors, changes in transport mechanism(s) for water which cause changes in the radial water uptake (root hydraulic conductivity, Lp_r), are related to the intensity of water flow and to the nature of forces applied to drive water across the root. The Ohm's law analogy of water flow (van den Honert, 1948) may not hold for roots in all cases. This law predicts that water flow across a root would linearly increase in proportion to the force driving the flow which is the gradient in water potential between soil and root xylem. One would think that the different components of water potential (hydrostatic and osmotic pressure, matric and gravitational potential) have the same effect on the flow. However, this is not the case. Depending on the species used, there are large differences in root hydraulic conductivity ('root Lp_r') observed either during osmotic (such as during conventional exudation of an excised root) or hydraulic water flow (such as in transpiring plants). Careful investigations have shown that the story is tied up with the question of how water uses the different pathways which are potentially available in roots (see below). Furthermore, the hydraulic resistance of roots usually decreases when the force (flow rate) increases, i.e. there is a non-linear relationship between forces and flows.

In the root cylinder there are three different pathways (Figure 1). The first is the apoplastic path around protoplasts. The second path is the symplastic which is through plasmodesmata. When travelling along this latter path, water stays within the 'cytoplasmic continuum'. Last but not least, we have the transcellular or vacuolar pathway which is across membranes. This latter route is something special for water. It is due to the high permeability of membranes to water. For ions and other solutes present in plant cells, the contribution of this component will be usually negligible. Although there have been attempts to use the pressure clamp-technique to work out symplastic and transcellular components of water flow (Murphy and Smith, 1998), there are, to date, no approaches to separate components in a way which is experimentally simple and straightforward. Therefore, they are summarized as a 'cell-to-cell' or 'protoplastic' component.

In the usual picture of water transport in roots which is dealt with in textbooks, the flow of water across the root cortex is largely apoplastic. This changes at the endodermis because of the Casparian band which interrupts the apoplastic path. There will be a transcellular transport step at the endodermis which is usually thought to rate-limit water transport across young roots. In the stele, the situation will be then similar to that in the cortex.

Detailed measurements of root hydraulics have shown that this simple picture has to be modified. Depending on the conditions, the relative contribution of pathways to the overall uptake or hydraulic conductivity may change substantially. A 'composite transport model of the root' has been established which takes into account the structure of roots (or of other tissue) and is based on irreversible thermodynamics (Figure 2; Steudle, 1989; 1994; 1997; Steudle and Frensch, 1996; Steudle and Peterson, 1998). The model shows that the different pathways may be used with different intensity which then results in the observed plasticity in root hydraulics. Besides the intensity of water flow, the physical nature of driving forces is crucial. In the presence of (hydrostatic) pressure gradients, flow is largely around protoplasts (apoplastic), because this path represents a low hydraulic resistance. The data indicate that there should be some apoplastic flow even across the endodermis, i.e. Casparian bands appear to be somewhat permeable to water (but not to ions) during state I or even later. This results in a high overall hydraulic conductivity of the root (root Lp_r) which is sometimes even larger than that of the plasma membrane of individual root cells. On the other hand, water flow in the presence of osmotic gadients is low as would be the case in the absence of transpiration and during phenomena such as root exudation. Osmotic driving forces only cause a water movement in the presence of membranes. Provided that osmotic gradients applied from outside do not create hydrostatic or matric forces within the root (Zimmermann and Steudle, 1998), an osmotic water flow across the root has to pass many membranes which results in an overall root Lp_r which is much less than that found during hydraulic water flow.

The composite transport model of the root

The composite transport model is based on detailed measurements of hydraulic conductivity and other transport properties (salt permeability and reflection coefficients) which have been obtained both at the level of individual cells and entire roots (see reviews: Steudle, 1989; 1994; 1997; Steudle and Frensch, 1996; Steudle and Peterson, 1998). An important feature of the model is that there are two parallel pathways present which exhibit a quite different 'passive selectivity' as expressed by their reflection coefficients (σ_s). To a first approximation, the cell-to-cell (protoplastic) path is semipermeable, i.e. it exhibits a σ_s^{cc} of close to unity. The apoplastic path, on the other hand, does not select between water and solutes (nutrient salts and others). The reflection coefficient here (σ_s^{apo}) is, therefore, close to zero. The two pathways interact with each other, and the interaction results in phenomena such as a circulation flow of water and a

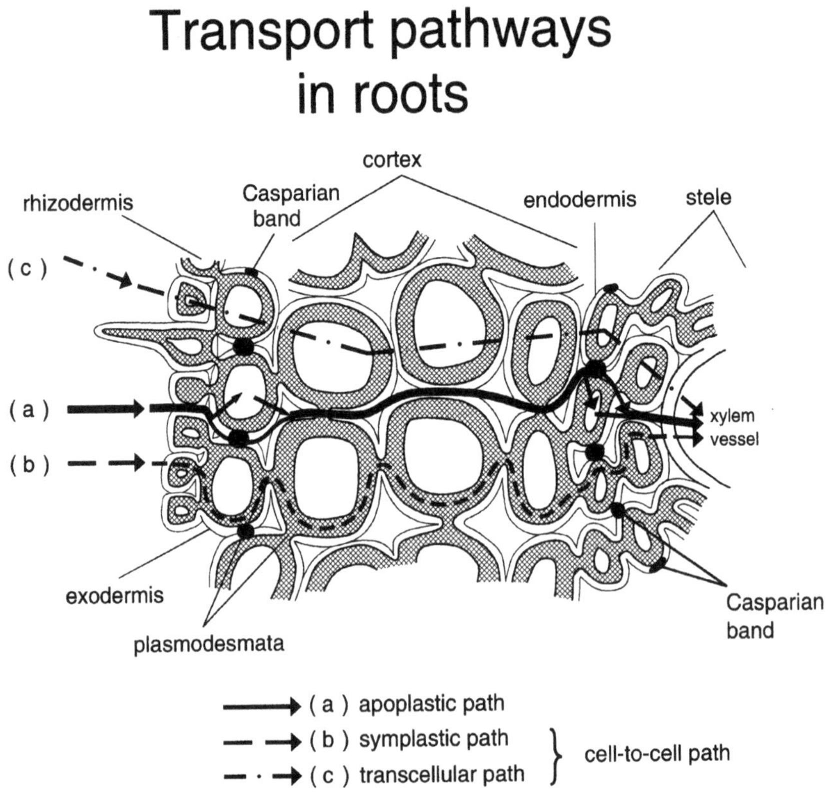

Figure 1. Pathways for the movement of water and solutes in roots. The apoplastic path (**a**) refers to the flow around protoplasts. The symplastic component defines flow from cell to cell via plasmodesmata (**b**). On the third route (**c**), water and solutes have to cross cell membranes (two membranes per cell layer; transcellular path). The transcellular path is important for water, but is of minor importance for solutes. For water, pathways (**b**) and (**c**) cannot be separated experimentally to date. Therefore, they are summarized as a cell-to-cell path. It is usually assumed that, in roots, the Casparian bands in the exo- and endodermis completely interrupt apoplastic transport. Here, it is, however, indicated that there may be an apoplastic component of water flow across Casparian bands.

low overall reflection coefficient of the root (as measured). An important consequence of the model is that the hydraulic resistance of roots of transpiring plants will be low, i.e. in the presence hydrostatic pressure gradients. Moreover, the supply from the root is adjusted according to the demands of the shoot. In the absence of a demand, there will be only osmotic gradients present due to the active uptake of solutes by the root. These, however, will cause a much smaller root Lp_r and water flow. The hydraulic resistance of the root will be high. As a consequence, losses of water to a dry or saline soil will be delayed under these conditions. Hence, composite transport provides some kind of a switching between transport models and a 'coarse regulation of water flow' across roots which is favorable for the plant.

The composite transport model readily explains the variability of root hydraulic properties in terms of changes in forces which cause a switching between the pathways used. Along the apoplast, a dependence of the hydraulic conductivity on the nature of the driving force has been discussed as well, to account for some of the changes in the overall root Lp_r. These changes in the hydraulics of the apoplast may be brought about by changes in the water content of cell walls and other things (Steudle and Frensch, 1996). Most interestingly, there are species in which root Lp_r is not as variable as in others. For example, in *Phaseolus coccineus* and barley, the hydraulic and osmotic Lp_r were similar (Table 1). This has been interpreted by a high membrane Lp (in *P. coccineus*) or a rather tight Casparian band (in barley).

Table 1. Root hydraulic conductivity (Lp_r), solute permeability (P_{sr}), and reflection coefficients (σ_{sr}) of roots of **herbaceous (A) and woody (B) species** as determined with the root pressure probe and other techniques. Where available, hydraulic conductivities of root cell membranes (cell Lp) are given for comparison. Because of the high cell Lp, there are no differences between osmotic and hydraulic water flow (Lp_r) in barley and *Phaseolus coccineus*. For maize and *Phaseolus vulgaris*, there are large differences. Differences between osmotic and hydraulic water flow are much larger for roots of trees than for those of herbs. Tree root Lp_r is, on average, smaller by an order of magnitude than that of herbaceous plants. Values of root σ_{sr} are significantly lower than unity for solutes for which cell membranes exhibit a σ_s of virtually unity. σ_{sr} values of tree roots are substantially smaller than those of roots of herbs. The findings are explained by the composite transport model of the root (see text). (n.m. = not measurable)

Species	Root $Lp_r \cdot 10^8$ ($m\ s^{-1}\ MPa^{-1}$) hydraulic	osmotic	Root permeability, $P_{sr}\ 10^9$ ($m\ s^{-1}$)	Root reflection coefficient, σ_{sr} (1)		Techniques	Ref.
A. Herbaceous plants							
Hordeum distichon, primary root	0.3 – 4.3 *Cell Lp: 12*	0.3 – 4.3	–	mannitol:	≈ 0.5	Cell and root pressure probe	(a)
Zea mays, primary root	1–46 *Cell Lp: 24*	0.1–5	sucrose: 3.0 NaCl: 6–14	mannitol: sucrose: NaCl, KCl:	04.–0.7 0.54 0.5–0.6	Cell and root pressure probe	(b) to (e)
Zea mays, root system	21	2.2	—	nutrients:	0.85	Stop-flow technique and osmotic flow	(f) and (g)
Allium cepa, primary root	14	0.02–2	$NaNO_3$: 0.7	KCl, mannitol, and NH_4NO_3:	$NaNO_3$, 0.35–0.88	Root pressure probe	(h) to (j)
Phaseolus coccineus, coccineus, primary root	2–8 *Cell Lp: 30-470*	3–7	mannitol; 0.15 NaCl:0.21 KCl: 0.7–0.9	mannitol: NaCl: KCl:	0.68 0.59 0.43–0.54		
Phaseolus vulgaris, root system	30	0.56	nutrients: 1.3	nutrients:	0.98	pressure chamber and osmotic flow	(f) and (k)
B. Woody plants							
Picea abies, root system	6.4	0.017	n.m.	Na_2SO_4, K_2SO_4,	$Ca(NO_3)_2$: 0.18–0.28		
Quercus robur, root system	0.5–4.8	0.003-0.062	n.m.	mannitol: NaCl, KCl:	0.19–0.43 0.12–0.35	Root pressure probe	(l) to (n)
Fagus sylvatica, root system	0.35–1.6	0.022–0.11	n.m.	mannitol: KCl: NaCl:	0.29–0.82 0.22–0.55 0.32–0.64		

(a) Steudle and Jeschke, 1983; (b) Steudle et al. 1987; (c) Steudle and Frensch 1989; (d) Zhu and Steudle, 1991; (e) Peterson et al., 1993; (f) Newman, 1973; (g) Miller, 1985; (h) Melchior and Steudle, 1993; (i) Melchior and Steudle, (1995); (j) Steudle and Brinckmann 1989; (k) Fiscus, 1986; (l) Rüdinger et al., 1994; (m) Steudle and Meshcheryakov, 1996; (n) Steudle and Heydt, 1997.

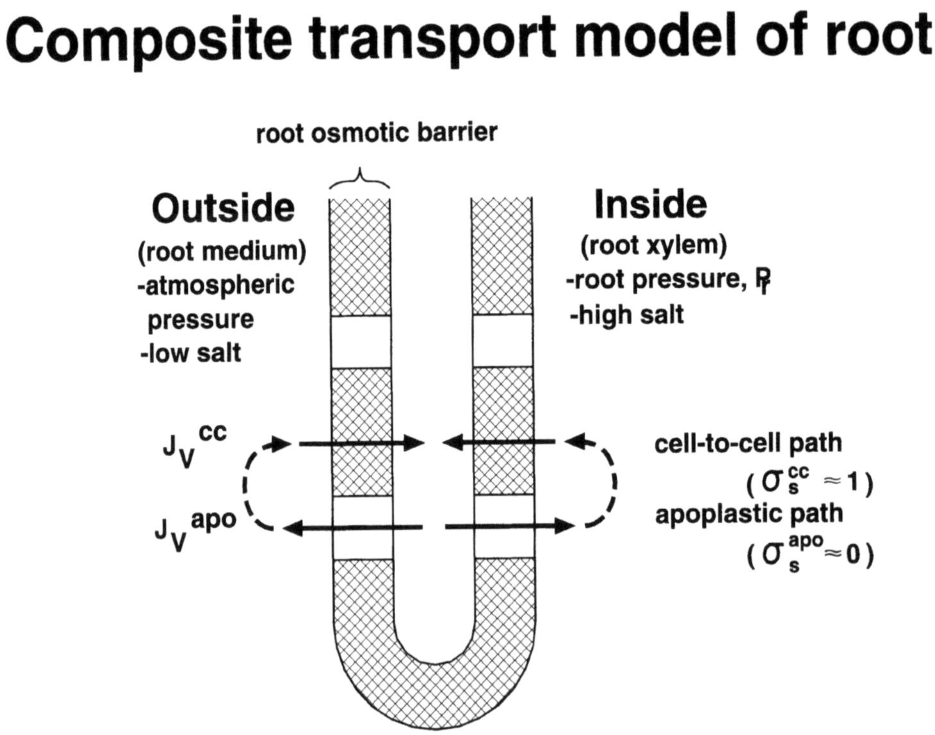

Figure 2. Composite transport model of root (schematical). The root osmotic barrier is comprised of cells (protoplasts) and the apoplast. The apoplastic path may be interrupted by Casparian bands in the endo- and exodermis (not shown for sake of simplicity). Water and solutes move along the two parallel pathways (cell-to-cell and apoplastic route). The cell-to-cell path has a high selectivity (reflection coefficient, $\sigma_s{}^{cc} \approx 1$), and the apoplastic path a very low selectivity ($\sigma_s{}^{apo} \approx 0$) for solutes. At low rates of transpiration, this results in a circulation flow of water in the root and in a low overall root σ_{sr} (as found experimentally). The model explains variable root hydraulic conductivity which depends on the nature of the driving force and other things. For further explanation, see text.

Interactions between water and solute flow

As solutes (nutrients) move across the root cylinder, osmotic concentrations within the apoplastic and protoplastic compartments change. This, in turn, causes a redistribution of water between pathways. The composite transport model accounts for these interactions. In the model, the diffusion along the apoplast is taken into account. Along the cell-to-cell path, the diffusive transport of solutes will be usually negligible because of the low membrane permeability of solutes (nutrients). Active transport may be important at low rates of transpiration. This may be accounted for in the model provided that the rates of uptake by cells in different layers is known.

Another important point which relates to water/solute interactions, is the fact that water uptake into the root xylem reduces the osmotic concentration of xylem sap. In turn, this modifies the force driving water uptake. As water uptake increases during transpiration, the concentration of xylem sap is reduced at a constant rate of active uptake of nutrients. At sufficiently large rates of uptake, the osmotic driving force will vanish. Fiscus (1975) analyzed this 'dilution effect' quantitatively. He found that the effect may account for some of the 'apparent' variability of root Lp_r reported in the older literature. This may be true. However, recent studies with excised root systems of maize have shown that most of the differences between osmotic and hydraulic Lp_r remain even when the dilution effect is accounted for (Zimmermann and Steudle, 1998). Thus, there are inherent differences in root Lp_r which are related to changes in the mechanisms of flow. As pointed out, the differences are physiologically important, namely, during the adaptation of plants to water stress.

Role of water channels (aquaporins) in roots: is there a fine regulation of water flow?

It has been shown that water channels or aquaporins contribute to most of the hydraulic conductivity (water permeability) of plant cell membranes (Chrispeels and Maurel, 1994; Hertel and Steudle, 1997; Maurel, 1997; Schäffner, 1998; Steudle, 1997; Steudle and Henzler, 1995; Schütz and Tyerman, 1997; Tyerman et al., 1999). Water channels are transport proteins of a molecular weight of about 30kDa. They belong to a highly conserved group of proteins called membrane integral proteins (MIPs). MIPs are fairly hydrophobic. Common to all MIPs are six transmembrane domains which span the membrane and two smaller hydrophobic loops on either side containing the asparagine-proline-alanine (NPA) motif (Park and Saier, 1996; Reizer et al. 1993). The NPA motif is thought to be the part of the pore region creating the selectivity for water. The entire structure resembles an hour glass (Jung et al., 1994; Walz et al., 1995). The aquaporin AQP1(CHIP28) forms water channels in the plasma membrane of red cells and epithelia (Nielson and Agre, 1995; Preston et al., 1992). It has been crystallized and structurally resolved (Jap and Li, 1995; Cheng et al., 1997; Walz et al., 1997). AQP1 forms tetramers but each monomer is thought to facilitate water flow (Verkman et al., 1996). The first MIP from plants was cloned and functionally expressed in 1993 (γ-TIP; Maurel et al., 1993). This has renewed interest in the biophysics of plant water (Chrispeels and Maurel 1994; Maurel, 1997; Schäffner, 1998; Steudle and Henzler, 1995). It has been shown that, although water channels are selective for water, there may be some slippage of small organic molecules (Henzler and Steudle, 1995; Hertel and Steudle, 1997). Mercurials such as mercuric chloride ($HgCl_2$) have been used to reversibly block the channel function. This has been demonstrated either in frog oocytes, where channel protein was expressed, or in *Chara* internodes (Chrispeels and Maurel, 1994; Schütz and Tyerman, 1997; Steudle and Henzler, 1995). During tissue transport such as in roots, water channel activity should affect root Lp_r and, hence water uptake. However, from what has been discussed so far, this should depend on the flow model, i.e. on the actual contribution of the cell-to-cell (protoplastic) path to overall water flow (root Lp_r).

It is interesting that water channel activity may be controlled by metabolism (e.g. by a phophorylation of aquaporins; Johansson et al., 1996) or may be triggered by environmental factors (Steudle and Henzler, 1995). In pea, water shortage increased the expression of water channels (Guerrero et al., 1990). In corn roots, high salinity reduced the permeability to water of both root cells and (to a lesser extent) whole roots (Azaizeh and Steudle, 1991; Azaizeh et al., 1992). Carvajal et al. (1996) have shown that depriving wheat plants of nutrients also decreased root Lp_r, apparently by affecting the activity of water channels. These findings are in line with the observation that treating roots with mercurials reduced root Lp_r reversibly (Maggio and Joly, 1995). At the level of individual internodes of *Chara*, the effect has been documented in great detail (Henzler and Steudle, 1995; Schÿtz and Tyerman, 1997; Steudle and Henzler, 1995; Tazawa et al., 1996; Wayne and Tazawa, 1990) In these experiments, water channel activity has been titrated with mercuric chloride or p-chloromercuriphenylsuolfonic acid (p-CMBS). Similarly, high external concentrations of permeating solutes have been employed to cause a closure of water channels in *Chara* (Steudle and Henzler, 1995). Higher plant cells have been also used in these latter studies which were performed with the aid of the cell pressure probe (Steudle, 1993). The results demonstrated that water and the small uncharged solutes used different pathways to cross the plama membrane, as postulated in the presence of selective pores for water which allow for most of the water flow. However, the quantitative treatment of interactions between water and solute flows showed that there was some slippage of the small test solutes across water channels which caused some drag of solvent (water). To date, there are no equivalent experiments with cells of plant tissue such as with root cortical cells. One would need to measure cell Lp in the presence and absence of functioning channels as well as the overall root Lp_r and compare the results. Changes in the expression of water channels in different root tissue would have to be demonstrated as well (Schäffner, 1998). These experiments are underway.

Fine regulation vs. coarse regulation

So far, the results with roots indicate that there is a regulation of water uptake along the cell-to-cell passage, namely, in older suberized roots where the apoplastic path is blocked. So, water channels may be looked at as a tool to provide some 'fine regulation of water uptake' in tissues when the apoplastic path cannot be used (Steudle, 1997; Steudle and Peterson,

1998). Regulation along the cell-to-cell path may be affected by external factors such as high salinity, nutrient deprivation, anoxia, and temperature (Azaizeh and Steudle, 1991; Azaizeh et al., 1992; Birner and Steudle, 1993; Carvajal et al., 1996; Henzler et al., 1998; Hertel and Steudle, 1997). It may be mediated by a phosphorylation of the transport proteins (Johansson et al., 1996). Future work has to show if this is true. This work has to imply both the *in situ* characterization of transport proteins and the measurement of its function at the cell and root levels. In terms of root physiology it would be most important to find out the mechanisms which trigger the opening and closing of water channels, i.e. the mechanisms of the gating of water channels (Tyerman et al., 1999).

Experimental techniques

At the level of individual root cells, water transport (hydraulic conductivity, Lp) has been measured using the cell pressure probe (Steudle, 1993; Steudle et al., 1987; Zhu and Steudle, 1991; Figure 3). The tip of the probe (diameter: 2 to 7 μm) was introduced into individual cells in different layers of the root which then allowed the monitoring of cell turgor. Cell Lp was measured after manipulating the equilibrium turgor to produce a water flow across the membrane which was monitored as a 'pressure relaxation'. The half-time of the process is a measure of Lp. In the same, or in a separate experiment, the root pressure probe has been employed to measure the overall hydraulic conductivity of roots (Steudle, 1993; Steudle and Jeschke, 1983; Steudle et al., 1987; Zhu and Steudle, 1991; Figure 4). Figure 4 also shows how root Lp_r has been measured by alternative, steady-state techniques. In the latter techniques, either pressure gradients have been applied (pressure chamber, vacuum), or the osmotic pressure of the medium was varied (for details, see legend to Figure 4 and Zimmermann and Steudle, 1998).

The root pressure probe has been used to separate the axial hydraulic resistance of xylem vessels from that related to flow across the root cylinder (Frensch and Steudle, 1989; Melchior and Steudle, 1993) and to measure the radial hydraulic resistance of individual root zones (Frensch et al., 1996).

Summary of recent results

Some results of root hydraulics are summarized in Table 1. The table separates between herbaceous and woody species. On average, roots from herbs have a root Lp_r which is larger by an order of magnitude than that of woody species. This is due to differences in root structure, namely, to the stronger degree of suberization of woody roots. As for the water, the permeability of woody roots to solutes (permeability coefficients of salts, sugars and the like) is substantially smaller than that of herbs. However, reflection coefficients of woody roots (which denote the 'passive selectivity' of a root) were smaller than those of herbs. In terms of the composite transport model this is understandable. In composite transport systems with parallel arrays, the components (cell-to-cell and apoplastic path) contribute to the overall reflection coefficient according to their hydraulic conductances. Since the permeability of woody roots is lower than that of herbs, the relative contribution of the apoplastic bypass is more significant for the former. In tree roots, differences of up to three orders of magnitude have been found between osmotic and hydraulic water flow (root Lp_r). This remarkable effect has been explained by the model but may, in addition, result from changes in the apoplastic hydraulic conductivity in the presence of pressure gradients (Steudle and Frensch, 1996).

It should be noted that, besides the water, the apoplastic component of transport could be also important for the stress hormone abscisic acid (ABA). When ABA was applied to the root medium, there was a considerable byass flow of ABA in young maize roots, i.e. ABA was transported by solvent drag with the transpiration stream imitated by a vacuum applied to the cut end of the excised roots (Freundl et al., 1998; 1999). In roots of sunflower, the effect was much smaller. The results indicate that the uptake of ABA present in the soil solution in low concentrations, could contribute to the xylem concentration of the hormone. Alternatively and more likely, ABA produced in cells of stressed roots may be delivered to the root apoplast and then transferred to the shoot with the transpiration stream. In both cases, the apoplastic flow of ABA will compensate for the dilution caused by water uptake. Hence, the root-to-shoot signal of ABA may have an apoplastic component which is directly coupled to water uptake, i.e. to transpiration.

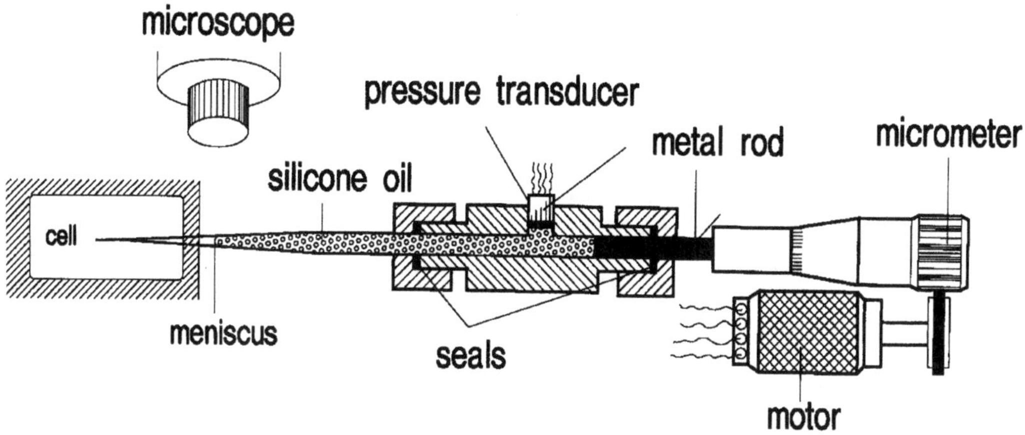

Figure 3. Cell pressure probe for measuring the hydraulic conductivity (water permeability) of plant cell membranes. The tip of the microcapillary is introduced into the cell so that turgor causes a shift of the oil/water meniscus in the tip which can be manipulated by moving the metal rod. Pressure relaxations are produced to work out the hydraulic conductivity of the membrane from half-times of the equilibration of turgor pressure. This requires also the measurement of the elasticity of the cell (also done with the probe) and cell dimensions (cell surface area and volume).

The integration of views

Composite transport provides an integration of views to explain the variability of root water uptake (hydraulic conductivity). Changes in root anatomy and structure have been taken into account such as the formation of apoplastic barriers (Casparian bands) in the endo- and exodermis and of suberin lamellae. The model may be used to integrate effects of water channels (aquaporins) which are much discussed at present. Water channel activity affects the cell-to-cell (protoplastic) rather than the apoplastic component of overall root Lp_r. The model may also account for changes in the driving forces by a dilution of xylem sap ('Fiscus-model'; Steudle and Jeschke, 1983; Steudle et al., 1987). As it stands, it considers parallel transport but serial array of tissues could be incorporated and axial transport in the xylem (Steudle and Peterson, 1998). Special components would be the transport across plasmodesmata (symplastic flow) which has been thought to substantially contribute to overall water flow across roots (Zhang and Tyerman, 1991). In some models, it has been speculated that this component may be affected by pressure gradients across roots (Passioura, 1988). There have been attempts to incorporate the interaction between active solute (nutrient) flow with the water (Steudle, 1989). However, this extension would require detailed knowledge of uptake rates and how this would affect osmotic gradients within and across the root cylinder.

The detailed data obtained so far for root hydraulics (cell and root level) are in agreement with the composite transport model. Pathways are coupled to each other, i.e. water and solutes are exchanged between pathways as they move across. The model explains the differences found between hydraulic and osmotic water flow and the variability of root Lp_r which changes in response to the nature of the driving force and the intensity of flow. Hence, it predicts that root Lp_r may vary according to external conditions (as found). It is proposed that the variability in root Lp_r provides a 'coarse regulation' of water uptake by roots. A 'fine regulation' may be brought about by the action of water channels along the cell-to-cell path. Present work concentrates on the role of water channels during water uptake, i.e., the activity of aquaporins is being mapped in different root tissues. The modes by which water channels are triggered by external and internal signals are currently being worked out (drought, salinity, temperature, nutritional state, heavy metals, etc.). The use of transgenic plants and of plants with reverse genetics play an important role in these studies (see Tyerman et al., 1999, for references). Another important issue is the separation of series hydraulic resistances in the root cylinder, namely, of those of the exodermis and endodermis, and how they would change during root development. Recent data from maize indicate that the development of an exodermis may strongly decrease the overall root Lp_r. The relative contributions of the cell-to-cell and apoplastic

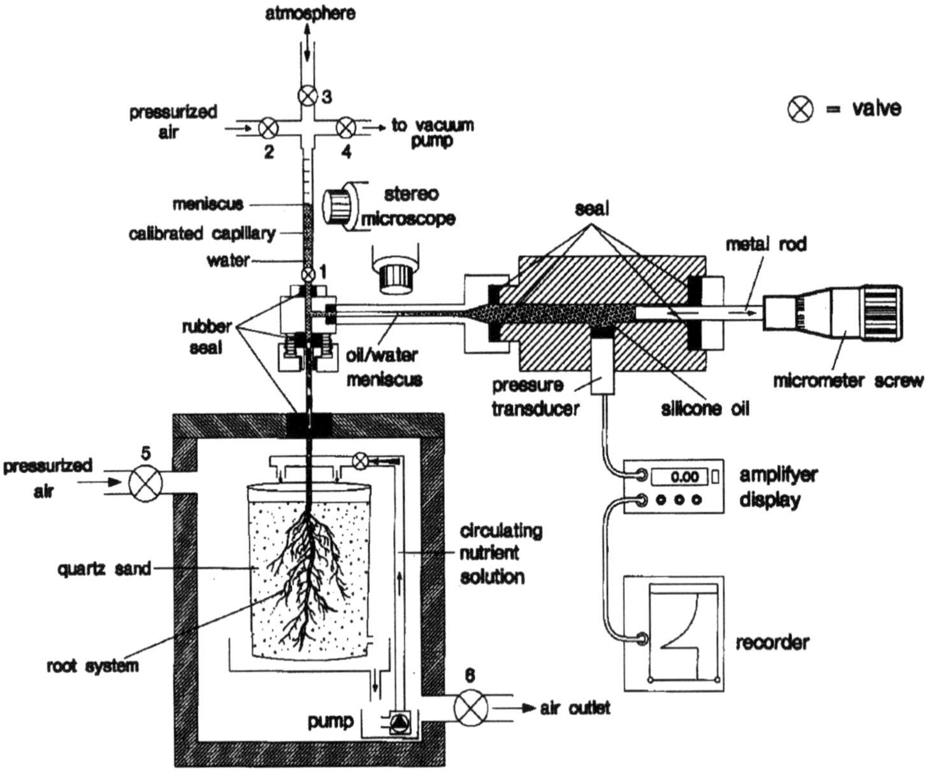

Figure 4. Experimental set-up employed to measure water transport and hydraulic conductivity of excised roots (root systems and individual roots). Different techniques may be used. In one procedure, root exudation is measured while pneumatic pressure is applied to the root system which is tightly sealed in a pressure chamber. The protruding root base is connected to a calibrated capillary. With valves 1 to 5 open and valve 6 closed, the steady water flow across the root (root system) is measured in response to a pneumatic pressure applied to the root. Alternatively, a pressure gradient may be set up by applying vacuum (valve 4 open) or pressurized air (valve 2 open) to the cut root base (xylem; valves 1, 5, 6 open; valve 3 closed). In the absence of pressure gradients (all valves open), water flow may be also measured in response to changes in the osmotic pressure of the root medium. With valve 1 closed and valves 5 and 6 open, the root pressure probe is employed to measure root Lp_r. In this type of experiment, root pressure builds up within the probe which is then changed to induce a water flow, analogous to the procedure used for cells (Figure 3). From the subsequent rate of equilibration of root pressure, root Lp_r is evaluated.

components of radial water flow and how its relative importance would change during root development are further quantified.

Acknowledgment

This work was supported by a grant from the Deutsche Forschungsgemeinschaft, Schwerpunktprogramm 'Apoplast'.

References

Azaizeh H, Gunse B and Steudle E 1992 Effects of NaCl and $CaCl_2$ on water transport across root cells of maize (*Zea mays* L.) seedlings. Plant Physiology 99, 886–894.

Azaizeh H and Steudle E. 1991 Effects of salinity on water transport of excised maize (*Zea mays* L.) roots. Plant Physiology 99, 1136–1145.

Birner T P and Steudle E 1993 Effects of anaerobic conditions on water and solute relations and active transport in root of maize (*Zea mays* L.). Planta 190, 474–483.

Brewig A 1937 Permeabilitätsänderungen der Wurzelgewebe, die vom Spross beeinflusst werden. Zeitschrift für Botanik 31, 481–540.

Brouwer R 1954 The regulating influence of transpiration and suction tension on the water and salt uptake by roots of intact *Vicia faba* plants. Acta Botanica Neerlandica 3, 264–312.

Carvajal M, Cooke D T and Clarkson D T 1996 Responses of wheat plants to nutrition deprivation may involve the regulation of water-channel function. Planta 199: 372–381.

Cheng A C, van Hoek A N, Yeager M, Verkman A S and Mitra A K 1997 Three-dimensional organization of a human water channel. Nature 387, 627–630.

Chrispeels M J and Maurel C 1994 Aquaporins: the molecular basis of facilitated water movement through living plant cells. Plant Physiology 105, 9–15.

Cruz R T, Jordan W R and Drew M C (1992 Structural changes and associated reduction of hydraulic conductance in roots of *Sorghum bicolor* L. following exposure to water deficit. Plant Physiology 99, 203–212.

Fiscus E L 1975 The interaction of osmotic- and pressure-induced water flow in plant roots. Plant Physiology 55: 917–922.

Fiscus E L 1986 Diurnal changes in volume and solute transport coefficients of *Phaseolus* roots. Plant Physiology 80, 752–759.

Frensch J and Steudle E 1989 Axial and radial hydraulic resistance to roots of maize (*Zea mays* L.). Plant Physiology 91, 719–726.

Frensch J, Hsiao T C and Steudle E 1996 Water and solute transport along developing maize roots. Planta 198, 348–355.

Freundl E, Steudle E and Hartung W 1998 Water uptake by roots of maize and sunflower affects the radial transport of abscisic acid and its concentration in the xylem. Planta 207, 8–19.

Freundl E, Steudle E and Harting W (1999) Apoplastic transport of abscisic acid through roots of maize: effects of the exodermis. Planta (*In press*).

Guerrero F D, Jones J T and Mullet J E 1990 Turgor-responsive gene transcription and RNA levels increase rapidly when pea shoots are wilted. Sequence and expression of three inducible genes. Plant Molecular Biology 15, 11–26.

van den Honert T H 1948 Water transport in plants as a catenary process. Discussions of the Faraday Society 3, 146–53.

Henzler T and Steudle E 1995 Reversible closing of water channels in *Chara* internodes provides evidence for a composite transport model of the plasma membrane. J Exper. Bot. 46, 199–209.

Henzler T, Carvajal M, Smyth A, Cooke D T and Clarkson D T 1998 Diurnal variation in root hydraulics and aquaporin gene-homologue expression in *Lotus japonicus*. J. Exper. Bot. 49 S: 10.

Hertel A and Steudle E 1997 The function of water channels in *Chara*: the temperature dependence of water and solute flows provides evidence for composite membrane transport and for a slippage of small organic solutes across water channels. Planta 202, 324–335.

Jap B K and Li H 1995 Structure of the osmoregulated H_2O-channel, AQP-CHIP, in projection at 3.5 Å resolution. J. Molec. Biol. 251, 413–420.

Johansson I, Larsson C, Ek B and Kjellbom P 1996 The major integral proteins of spinach leaf plasma membranes are putative aquaporins and are phoshorylated in response to Ca^{2+} and apoplastic water potential. Plant Cell 8, 1181–1191

Jones H, Tomos A D, Leigh R A and Wyn Jones R G 1983 Water-relation parameters of epidermal and cortical cells in the primary root of *Triticum aestivum* L. Planta 158, 230–6.

Jung J S, Preston G M, Smith B L, Guggino W B and Agre P 1994 Molecular structure of the channel activity of the seed specific aquaporin α-TIP. EMBO Journal 14, 3028–35.

Kramer P J and Boyer J S 1995 Water relations of plants and soils. Academic Press, Orlando.

Maggio A and Joly R J 1995 Effects of mercuric chloride on the hydraulic conductivity of tomato root systems: evidence for a channel-mediated pathway. Plant Physiology 109, 331–335.

Maurel C, Reizer J, Schroeder J L and Chrispeels M J 1993 The vacuolar membrane protein γ-TIP creates water specific channels in *Xenopus* oocytes. EMBO Journal 12, 2241–2247.

Maurel C 1997 Aquaporins and water permeability of plant membranes. Annu. Rev. Plant Physiol. Plant Molec. Biol. 48, 399–429.

Melchior W and Steudle E 1993 Water transport in onion (*Allium cepa* L.) roots. Changes of axial and radial hydraulic conductivities during root development. Plant Physiol. 101, 1305–1315.

Melchior W and Steudle E 1995 Hydrostatic and osmotic hydraulic conductivities and reflection coefficients of onion (*Allium cepa* L.) roots. *In* Structure and Function of Roots. Eds. Baluska F, Ciamporova M, Gasparikova O, Barlow P W. pp 209–213. Kluwer Academic Press Publ., Dordrecht, The Netherlands,

Miller D M 1985 Studies of root function in *Zea mays*. III. Xylem sap composition at maximum root pressure provides evidence of active transport into the xylem and a measurement of the reflection coefficient of the root. Plant Physiol. 77, 162–167.

Murphy R and Smith J A C 1998 Determination of cell water-relation parameters using the presure probe: extended theory and practice of the pressure-clamp technique. Plant Cell Environ. 21, 637–657.

Newman E I 1973 Permeability to water of five herbaceous species. New Phytologist 72, 547–555.

Nielson S and Agre P 1995 The aquaporin family of water channels in kidney. Annual Kidney International 48, 1057–1068.

North G B and Nobel P S 1991 Changes in hydraulic conductivity and anatomy caused by drying and rewetting roots of *Agave desertii* (Agavaceae). Am. J. Bot. 78, 906–915.

Park J H and Saier M H 1996 Phyolgentic characterization of the MIP family of transmembrane channel proteins. J. Membrane Biol. 153, 171–180.

Passioura J B 1988 Water transport in and to roots. Annu. Rev. Plant Physiol. Plant Molec. Biol. 39, 245–56.

Peterson C A and Enstone D E 1996 Functions of passage cells in the endodermis and exodermis of roots. Physiologia Plantarum 97, 592–598.

Peterson C A, Murrmann M and Steudle E 1993 Location of major barriers to water and ion movement in young roots of *Zea mays* L. Planta 190, 127–136.

Peyrano G, Taleisnik E, Quiroga M, de Forchetti S M and Tigier H 1997 Salinity effects on hydraulic conductance, lignin content and peroxidase activity in tomato roots. Plant Physiol. Biochem. 35, 387–393.

Preston G M, Carroll T P, Guggino W B and Agre P 1992 Appearance of water channels in *Xenopus* oocytes expressing red cell CHIP28 protein. Science 256, 385–387.

Reizer J, Reizer A and Saier M H 1993 The MIP family of integral membrane channel proteins: sequence comparisons, evolutionary relationships, reconstructed pathway evolution, and proposed functional differentation of the two repeated halves of the proteins. Critical Reviews in Biochemistry and Molecular Biology 28, 235–257.

Rüdinger M, Hallgren S W, Steudle E and Schulze E D 1994 Hydraulic and osmotic püroperties of spruce roots. J. Exper. Bot. 45, 1413–1425.

Schäffner A R 1998 Aquaporin function, structure and expression: are there more surprises to surface in plant water relations. Planta 204, 131–9.

Schütz K and Tyerman S D 1997 Water channels in *Chara corallina*. J. Exper. Bot. 48, 1511–1518.

Stavosky E and Peterson C A 1993 Effects of drought and subsequent rehydration on the structure, vitality and permeability of *Allium cepa* adventitious roots. Can. J. B. 71, 700–7.

Steudle E 1989 Water flow in plants and its coupling to other processes: an overview. Methods of Enzymology 174: 183–225.

Steudle E 1993 Pressure ropbe techniques: basic principles and application to studies of water and solute relations at the cell, tissue and organ level. *In* Water deficits: plant responses from cell to community. Eds. J A C Smith and H Griffiths. pp 5–36. Bios Scientific Publ. Ltd., Oxford.

Steudle E 1994 Water transport across roots. Plant Soil 167, 79–90.

Steudle E 1995 Trees under tension. Nature 378, 663–4.

Steudle E 1997 Water transport across plant tissue: role of water channels. Biology of the Cell 89, 259–273.

Steudle E and Brinckmann E 1989 The osmometer model of the root: water and solute relations of *Phaseolus coccineus*. Botanica Acta 102, 85–95.

Steudle E and Frensch J 1989 Osmotic responses of maize roots. Water and solute relations. Planta 177, 281–295.

Steudle E and Frensch J 1996 Water transport in plants: role of the apoplast. Plant Soil 187, 67–79.

Steudle E and Henzler T 1995 Water channels in plants: do basic concepts of water transport change? J. Exper. Bot. 46, 1067–1076.

Steudle E and Heydt H 1997 Water transport across tree roots. *In* Trees – Contributions to Modern Tree Physiology. Eds. Rennenberg H, Eschrich W, Ziegler H. pp 239–255. Backhuys Publishers, Leiden, The Netherlands.

Steudle E and Jeschke W D 1983 Water transport in barley roots. Planta 158, 237–248.

Steudle E and Meshcheryakov A B 1996 Hydraulic and osmotic properties of oak roots. J. Exper. Bot. 47, 387–401.

Steudle E, Murrmann M and Peterson C A 1993 Transport of water and solutes across maize roots modified by puncturing the endodermis. Further evidence for the composite transport model of the root. Plant Physiology 103, 335–349.

Steudle E, Oren R and Schulze 1987 Water transport in maize roots. Plant Physiol. 84, 1220–1232.

Steudle E and Peterson C A 1998 How does water get through roots? J. Exper. Bot. 49, 775–788.

Tazawa M, Asai K and Iwasaki N 1996 Characteristics of Hg- and Zn-sensitive water channels in the plasmamembrane of *Chara* cells. Botanica Acta 109, 388–96.

Tyerman, S D, Bohnert H, Maurel C, Steudle E and Smith J A C 1999 Plant water channels: molecular biology meets biophysics. J. Exper. Bot. (*In press*).

Tyree M T 1977 The cohesion-tension theory of sap ascent: current controversies. J. Exper. Bot. 48, 1753–65.

Verkman A S, van Hoek A N, Ma T, Frigeri A and Skach W R 1996 Water transport across mammalian cell membranes. Am. J. Physiol. 270, C12–C30.

Walz T, Typke D, Smith B L, Agre P and Engel A 1995 Projection map of aquaporin-1 determined by electron crystallography. Nature Structural Biology 2, 730–32.

Walz T, Hirai T, Murata K, Heymann J B, Mitsuoka K, Fujiyioshi Y, Smith B L, Agre P and Engel A 1997 The three-dimensional structure of aquaporin-1. Nature 387, 624–627.

Wayne R and Tazawa M 1990 Nature of the water channels in the internodal cells of *Nitellopsis*. J. Membrane Biol. 116, 31–39.

Weatherley P E 1982 Water uptake and flow into roots. *In* Encyclopedia of Plant Physiology Vol 12B. Eds. Lange O L, Nobel P S, Osmond C B, Ziegler H pp 79–109. Springer-Verlag, Berlin.

Zhang W H and Tyerman S D 1991 Effect of low O_2 concentration and azide on hydraulic conductivity and osmotic volume of the cortical cells of wheat roots. Australian J. Plant Physiol. 18, 603–613.

Zhu GL and Steudle E 1991 Water transport across maize roots. Plant Physiol. 95, 305–315.

Zimmermann H M and Steudle E 1998 Apoplastic transport across young maize roots: effect of the exodermis. Planta 206, 7–19.

Section editor: T J Flowers

Recent Advances of Plant Root Structure and Function, 83–95
Edited by O. Gašparíková, M. Čiamporová, I. Mistrík and F. Baluška
© 2001 *Kluwer Academic Publishers.*

Root-shoot interactions in mineral nutrition

W. Dieter Jeschke* and Wolfram Hartung
Julius-von-Sachs-Institut für Biowissenschaften, University of Würzburg, Julius-von-Sachs-Platz 2, D 97082 Würzburg, Germany

Key words: ABA, K/Na selectivity, nitrate reduction, nitrogen, phloem, phosphate, root-shoot interactions, salinity, xylem

Abstract

In this paper four classes of co-operative root-shoot interations are addressed. (I) Nitrogen concentrations in the xylem sap originating from the root and in the phloem sap as exported from source leaves are much lower than those required for growth by apices and developing organs. Enrichment of xylem sap N is achieved by xylem to xylem (X-X) transfer, by which reduced N, but not nitrate, is abstracted from the xylem of leaf traces and loaded into xylem vessels serving the shoot apex. Nitrogen enrichment of phloem sap from source leaves is enacted by transfer of reduced N from xylem to phloem (X-P transfer). Quantitative data for the extent of the contribution of X-X and X-P transfer to the nutrition of young organs of *Ricinus communis* L. and for their change with time are presented. (II) Shoot and root cooperate in nitrate reduction and assimilation. The partitioning of this process between shoot and root is shifted towards the root under conditions of nitrate- and K-deficiency and under salt stress, while P deficiency shifts nitrate reduction almost totally to the shoot. All four changes in partitioning can be attributed to the need for cation-anion balance during xylem transport and the change in electrical charge occurring with nitrate reduction. (III) Even maintenance of the specificity of ion uptake by the root may – in addition to its need for energy – require a shoot-root interaction. This is shown to be needed in the case of the maintenance of K/Na selectivity under the highly adverse condition of salt stress and absence of K supply from the soil. (IV) Hormonal root to shoot interactions are required in the whole plant for sensing mineral imbalances in the soil. This is shown and addressed for conditions of salt stress and of P deficiency, both of which lead to a strong ABA signalling from root to shoot but result in different patterns of response in the shoot.

Abbreviations: ABA – abscisic acid; DAS – days after sowing; NRA – nitrate reductase activity; X-P transfer – xylem to phloem transfer; X-X transfer – xylem to xylem transfer.

Introduction

Supply of plant roots with mineral nutrients has a strong influence on vegetative and reproductive development of the shoots. Correspondingly, nutrient uptake needs to be regulated in response to shoot demand (Engels and Marschner, 1992; Marschner, 1995) and this has attracted particular interest in the case of nitrogen nutrition (Engels et al., 1992; Imsande and Touraine, 1994). In addition to the need for regulation other sorts of co-operative interactions between root and shoot are required. The root is the unique higher plant organ responsible, amongst other functions, for mineral nutrition and the xylem is the principal pathway for supplying the shoot with the mineral elements essential for growth. However, the xylem *per se* is inadequate for this very function and a subtle co-operation between root and shoot, i.e. between root-sourced xylem and leaf-sourced phloem is needed for supplying growing organs with the required nutrients as has been especially shown for N (Pate et al., 1975). A similar inadequacy exists for the phloem sap (Pate et al., 1975) The required enrichment of xylem and phloem sap with N is achieved by transfers between xylem and xylem and between xylem and phloem (Pate and Jeschke, 1995). In this paper some

* FAX No: 0931 8886158. E-mail: w.jeschke@mail.uni-wuerzburg.de
Dedicated to Professor Wilhelm Simonis on the occasion of his 90th birthday.

quantitative data for the interactions between xylem and phloem and between leaf- and apex-bound xylem streams will be given and the changes of these transfers with time will be addressed.

Root and shoot co-operate in the assimilation of mineral nutrients into organic matter as for instance in nitrate reduction which can occur in both organs (Andrews, 1986). Under conditions of mineral imbalances sensed by the root the partitioning of nitrate reduction between shoot and root is altered (Andrews, 1986; Cramer et al., 1995). These changes will be addressed and compared with the partitioning of phosphate assimilation. Even the maintenance of the specificity of ion uptake by root cells, a membrane transport process, may require the co-operation between shoot and root as will be discussed for the selective uptake of potassium in favour of sodium under conditions of salinity and potassium deficiency.

Communication between root and shoot may be direct or indirect, mediated by the nutrients themselves or by hormonal signals translocated in the long distance vessels to the developing tissues or organs. Hormones such as ABA are synthesised in roots in response to perturbations of nutrient supply and transported to target tissues and cells where developmental processes can be initiated. Hormonal signals from root to shoot induced by mineral imbalances have been postulated and investigated particularly in the case of N deficiency. Often a slight stimulation of ABA accumulation in leaves of nitrate-deficient plants has been reported (for refs. see Peuke et al., 1994). In that paper, an increase in ABA concentrations in the xylem sap but not in leaves has been observed in *Ricinus* plants grown under low supply of nitrate. However, in *Ricinus* sufficiently supplied with N in the form of ammonium led to severe increases in leaf and xylem sap ABA compared to nitrate. Similar data were reported by Brewitz et al. (1995) in barley plants and by Peuke et al. (1998) in *Ricinus* plants with nitrate-sprayed leaves. The present contribution deals with abscisic acid signals from root to shoot in *Ricinus* plants as initiated by salt stress and by phosphorus deficiency.

Co-operative root-shoot interactions in the supply of nutrients to growing shoot organs

It has long been known that, since flows of water and of nutrients are coupled both in the xylem and in the phloem, scarcely transpiring organs like developing fruits in certain species tend to be under-supplied with nutrients such as Ca, which are phloem-immobile and predominantly supplied by the xylem (Marschner, 1995). However, similar limitations of nutrient supply to growing organs can arise also for phloem-mobile elements like nitrogen, since the import via xylem is dominated by the intake of water and phloem import is linked to the import of assimilates. Therefore growing organs could be in danger of being under-supplied with N if nitrogen concentrations in xylem or phloem sap were lower than required. This is underlined by the data of Table 1 showing that N concentrations in xylem sap as supplied by the root are too low and C:N ratios in phloem sap as exported by source leaves too high to meet the demand of a developing white lupin fruit or *Ricinus* leaf. In the whole plant, therefore, both xylem and phloem sap need to be enriched in nitrogen and this is achieved by transfers of the originally root-sourced nitrogen from xylem to phloem in leaves and in the stem (Pate et al., 1979), from phloem to xylem in the root and in lower stem segments and specifically also by a transfer from xylem to xylem (Jeschke and Pate, 1991a; Layzell et al., 1981), as shown for example by the pattern of N flows in *Ricinus communis* in Figure 1.

The first process to be addressed is the xylem to xylem transfer (X-X transfer) indicated by the looped pathways in Figure 1, in which reduced nitrogen and not nitrate (Jeschke and Pate, 1991b) is abstracted from vessels of leaf traces and reloaded into xylem vessels passing further up the stem. This transfer occurs in a cascading way successively in each of the lower stem segments and functions as a nitrogen concentrating mechanism through which the high demand of young growing organs is met while the xylem sap N concentration reaching mature leaves is decreased (Table 1). More than three-fold increases in the xylem sap N concentration are needed (Table 1) and have experimentally been found by analysing xylem sap collected at various stations along the stem axis of *Lupinus albus* (Layzell et al, 1981) and *Ricinus communis* (Jeschke and Pate, 1991a).

Node-based transfer cells surrounding all xylem vessels of leaf traces are involved in this transfer as evidenced by the observation that the development of wall ingrowths in these transfer cells coincides with the onset of X-X transfer in a stem segment (Jeschke and Pate, 1991b). In the youngest stem segments X-X transfer was not observed (Figure 1) and indeed, throughout the life span of a leaf X-X transfer in the nodal segment supporting it changed dramatically (Figure 2B). The transfer commenced about 12 days

Table 1. Nitrogen demand and supply during fruit and leaf development. a) Concentrations of total nitrogen in the xylem sap as required by developing fruits of *Lupinus albus* or leaves of *Ricinus communis* under the prevailing transpiration and as supplied by the root at the stem base. b) Concentrations of C relative to N (ratios of C:N) in the phloem sap as required by fruits and leaves at the current assimilate import and as supplied by the source leaf

Plant, organ	a) Xylem sap mM N		b) Phloem sap C:N ratio		Reference
	as required[1)]	as supplied by root[3)]	as required[2)]	as supplied by source leaf[4)]	
Lupinus albus					
developing fruit	21±2	7.2	19±2.3	60±3.4	Pate et al., 1975
Ricinus communis					
young leaf	17	5	17	35	Jeschke & Pate, 1991a
Ricinus communis					
mature leaf	1.8	2.7[5)]	n.a.[6)]	–	

[1)]Estimated from xylem import of N and of water into the organ.
[2)]Estimated from phloem import of C and N into the organ.
[3)]Measured N concentration in stem base xylem sap.
[4)]Measured C:N ratio in phloem sap collected from petiole of source leaf.
[5)]Lower concentration in xylem sap, since external nitrate concentration was lowered.
[6)]Not applicable, since mature leaf is exporting via phloem.

after leaf emergence and then rose to very substantial proportions of xylem import into the leaf, thereby drastically lowering the concentrations of reduced N in the xylem sap serving the leaf (Figure 2C). The extent to which X-X transfer contributes to the nitrogen imported into a particular apical organ is not directly evident from flow patterns like Figure 1, but it can be derived from flow rates and xylem sap concentrations. By evaluating the cascading, repetitive additions to the xylem sap it was found that even more than 100% of the reduced N had been transferred and this amounted to 85% of total N in the xylem sap (Figure 3). Owing to effective transpiration of leaves xylem-to-xylem-shuttled N contributed very substantially (60 to 71%) to the N nutrition of young leaves, more than to that of the mainly phloem-fed apical and virtually non-transpiring bud.

So far xylem to xylem shuttling has primarily been described for nitrogen, but also K has been shown to be transferred from the xylem of leaf traces to xylem vessels serving apical organs in barley (Wolf et al., 1991) and in *Ricinus* (Pate and Jeschke, 1995). By means of this transfer the high demand of developing organs for K is met. Interestingly an opposing transfer from xylem vessels passing further up the stem to leaf traces has been found in the case of Na in salt-treated barley (Wolf et al., 1991). In these plants xylem strands serving apical tissues were depleted of Na by means of this transfer and growing organs protected from this potentially damaging ion.

The second equally important transfer process occurring in the stem and contributing to the adequate supply of mineral nutrients towards apical organs is the xylem to phloem transfer (marked with crosses in Figure 1) which occurs prominently in upper segments of the stem. By this process nitrogenous compounds are extracted from N-rich upper xylem streams and loaded into phloem strands serving the shoot apex and developing leaves or fruits. Only with the aid of this redirection of nitrogen from xylem to phloem is sufficient supply of root-sourced nitrogen to phloem-fed sink organs warranted. In nitrogen-fixing *Lupinus albus* the xylem carries amino acids as the principal nitrogenous compounds and these are transferred (Pate et al., 1979), whilst in *Ricinus* the xylem translocates both amino acids and nitrate, but transfer is specifically restricted to amino acids (Jeschke and Pate, 1991b). Since the compounds transferred, such as asparagine and glutamine in white lupin (Pate et al., 1979) and principally glutamine in *Ricinus* (Jeschke and Pate, 1991b), have low C:N ratios of 2 or 2.5, respectively, the process leads to the required substantial decrease in the C:N ratio (Table 1). As opposed to X-X transfer, the transfer from xylem to phloem commences early within a stem internode and decreases with time in relative terms to the magnitude of

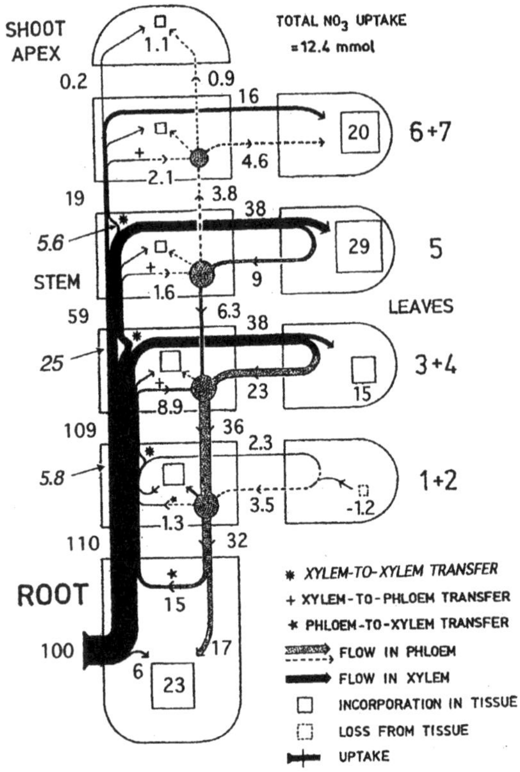

Figure 1. Flow profiles for uptake, transport and utilisation of nitrogen in whole plants of *Ricinus communis* over the period 44–53 days after sowing, depicting in particular the transfer processes from xylem to xylem (∗), from xylem to phloem (+) and from phloem to xylem (⋆) by which the concentration of reduced nitrogen in xylem and phloem is elevated to meet the demand of developing sink organs such as the terminal bud and the developing leaves No. 6 and 7. The numbers denote the rates of flow or deposition of nitrogen as percentage of the total uptake of N as nitrate of 12.37 mmoles during the study period of nine days. Redrawn from Jeschke and Pate, 1991a.

phloem flow within the corresponding petiole (Figure 2D). The impact of xylem to phloem transfer hence is particularly large in the N supply to the apical bud and it diminishes as developing leaves gain in size and maturity (Figure 3). The slowly declining impact X-P transfer is clearly reflected in the rising but low C:N ratios of phloem sap entering a leaf during the phase of assimilate import (Figure 2E, 0–12 days after leaf emergence).

Xylem to phloem transfer is not limited to nitrogen but has been found also for K in white lupin (Pate and Jeschke, 1995) and barley (Wolf et al., 1991) and for P in *Ricinus* (Jeschke et al., 1997a).

Root-shoot interactions in the assimilation of mineral nutrients into organic matter; nitrate reduction and phosphate assimilation

Nitrate reduction and assimilation can occur in the root as well as in the shoot (Andrews, 1986) and the partitioning of this activity between root and shoot varies greatly amongst species (Pate, 1973) and with nutritional conditions (Andrews, 1986). In this chapter the impact of imbalances in mineral nutrition such as nutrient deficiencies and salinity stress on the partitioning of nitrate assimilation will be considered. It has long been known that nitrate reduction is shifted towards the root under N-limiting conditions (Andrews, 1986, see Table 2). A comparable change was observed also under conditions of K deficiency, whilst severe P deficiency almost abolished nitrate reduction in the root and led to a shift of this activity towards the shoot (Table 2). NaCl salinity not only inhibits nitrate uptake but it leads to an increase in NRA in the root (Cramer et al., 1995) and again nitrate reduction is shifted from the shoot to the root (Table 2).

The question arises, whether the varied co-operation between root and shoot in nitrate assimilation is due to some signal between the two and/or whether there is a common link between the effects of the four quite different classes of mineral imbalances. In the case of low and high nitrate supply the partitioning between root and shoot has been attributed to a limited capacity of the root for nitrate reduction (Andrews, 1986). Once this capacity is exceeded at higher nitrate supply more nitrate will be transported to the shoot, as was measured (Peuke et al., 1996), and hence must be reduced there. However, there is another important set of conditions, which inevitably will contribute to the partitioning of nitrate reduction and may be the common basis of regulation. These are the prominent role of nitrate as a counter-anion for cations, notably K, during xylem transport, the strict need for cation-anion balance during xylem transport and the fact that nitrate taken up by the root can be translocated either as an anion or – after reduction – in the form of to some proportion only charged amino acids. Given these conditions any major change in the cation-anion balance would have an impact on the site of nitrate reduction (Rufty et al., 1981) and this applies to all four classes of mineral imbalance: (a) When experimentally imposing a low nitrate supply without also causing a shortage in K, nitrate is in part substituted by another anion, usually chloride, and then chloride replaces nitrate in its function as a counter-

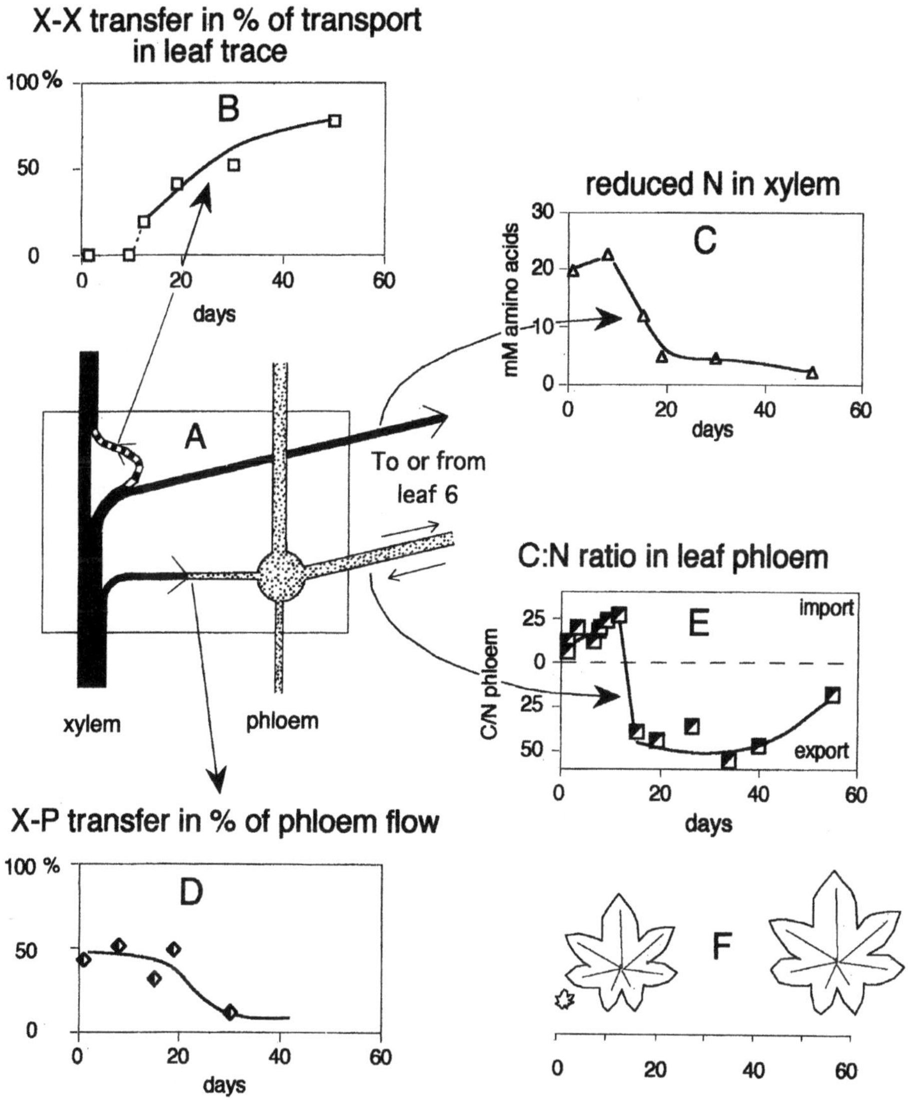

Figure 2. Changes in the extent of xylem to xylem (X-X) and of xylem to phloem (X-P) transfer of reduced nitrogen within the nodal stem segment supporting leaf No. 6 of *Ricinus communis* during its 60 days life span; the x-axis is given in days after leaf emergence. A) Diagram depicting the two transfer processes; the loop designates transfer cell-mediated X-X transfer and the horizontal arrow indicates X-P transfer. B) Changes in the magnitude of X-X transfer of reduced N from leaf trace to stem-own vessels in relation to transport in the leaf trace. C) Time-dependent changes in the concentration of reduced N in the xylem sap of the leaf trace. Note the massive decrease starting with the onset of X-X transfer. D) Changes in the magnitude of xylem to phloem transfer of nitrogen expressed in relation to the flow of N toward or from leaf 6. E) C:N ratio in the phloem sap in the petiole of leaf 6. Positive values indicate import and negative values export of C and N. Note low C:N ratios during import into the developing leaf. F) Schematic representation of the leaf size of leaf 6. Estimated from the data obtained in the experiments published in Jeschke and Pate, 1992.

Table 2. Partitioning of A) nitrate reduction and B) phosphate assimilation between root and shoot as affected by mineral imbalances. A) The contribution of the root to nitrate reduction in the whole plant was measured either on the basis of the balances of flows and deposition of reduced nitrogen and of nitrate within the plant 'M' (empirical modelling of flows) (Jeschke and Pate, 1991b) and confirmed by measurements of in vitro nitrate reductase 'NRA' or on the basis of the ratio of reduced:total N in the xylem sap 'xyl' (Pate, 1973). B) the distribution of phosphate assimilation was similarly measured on the basis of the flows and deposition of PO_4 and organic phosphates 'Mp' (modelling of phosphate flows) (Jeschke et al., 1997)

Species and conditions	A) Nitrate reduction % in root	Method	B) Phosphate assimilation % in root	Method	Reference
Ricinus communis					
control[a]	41	M, xyl	21	Mp	Jeschke et al., 1997
128 m*M* NaCl[a]	51	M, xyl			Jeschke & Pate, 1991b
P deficiency[a),b)]	12	M, xyl	26	Mp	Jeschke et al., 1997
Ricinus communis					
control[c]	33	M, xyl, NRA			Peuke et al., 1996
40 m*M* NaCl[c]	58	M, xyl			
N deficiency[d]	66	M, xyl, NRA			
Lycopersicum esculentum					
control	23[f]	xyl			Cramer et al., 1995
100 m*M* NaCl	60[f]	xyl			
200 m*M* NaCl	57[f]	xyl			
Hordeum vulgare					
control[e]	30	xyl			Förster & Jeschke, 1993
minus K[e]	52	xyl			

[a)] *Ricinus* plants grown in quartz sand culture with 12 m*M* nitrate as N source.
[b)] External phosphate concentration 5 μM.
[c)] *Ricinus* grown in quartz sand culture with 1 m*M* nitrate as N source.
[d)] Plants grown in quartz sand culture with 0.2 m*M* nitrate as the N source.
[e)] *Hordeum* plants grown with split roots in quartz sand culture with 8 m*M* nitrate as N source. Controls received 4 m*M* K on both sides, 'minus K roots' were exposed to an external medium without added K.
[f)] Calculated from the xylem sap concentrations published in Cramer et al., 1995.

anion in xylem transport (Peuke and Jeschke, 1995) and a higher percentage of nitrate needs to be reduced in the root. (b) In presence of saline concentrations of NaCl nitrate uptake is inhibited (Cramer et al., 1995) and chloride competes for nitrate as a counter-anion in xylem transport and hence nitrate reduction is again shifted towards the root. (c) In the case of K deficiency potassium is largely missing as a counter-cation for nitrate transport in the xylem and not to a large extent replaced by other cations (Förster and Jeschke, 1993). The resulting cation deficit limits the possibility of nitrate transport in the xylem and thereby enforces reduction of nitrate in the root and transport of N in the form of amino acids. (d) In the case of P deficiency the strong decrease in root nitrate reduction (Table 2) may be attributed to an anion deficit (Jeschke et al., 1997a), which arises both from the lack in phosphate and from a strong depression of nitrate uptake (Heuwinkel et al., 1992; Pilbeam et al., 1993) under P-deficient conditions and which was apparently not fully compensated by organic anions (Jeschke et al., 1997a). The anion deficit will affect xylem loading and requires a higher allocation of nitrate to transport and results in a lack of nitrate as a substrate for reduction in the root (Jeschke et al., 1997a).

So it appears that the strict requirement for maintenance of cation-anion balance in the xylem can account for the shifts in the partitioning of nitrate reduction under quite diverse classes of imbalances in mineral nutrition and this may hence be the common basis of regulation. The question remains, however, which signals are sensed by root cells and translated into retarded or accelerated nitrate reduction in the incidence of incipient anion or cation deficits.

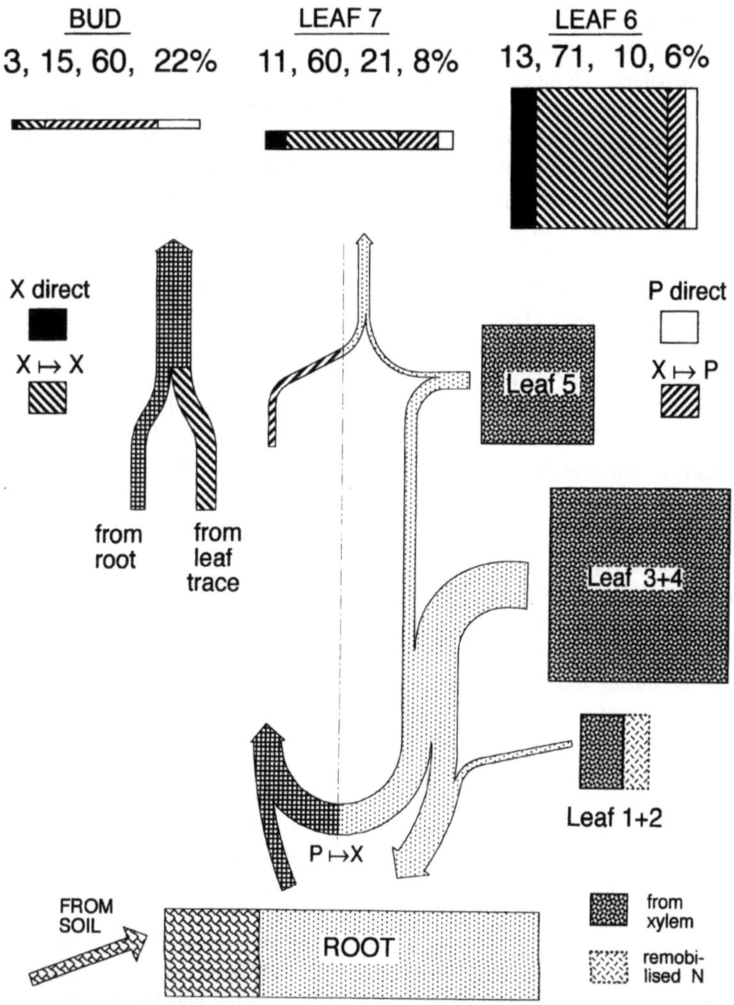

Figure 3. Schematic representation of the supply of nitrogen to the sink organs, the terminal bud, and the developing leaves No. 6 and 7 in the whole plant of Ricinus during the period 44 to 53 days after sowing. The size of the rectangles is drawn in proportion to molar quantities of N incorporated and the numbers denote the sources of nitrogen, from left to right: ■ directly from the root-sourced xylem, ▧ from the xylem after X-X transfer, ▨ from the phloem after X-P transfer or □ from the phloem as exported from the source leaf. The squares similarly are drawn in proportion to the quantities of N exported from leaves 1 to 5, either after transfer from the xylem ■ or after remobilisation ▒. The dotted arrows on the right hand side denote flows in the phloem and the thickness of arrows is drawn in proportion to the flow rates. The crosshatched arrows on the left hand side denote the flows in the xylem (not to scale). The rectangle at the bottom represents the N incorporated into the root and the patterns indicate the quantities taken up from the soil ■ and those imported from the phloem □. Estimated from the data published in Jeschke and Pate, 1991a and Pate and Jeschke, 1995.

In relation to the importance of X-X transfer discussed above and the fact that amino acids and not nitrate are transferred in this process the very necessity for X-X transfer, see above, may be one reason why the costly process of nitrate reduction in the root is of vital importance for the plant, since it is one source for xylem sap amino acids, the substrate for this transfer. The other source is the massive transfer of amino acids from phloem to xylem occurring in the root (Jeschke and Pate, 1991b and see Figure 1 and 3: all of the cycling of N through the root occurs in the form of amino acids).

If the above interpretations for the shifts in nitrate reduction are correct, then the partitioning of phosphate assimilation might *not* be changed by ion imbalances in the xylem, since assimilation of the anion phosphate results in negatively charged organic phosphates and the assimilation should therefore *not* be affected by the charge balance. Compared to nitrate the site of phosphate assimilation so far has attracted little attention, but recently we have been able to assess its distribution within the whole plant of *Ricinus* (Jeschke et al., 1997a) and found a particularly high involvement of young, growing organs in phosphate assimilation. Control roots contributed 21% to total phosphate assimilation and indeed this share was not much altered under P deficiency (Table 2). The slightly larger contribution of the root may be related to relatively favoured root growth in P-deficient plants.

Shoot-root interactions in the selective uptake of K vs. Na by the root

Most glycophytes have a high capacity to discriminate between K and Na and maintain high K and low Na concentrations in the cytoplasm of root cells, in the xylem sap and hence in the shoot. This K/Na selectivity is primarily achieved during membrane transport across the plasmalemma of root cells. Here univalent cation uptake is highly selective for K (Rains and Epstein, 1967; Rains, 1972) and may be achieved by K inwardly rectifying channels (Amtmann et al., 1997; Schroeder et al., 1994). Particularly under saline conditions Na passively enters into the cytoplasm and is then pumped out of the cytoplasm in the vacuole (Barkla and Pantoja, 1996; Dupont, 1992; Jeschke, 1984) or by extrusion to the external medium by Na^+/H^+ antiport (Allen et al., 1995; Jeschke, 1983, 1984; Ratner and Jacoby, 1976). The extrusion of Na across the plasmalemma, requires the presence of external K and proceeds together with K uptake in essence as an exchange of cytoplasmic Na for external K (Jeschke, 1973, 1984) and the overall process will be termed a proton-mediated K/Na exchange. The K/Na selective properties of the root and the maintenance of sufficiently high cytoplasmic K levels are of high importance for salt tolerance (Greenway and Munns, 1980; Jeschke, 1984; see also Tyerman et al., 1997).

Apart from a dependence of these root activities on energy supply by the shoot and apart from a regulation of K uptake in accordance with shoot demand (Engels and Marschner, 1992) maintenance of K/Na selectivity would appear to be unique for the root and independent of the shoot. However, even this root-specific process may be exclusively dependent on the shoot, if external K is missing, as shown by Jeschke and Wolf, 1988. In these experiments roots of *Ricinus* were exposed to a medium without added external K and to a salinity of 100 m*M* NaCl. In order to avoid a severe K deficiency under these conditions, the experiments were performed using a reciprocal split-root system (Jeschke and Wolf, 1988) in which only one half of the root system was exposed to a solution devoid of K but with nitrate whilst the other half was supplied with K but not with nitrate; both root halves were exposed to salinity. Under the highly adverse conditions of both, long-term lack of external K (23 d) and salinity stress (12 d) some K was lost from the low-K roots and a stable external concentration of 10 μM K was thereby established. Despite the absence of added external K and an external Na:K ratio of 10 000:1 the roots stayed healthy, grew vigorously and were able to maintain an almost unchanged and sufficiently high cytoplasmic K and fairly low Na concentration (Table 3). Since external K was missing, this can be attributed only to a co-operation between shoot and root and to an import of K from the phloem. Figure 4 schematically depicts the processes governing K/Na selectivity in these roots. As indicated, phloem import is the sole source for K in these roots. Based on continuous import of K together with assimilates and amino acids K-rich meristematic cells can be formed and then also be elongated at the expense of phloem-borne K, as evidenced by high vacuolar K concentration in newly expanded cells (2.6 mm from the tip, Figure 4). K/Na selectivity at the plasmalemma and maintenance of a low cytoplasmic Na concentration is depicted to be maintained by proton-mediated K/Na exchange (Na^+/H^+ antiport (Allen et al., 1995; Jeschke, 1983, 1984; Ratner and Jacoby, 1976;)) and Na uptake into the vacuole in exchange for K previously taken up during cell expansion. In the present case, where no external K is added, it is suggested that the exchange across the plasmalemma is facilitated by a recycling of the K that is lost from the cytoplasm to the apoplast, whereby the cell wall acts as a reservoir from which previously lost K is retrieved. Alternatively, the proton-mediated exchange of cytoplasmic Na for external K across the plasmalemma may operate on the basis of the 10 μM residual external K (see above). At this concentration this process operates at one quarter of its V_{max} (K_m=20 μM, Jeschke, 1984). The K uptake at this low external K concentration possibly is mediated by a H^+/K^+ symporter (Maathuis and Sanders, 1997).

Clearly these conditions are extremely adverse, but the findings are consistent with a basic function of K

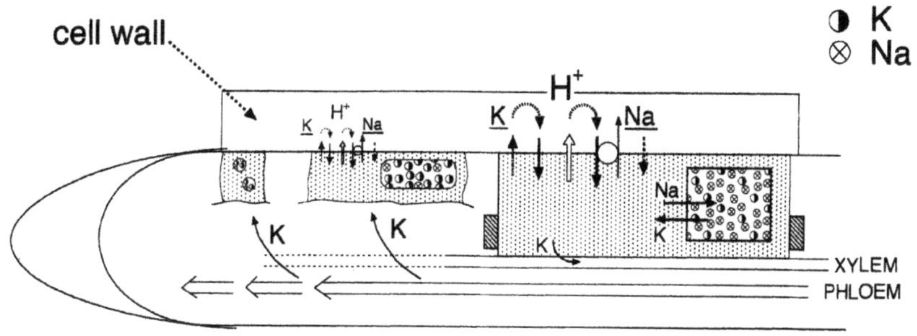

mm from tip:		2.6	8
vacuolar concentration: (mM)	K	96	27
	Na	101	143

Figure 4. Schematic diagram of a root of *Ricinus communis* indicating the processes, which result in the maintenance of K/Na selectivity and allow root growth in an external medium without added K but containing 100 m*M* NaCl. Both cell division and formation of new cells with adequate cytoplasmic K concentration and cell expansion are indicated to proceed at the expense of K imported from the phloem. As shown by the table at the bottom, cell expansion with phloem-borne K results in cells with a high vacuolar K concentration (at 2.6 mm distance from the root tip); in maturing cells (8 mm from tip) this K is exchanged for Na from the external medium. Newly drawn model based on the data of Jeschke and Wolf, 1988 and an earlier scheme published in Jeschke, 1984.

retranslocation in salinity tolerance in the whole plant (Jeschke and Pate, 1991c, Jeschke et al., 1995) and they show that *Ricinus* is capable of growing roots as integral parts of the plant in agreement with the concept of shoot-borne roots (Barlow, 1994). This can be of great ecological advantage, since high salinity can be associated with local depletion of K but the plant is nevertheless capable of promoting root growth. Similar shoot-root co-operations might occur also with other phloem-mobile nutrients.

Hormonal root to shoot signalling in response to mineral imbalances

Under many circumstances ABA is related to stomatal reactions even when leaf water potential is not directly affected by stress (reviews of Davies and Zhang 1991; Hartung and Davies, 1991). This root to shoot signalling not only can occur in case of inadequate soil moisture but also in response to soil mineral imbalances sensed by the root in well watered soils, such as salinity and deficiencies of the soil in essential nutrients (see the review by Hartung and Jeschke, 1999). Mineral deficiencies occur in soils of extreme habitats, which often have high concentrations of different salts and can be very alkaline. Under these conditions nutrients such as phosphate and iron can not be utilised. Therefore salt stress and phosphate deficiency are compared in this chapter.

Salt stress was shown to cause 3–6 fold increases in ABA concentrations in leaf and root tissues of several plants (for references see Hartung and Jeschke, 1999). In seriously salt-stressed white lupins there were dramatic increases in ABA synthesis in the root (20-fold), in xylem transport to the shoot (10-fold) and also in phloem transport of ABA (5-fold) (Wolf et al., 1990). However, in this salt-sensitive plant ABA concentrations in the mature leaves were decreased by salt stress despite the strong increase in the circulation of ABA via phloem and xylem.

Figures 5 and 6 present flow patterns of ABA in intact *Ricinus* plants in response to salt stress and P deficiency. The models have been calculated on the basis of previously measured flow patterns for C or N (see Figure 1) and the ratios of ABA:C or ABA:N in xylem and phloem sap (Wolf et al., 1990).

Under control conditions (Figure 5A) major sites of ABA synthesis are mature leaves with large amounts of ABA transported in the phloem to the sinks, the root and shoot meristems. A substantial part of the ABA imported to the root is recirculated to the shoot.

Table 3. Concentrations of K^+ and Na^+ in the cytoplasm and vacuole of root tissues in *Ricinus communis* grown with reciprocal split roots and in presence of 100 m*M* NaCl salinity. In control plants both root halves obtained the same solution containing 4 m*M* K^+ and 12 m*M* NO_3^-; in reciprocal split root plants one root half '+K–N' was supplied with K^+ but not with NO_3^-, while the other half '−K+N' obtained NO_3^- but no K was added; due to some loss of K from the roots the external K^+ concentration was then 0.01 m*M*. The cytoplasmic and vacuolar K and Na concentrations were obtained by measuring longitudinal ion profiles along the roots, for details see Jeschke and Wolf, 1988

Roots	K^+			Na^+		
	External solution	Cytoplasm	Vacuole	External solution	Cytoplasm	Vacuole
Control	4	143	75	100	17	82
+K–N	4	133	66	100	13	121
−K+N	0.01	130	27	100	46	143

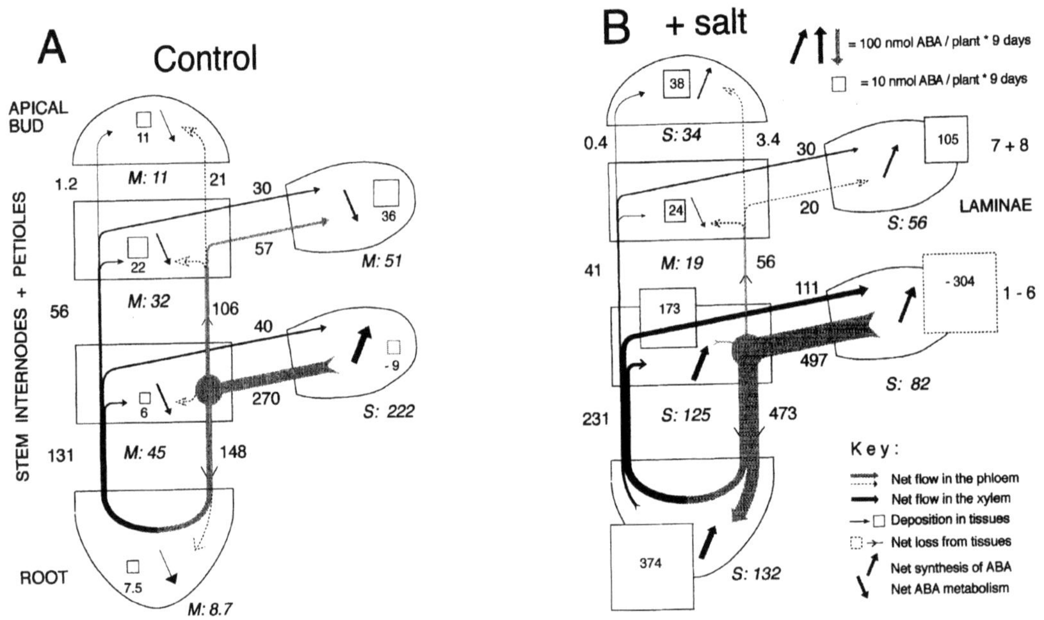

Figure 5. Flow profiles for transport, utilisation and net synthesis or metabolism of ABA in A: control and B: salt-treated plants of *Ricinus communis* L. over the study period 44–53 DAS. The plants were fed 0.5 m*M* PO_4; salt treatment (128 m*M*) commenced 35 DAS. The numbers indicate net rates of flow, deposition or synthesis in nmoles ABA $plant^{-1}$ $(9d)^{-1}$. The width of arrows and areas of squares (see Key) are drawn in proportion to rates, see scale. Redrawn and recalculated for two leaf strata – with omission of glucose esters of ABA – from Jeschke et al., 1997b.

Moderate salt stress (128 m*M* NaCl) led to massive increases in ABA biosynthesis and deposition in the root and to higher transport of ABA from root to shoot, all pointing to the root as the source of an ABA signal. However, phloem import from mature leaves to the root and the ABA recirculation to the shoot were also increased, pointing to a co-operative root-shoot interaction. As a result ABA concentrations in mature leaves were strongly increased in this fairly salt-tolerant species and transpiration was reduced by 82% in terms of the whole plant or by 31% on a leaf area basis (Table 4). (The net loss of ABA from salt-stressed leaves 1–6 (Figure 5) occurred mainly in the senescing lower leaves, see Jeschke et al., 1997b).

It is interesting to note that broadly similar salt stress responses of ABA synthesis in the salt-exposed root and of ABA flows from root to shoot are differently effective in salt-tolerant *Ricinus* and the salt-sensitive *Lupinus albus*. In the latter ABA concentrations of leaves decreased in presence of salt despite

Table 4. Mean ABA concentrations in tissues of mature leaves and the root of *Ricinus communis* during the study period 44–53 days after sowing and transpirational water loss by the plants over this study period

Conditions	ABA concentration[a)] [nmol g^{-1} FW]		Transpirational water loss by the leaves[b)]	
	Mature leaves	Root system	g H_2O $plant^{-1}$ $(9d)^{-1}$	kg H_2O m^{-2} $(9d)^{-1}$
Control	0.7±0.13	0.13±0.08	5534	23.4
P-deficient (5 μM PO_4)	1.2±0.23	0.062±0.007	977	15.2
Salt (128 mM NaCl)	13±3.5	7.2	1012	16.2

[a)] Means of the data for the initial and final harvest as given in Jeschke et al., 1997.
[b)] Transpiration was measured by daily weighing of the plants during the study period and related to the mean leaf area of the whole plant.

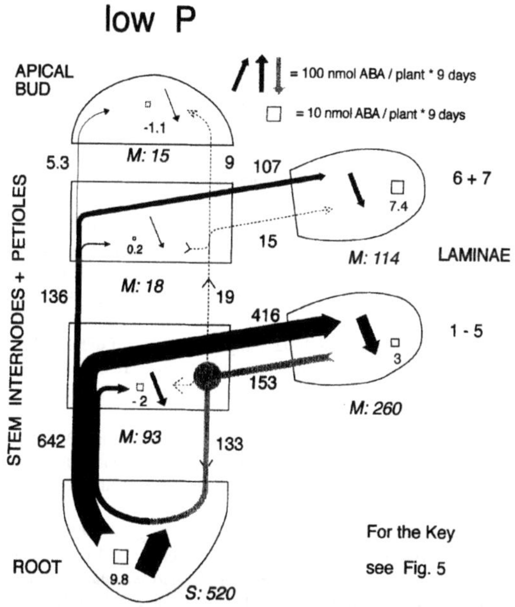

Figure 6. Flow profiles for transport, utilisation and net synthesis or metabolism of ABA in P-deficient plants of *Ricinus communis* L. over the study period 44–53 DAS. PO_4 had been withdrawn at 29 DAS and 5 μM PO_4 was then given starting at 41 DAS. Note that the scale for arrows and squares is the same as in Figure 5, highlighting the substantial increase in xylem transport of ABA in the case of P deficiency. In these plants the youngest leaves were leaf 6 and 7. Other details as in Figure 5. Redrawn and recalculated for two leaf strata – with omission of glucose esters of ABA – from Jeschke et al., 1997b.

much higher transport from root to shoot (Wolf et al., 1990) and leaves showed symptoms of wilting (Jeschke et al., 1986).

As evident from Figure 6, *Ricinus* roots responded in a way similarly to severe phosphate deficiency in the soil (5 μM PO_4 in the watering solution) as to salinity and increased ABA biosynthesis substantially. However, although most of this newly synthesised ABA was loaded into the xylem and transported to the leaves, ABA concentrations in leaf tissues increased only slightly (Table 4) and the modelling predicted that most of the ABA imported via xylem was degraded by metabolism and phloem transport was not increased. Nevertheless, leaf transpiration was similarly decreased as by salt stress (Table 4). The data suggest that xylem-imported ABA acts on the stomata before it is metabolised in mesophyll cells. These findings explain the data published by Radin (1984). In his study leaves of P-deficient cotton plants showed only slightly increased ABA concentrations. The responsiveness of stomata to drought stress was, however, markedly enhanced under P deficiency.

Concluding remarks

Root to shoot interaction in relation to mineral nutrition include complex interactions and transfers between xylem vessels serving the leaves and those passing further up the stem and between xylem and phloem within the stem. Diverse nutritional imbalances affect the partitioning of nitrate assimilation between shoot and root, while this was not the case for phosphate assimilation. Both of these effects can be attributed to the strict requirement for anion-cation balance during xylem transport. An interaction between shoot and root in the form of potassium recirculation via phloem is required for the maintenance of root growth and adequate symplastic K concentrations in root cells under K-deficient conditions. Abscisic acid plays an important role as hormonal root to shoot signal under mineral deficiency and salinity to regulate growth, development and stomatal activity in shoot meristems and leaves.

Acknowledgements

We are grateful to the Deutsche Forschungsgemeinschaft (SFB 2512, TP A3 and A5) for financial support and to Mrs Andrea Hilpert and Barbara Dierich for expert technical assistance.

References

Allen G J, Wyn Jones R G and Leigh R A 1995 Sodium transport in plasma membrane vesicles isolated from wheat genotypes with different K^+/Na^+ discrimination traits. Plant Cell Environ. 18, 105–115.

Amtmann A, Laurie S, Leigh R and Sanders D 1997 Multiple inward channels provide flexibility in Na^+/K^+ discrimination at the plasma membrane of barley suspension culture cells. J. Exp. Bot. 48, 481–497.

Andrews M 1986. The partitioning of nitrate assimilation between root and shoot of higher plants. Plant Cell Environ. 9, 511–519.

Barkla B J and Pantoja O 1996 Physiology of ion transport across the tonoplast of higher plants. Annu. Rev. Plant Physiol. Mol. Biol. 47, 159–184.

Barlow P W 1994 The origin , diversity and biology of shoot-borne roots. *In* The biology of adventitious root formation. Eds. T D Davies and B E Haissig. pp 1–23. Plenum Press, New York.

Brewitz E, Larsson C M and Larsson M 1995. Influence of nitrate supply on concentrations and translocation of abscisic acid in barley (*Hordeum vulgare*). Physiol. Plant. 95, 499–506.

Cramer M D, Schierholt A, Wang Y Z and Lips S H 1995. The influence of salinity on the utilization of root anaplerotic carbon and nitrogen metabolism in tomato roots. J. Exp. Bot., 46, 1569–1577.

Davies W J and Zhang J 1991 Root signals and the development of plants growing in drying soil. Annu. Rev. Plant Physiol. Mol. Biol. 42, 55–76.

Dupont F M 1992 Salt-induced changes in ion transport: regulation of primary pumps and secondary transporters. *In* Transport and receptor proteins of plant membranes. Molecular structure and function. Eds. D T Clarkson and D T Cook. pp 91–100. Plenum Press, New York.

Engels C and Marschner H 1992 Adaptation of potassium translocation into the shoot of maize (*Zea mays*) to shoot demand: Evidence for xylem loading as a regulating step. Pysiol. Plant. 86, 263–268.

Engels C, Münkle L and Marschner H 1992 Effect of root zone temperature and shoot demand on uptake and xylem transport of macro nutrients in maize (*Zea mays* L.). J. Exp. Bot. 43, 537–547.

Greenway H and Munns R 1980 Mechanisms of salt tolerance in non-halophytes. Annu Rev. Plant Phsysiol. 31, 149–190.

Hartung W and Davies W J 1991 Drought induced changes in physiology and ABA. *In* Abscisic Acid. Physiology and Biochemistry. Eds. W J Davies and H G Jones. pp 63–79. Bios Scientific Publishers, Oxford.

Hartung W and Jeschke W D 1999 Abscisic acid: A long distance stress signal in salt-stressed plants. *In* Plant response to environmental stresses. Ed. H R Lerner. pp 333–348 Marcel Dekker, New York. (*In press*).

Heuwinkel H, Kirkby E A, Le Bot J and Marschner H 1992 Phosphorus deficiency enhances molybdenum uptake by tomato plants. J. Plant Nutr. 15, 549–568.

Imsande J and Touraine B 1994 N demand and the regulation of nitrate uptake. Plant Physiol. 105, 3–7.

Jeschke W D 1973 K^+-stimulated Na^+ efflux and selective transport in barley roots. *In* Ion transport in plants. Ed. W P Anderson. pp 285–296. Academic Press, New York.

Jeschke W D 1980 Involvement of proton fluxes in K^+-Na^+ selectivity at the plasmalemma; K^+-dependent net extrusion of sodium in barley roots and the effect of anions and pH on sodium fluxes. Z. Pflanzenphysiol. 98, 155-175.

Jeschke W D 1984 K^+–Na^+ exchange at cellular membranes, intracellular compartmentation of cations, and salt tolerance. *In* Salinity tolerance in plants: Strategies for crop improvement. Eds. J D Staples and G H Toenniessen. pp. 37–66. Wiley, New York.

Jeschke W D, Klagges S, Hilpert A, Bhatti A S and Sarvar G 1995 Partitioning and flows of ions and nutrients in salt-treated plants of *Leptochloa fusca* L. Kunth. New Phytol. 130, 23–35.

Jeschke W D, Pate J S and Atkins C A 1986 Effects of NaCl salinity on growth, development, ion transport and ion storage in white lupin (*Lupinus albus* L., cv. Ultra). J. Plant Physiol. 124, 257–274.

Jeschke W D and Pate J S 1991a Modelling the uptake, flow and utilization of C, N and H_2O within whole plants of *Ricinus communis* L. based on empirical data. J. Exp. Bot. 137, 488–498.

Jeschke W D and Pate J S 1991b Modelling the partitioning, assimilation and storage of nitrate within the root and shoot organs of castor bean (*Ricinus communis* L.). J. Exp. Bot. 42, 1091–1103.

Jeschke W D and Pate J S 1991c Cation and chloride partitioning through xylem and phloem within whole plants of *Ricinus communis* L. under conditions of salt stress. J. Exp. Bot. 42, 1105–1116.

Jeschke W D and Pate J S 1992 Temporal patterns of uptake, flow and utilization of nitrate, reduced nitrogen and carbon in a leaf of salt-treated castor bean (*Ricinus communis* L.). J. Exp. Bot. 43, 392–402.

Jeschke W D, Kirkby E A, Peuke A, Pate J S and Hartung W 1997a Assimilation and transport of nitrate and of phosphate in whole plants of castor bean (*Ricinus communis* L.) as affected by P deficiency. J. Exp. Bot. 48, 75–91.

Jeschke W D, Peuke A, Pate J S and Hartung W 1997b Synthesis, Transport and Catabolism of Abscisic Acid in a single leaf and whole plant of castor bean (*Ricinus communis* L.) under moderate salinity and phosphate deficiency. J. Exp. Bot. 48, 1737–1747.

Jeschke W D and Wolf O 1988 External potassium supply is not required for root growth in saline conditions: Experiments with *Ricinus communis* L. grown in a reciprocal split-root system. J. Exp. Bot. 39, 1149–1167.

Layzell D B, Pate J S, Atkins C A and Canvin D T 1981 Partitioning of carbon and nitrogen and the nutrition of root and shoot apex in a nodulated legume. Plant Physiol. 67, 30–36.

Maathuis F J M and Sanders D 1997 Regulation of K^+ absorption in plant root cells by external K^+: interplay of different plasma membrane K^+ transporters. J. Exp. Bot. 48, 451–458.

Marschner H 1995 Mineral nutrition of higher plants. Second Edition. Academic Press, London. pp 101–105.

Pate J S 1973 Uptake, assimilation and transport of nitrogen compounds by plants. Soil Biol. Biochem. 5, 109–119.

Pate J S and Jeschke W D 1995 Role of stems in transport, storage, and circulation of ions and metabolites by the whole plant. *In* Plant stems, physiology and functional morphology. Ed. B L Gardner. pp 177–204. Academic Press, San Diego.

Pate J S, Layzell D B and McNeil D L 1979 Modelling the transport and utilization of carbon in an nodulated legume. Plant Physiol. 62, 730–738.

Pate J S, Sharkey P J and Lewis O A M 1975 Xylem to phloem transfer of solutes in fruiting shoots of legumes, studied by a phloem bleeding technique. Planta 122, 11–26.

Peuke A D and Jeschke W D 1995 Effect of nitrogen source, nitrate concentration and salt stress on element and ion concentration in transport fluids on C and N flows in *Ricinus communis* L. *In* Structure and function of roots: Eds. Baluska F, Ciamporová M, Gasparíková O M and Barlow P W. pp 229–236. Kluwer Academic Publishers, Dordrecht.

Peuke A D, Jeschke W D and Hartung W 1994 The uptake and flow of C, N and ions between roots and shoots in *Ricinus communis* L. III. Long distance transport of abscisic acid depending on nitrogen nutrition and salt stress. J. Exp. Bot. 45, 741–747.

Peuke A D, Jeschke W D and Hartung W 1998 Foliar application of nitrate or ammonium as the sole nitrogen supply in *Ricinus communis*. II. The flow of cations, chloride and abscisic acid. New Phytol. 140, 625–636.

Peuke A D, Glaab J, Kaiser W M and Jeschke W D 1996 The uptake and flow of C, N and ions between roots and shoots in *Ricinus communis* L. IV. Flow and metabolism of inorganic nitrogen and malate depending on nitrogen nutrition and salt treatment. J. Exp. Bot. 47, 377–385.

Pilbeam D J, Cakmak I, Marschner H and Kirkby E A 1993 Effect of withdrawal of phosphorus on nitrate assimilation and PEP carboxylase activity in tomato. Plant Soil 154, 111–117.

Radin J W 1984 Stomatal responses to water stress and to abscisic acid in phosphorus deficient cotton plants. Plant Physiol. 76, 392–394.

Rains D W and Epstein E 1967 Sodium absorption by barley roots: role of dual mechanisms of alkali cation transport. Plant Physiol. 42, 314–318.

Rains D W 1972 Salt transport by plants in relation to salinity. Annu. Rev. Plant Physiol. 23, 367–3881.

Ratner A and Jacoby B 1976 Effect of K^+, its counter anion, and pH on sodium efflux from barley root tips. J. Exp. Bot. 27, 843–852.

Rufty T W Jr, MacKown C and Volk R J 1989 Effects of altered carbohydrate availability on whole-plant assimilation of $^{15}NO_3^-$. Plant Physiol. 89, 457–463.

Schroeder J I, Ward J M and Gassmann W 1994 Perspectives on the physiology and structure of inward rectifying K^+ channels in higher plants: biophysical implications for K^+ uptake. Annu. Rev. Biophys. Biomol. Struct. 23, 441–471.

Tyermann S D, Skerrett M, Garrill A, Findlay G P and Leigh R A 1997 Pathways for the permeation of Na^+ and Cl^- into protoplasts derived from the cortex of wheat roots. J. Exp. Bot. 48, 459–480.

Wolf O, Jeschke W D and Hartung W 1990 Long distance transport of abscisic acid in salt stressed Lupinus albus plants. J. Exp. Bot. 41, 593–600.

Wolf O, Munns R, Tonnet M L and Jeschke W D 1991 The role of the stem in the partitioning of Na^+ and K^+ in salt-treated barley. J. Exp. Bot. 42, 697–704.

Section editor: H Lambers

Recent Advances of Plant Root Structure and Function, 97–102
Edited by O. Gašparíková, M. Čiamporová, I. Mistrík and F. Baluška
© 2001 *Kluwer Academic Publishers.*

Calcium channels in the plasma membrane of root cells: Their roles in mineral nutrition and cell signalling

P.J. White[1,*] & M.S. Ridout[2]
[1]*Department of Cell Physiology, Horticulture Research International, Wellesbourne, Warwick CV35 9EF and*
[2]*Department of Biometrics, Horticulture Research International, East Malling, West Malling, Kent ME19 6BJ*

Key words: Calcium (Ca^{2+}), chilling, endodermis, kinetics, low temperature, permeation

Abstract

Calcium channels in the plasma membrane of root cells facilitate the movement of Ca^{2+} to the xylem, by providing a symplastic bypass to the Casparian band, and regulate cellular activities through changes in cytoplasmic Ca^{2+} concentration. In this paper, the biophysical properties and physiological functions of Ca^{2+} channels in the plasma membrane of root cells are reviewed. Acclimation to low temperatures is given as an example to illustrate how kinetic and permeation models can be used to understand the physiology of Ca^{2+} channels and to predict Ca^{2+} fluxes under physiological conditions.

Abbreviations: $[Ca^{2+}]_{cyt}$ – cytoplasmic Ca^{2+} concentration; E_{rev} – zero current (reversal) potential; GFP – green fluorescent protein; PLB – planar lipid bilayer; SAP – slow action potential

Introduction

Calcium is an essential and major plant nutrient. It is required in its ionic (Ca^{2+}) form for a variety of extracellular structures, as a cytoplasmic secondary messenger and as a countercation in the vacuole (Marschner, 1995). The delivery of calcium to the shoot requires calcium movement through the root symplasm and Ca^{2+} influx to root cells is also implicated in cell signalling (White, 1998). The entry of Ca^{2+} into root cells is likely to be mediated by ion channels, which allow the rapid movement of Ca^{2+} down its electrochemical gradient. There is a considerable electrochemical gradient for the entry of Ca^{2+} into root cells. Not only is there a cytoplasm-negative voltage difference across the plasma membrane but there is also a steep Ca^{2+} concentration gradient between the apoplast (millimolar) and cytoplasm (submicromolar). In this article, the physiological roles and properties of Ca^{2+} channels in the plasma membrane of root cells are discussed. A plasma membrane 'Ca^{2+} channel' is defined as a channel permeable to Ca^{2+} whose primary physiological role is to mediate Ca^{2+} influx, either to facilitate the movement of Ca^{2+} to the xylem or to regulate cellular activities through changes in $[Ca^{2+}]_{cyt}$. Thus, although several outward-rectifying K^+ channels in plant roots are permeable to Ca^{2+} (Gaymard et al., 1998), facilitating Ca^{2+} influx is unlikely to be their main physiological role and they are not considered here.

Physiological roles of root plasma membrane Ca^{2+} channels

Calcium must be acquired by the root from the soil solution and delivered to the xylem at rates sufficient for shoot growth. Since the movement of calcium through the symplasm may be restricted by the low cytoplasmic Ca^{2+} concentration ($[Ca^{2+}]_{cyt}$), it is often assumed that radial Ca^{2+} transport across the root cortex occurs primarily by an apoplastic route (Clarkson, 1993; White, 1998). However, since the development of both the Casparian band and a functional xylem appear to be coincident, and since the Casparian band prevents apoplastic ion transport, it has been proposed that Ca^{2+} destined for the xylem must enter the root symplasm at the endodermis (Clarkson,

* FAX No: (01789) 470552. E-mail: philip-j.white@hri.ac.uk

1993). To transfer Ca^{2+} across the Casparian band it is envisaged that catalytically active Ca^{2+} channels are located on the cortical side of endodermal cells and that transporters mediating Ca^{2+} efflux are located in cells within the stele. This scheme defines the role of Ca^{2+} channels in plant nutrition.

In addition to their role in plant nutrition, Ca^{2+} channels in the plasma membrane of root cells perform other vital physiological functions (reviewed by White, 1998). In expanding cells, such as cells in the elongation zone or root hair cells, elevated $[Ca^{2+}]_{cyt}$ is required both to maintain growth and to establish the direction of growth. The influx of Ca^{2+} through (mechanosensitive) Ca^{2+} channels has been implicated in this process. All root cells must respond to environmental signals and many must also respond to specific developmental cues. Calcium influx has been implicated in signalling during gravitropism and thigmotropism, and in the responses of plants to pathogens, wounding and low temperatures. It is expected that ligand-modulated Ca^{2+} channels in cells competent to respond to specific developmental and pathological stimuli initiate physiological responses, as they do for example in other tissues (Blumwald et al., 1998), and that voltage-dependent Ca^{2+} channels will provide a generalised signalling mechanism.

Ca^{2+} channels present in the plasma membrane of root cells

Calcium channels in the plasma membrane of plant roots have been characterised both from ^{45}Ca flux measurements into isolated vesicles (e.g. Huang et al., 1994; Marshall et al., 1994) and electrically, either after incorporating vesicles into planar lipid bilayers (PLB; reviewed by White, 1998) or by patch-clamping root cell protoplasts (Kiegle et al., 1998; Thion et al., 1998). All studies indicate the presence of Ca^{2+} channels which activate at (depolarised) membrane potentials more positive than the resting potential of the root cell. However, at least three types of depolarisation-activated Ca^{2+} channels can be distinguished on the basis of their differential sensitivities to La^{3+}, Gd^{3+} and verapamil (White, 1998).

Two depolarisation-activated Ca^{2+} channels have been characterised following the incorporation of plasma membrane vesicles from cereal roots into PLB. One was termed the maxi cation-channel because it has a high unitary conductance (450 pS in 100 m*M* KCl and 135 pS in 100 m*M* $CaCl_2$; White, 1993). The other has been called voltage-dependent cation channel two (VDCC2) in rye (White, 1994, 1998) and the root calcium (*rca*) channel in wheat (Piñeros and Tester, 1995, 1997). The VDCC2/*rca* channel has a lower unitary conductance than the maxi cation channel, approximating 170 pS in 100 m*M* KCl and 25 to 40 pS in 100 m*M* $CaCl_2$. Both channels are permeable to a wide variety of monovalent and divalent cations and models based on free energy profiles have been developed for K^+, Na^+ and Ca^{2+} permeation (White and Ridout, 1999; White and Piñeros, unpublished). Under physiological conditions both channels are likely to mediate Ca^{2+} influx and net K^+ efflux. However, since they are relatively non-selective, they may also contribute to the uptake of other cations and to cation toxicities. Both channels are voltage-regulated and inactivate rapidly at high negative membrane potentials under physiological ionic conditions. Both would be inactive at the normal resting membrane potential of root cells. However, they would open upon membrane depolarisation (which could occur in response to one of a number of environmental, pathological or hormonal stimuli) to allow Ca^{2+} into the cytoplasm. Since the $[Ca^{2+}]_{cyt}$, is of critical importance for the physiological poise of a plant cell, this could act as a signal to initiate physiological or developmental changes, and a role in cell signalling has been proposed for these channels.

When protoplasts from roots of *Arabidopsis* were studied by patch-clamp techniques a depolarisation-activated Ca^{2+}-current was also reported (Figure 1A; Thion et al., 1998). This current was stabilised by pharmaceuticals (oryzalin and colcicine) that disrupt microtubule organisation and was persistent in protoplasts from roots of the *Arabidopsis ton2* mutant, which has aberrant microtubule organisation. These observations strongly suggest that the underlying Ca^{2+} channels are regulated by interactions with the microtubule network. It has been suggested that cortical microtubules provide spatial information for cell wall deposition and are part of the sensing mechanism for mechanical stress (Wymer et al., 1996). Unfortunately, the anatomical origin of the protoplasts used in the study of Thion et al. (1998) is unknown.

Recently, a molecular-genetic cell-marking approach was described that enables electrophysiologists to identify specific cell-types for subsequent patch-clamp studies (Maathuis et al., 1998). In principle, the expression of green fluorescent protein (GFP) is driven by a cell-type specific promoter in roots of transgenic plants. These are digested and fluorescent

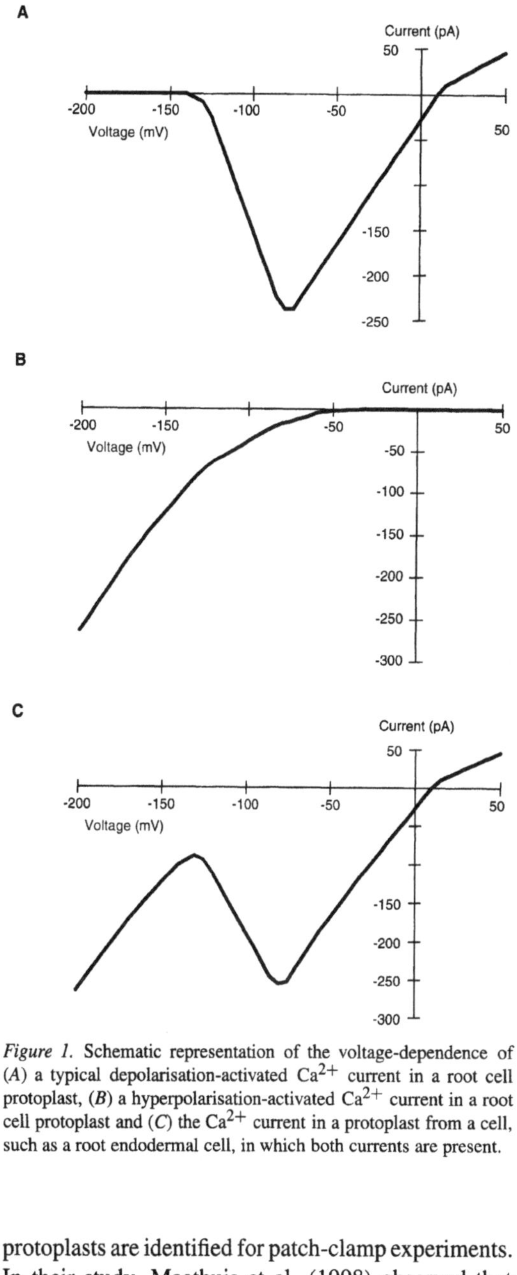

Figure 1. Schematic representation of the voltage-dependence of (*A*) a typical depolarisation-activated Ca^{2+} current in a root cell protoplast, (*B*) a hyperpolarisation-activated Ca^{2+} current in a root cell protoplast and (*C*) the Ca^{2+} current in a protoplast from a cell, such as a root endodermal cell, in which both currents are present.

protoplasts are identified for patch-clamp experiments. In their study, Maathuis et al. (1998) observed that GFP did not affect either the occurrence or biophysical characteristics of the dominant K^+ currents in root cells. This cell-marking approach is enabling the comparison of Ca^{2+} currents across the plasma membrane of various cell types from *Arabidopsis* roots (Kiegle et al., 1998). In addition to an ubiquitous depolarisation-activated Ca^{2+}-current (Figure 1A), a hyperpolarisation-activated Ca^{2+} current (Figure 1B) was observed in certain cell types (Kiegle et al., 1998). Since the depolarisation activated Ca^{2+} current occurred in all cell types, but the hyperpolarisation activated Ca^{2+} current occurred predominantly in endodermal cells, it is tempting to suggest that depolarisation activated Ca^{2+} channels may be involved in cell signalling, whereas the hyperpolarisation activated Ca^{2+} channels are involved in mineral nutrition.

Models of Ca^{2+} channel behaviour: from biophysics to physiology

The acclimation of chilling-resistant plants to growth at low temperatures is thought to be initiated in part by Ca^{2+} influx across the plasma membrane and mediated by the consequent increase in $[Ca^{2+}]_{cyt}$.

Rapid cooling of plant roots to low non-freezing temperatures evokes an electrical response termed the slow action potential (SAP; Minorsky, 1989). This consists of a rapid membrane depolarisation followed by repolarisation and restoration of the resting membrane potential (Figure 3B). It is believed, but not verified, that the SAP results from the successive opening of depolarisation-activated Ca^{2+} channels, Cl^- channels and outward-rectifying K^+ channels in the plasma membrane. It is thought that the opening of Ca^{2+} channels results in the transient increase in $[Ca^{2+}]_{cyt}$ observed experimentally (eg. Knight et al., 1996; Lewis et al., 1997). The kinetics of the changes in both membrane potential and $[Ca^{2+}]_{cyt}$ are related to the rate, magnitude, duration and repetitiveness of root cooling and both the SAP and the increase in $[Ca^{2+}]_{cyt}$ upon cooling are inhibited by La^{3+} but not by diltiazem, verapamil and nifedipine (White, 1998). This is the pharmacology of the maxi cation channel (White, 1996). To investigate a possible role for the maxi cation channel in low-temperature signalling, its responses to the changes observed in cell membrane potential during chilling were predicted using models of channel gating and permeation (White and Ridout, 1999).

Kinetic model. The kinetics of the maxi cation channel are dominated by inactivation at extreme voltages (White, 1993). The rates of transitions between the active (O) and inactive (I) states of the channel are denoted by the voltage-dependent rate constants k_1 and

k_{-1} (both expressed s^{-1}):

$$O \underset{k_{-1}}{\overset{k_1}{\rightleftarrows}} I$$

For simplicity it was assumed that the voltage-dependence of both k_1 and k_{-1} varied in parallel with changes in the zero current (reversal) potential (E_{rev}) for the channel (cf. White, 1993), and, where V is the voltage displacement from E_{rev}, these equations apply (White and Ridout, 1999):

$$k_1 = e^{(-20.72-(0.173V))}$$

$$k_{-1} = e^{(13.82+(0.210V))}$$

The probability of finding an active channel (P_a) at time (t) can be determined from the relationships:

$$P_a(t) = P_a(\infty) + [P_a(0) - P_a(\infty)]e^{-t/\tau}$$

$$P_a(\infty) = k_{-1}/(k_1 + k_{-1})$$

and

$$\tau = 1/(k_1 + k_{-1})$$

Again, for simplicity, a value of unity for the open probability of the active channel was taken. This is not precisely correct, but results in minor errors.

Permeation model. A model to describe the permeation of K^+, Na^+ and Ca^{2+} through the pore of the maxi cation channel was developed using the computer programme AJUSTE (Alvarez et al., 1992; White and Ridout, 1999). The model was composed of three energy-barriers and two ion binding sites (a 3B2S model). It assumed single-file permeation and allowed cations to occupy both binding-sites simultaneously and interact within the pore. The model (Figure 2A; White and Ridout, 1999) suggested an asymmetrical free energy profile, with differences in permeation between cations being attributed primarily to differences in their free energy profiles in the regions of the pore adjacent to the extracellular solution. Different affinities for ion binding sites and contrasting heights of the central free energy barrier between cations were implied. There was repulsion between cations within the pore, and the vestibules of the pore appeared to be negatively charged.

The model adequately fitted the diverse current/voltage relationships obtained over a wide variety of experimental conditions. It described the phenomena of non-Michaelian unitary-conductance versus activity relationships for K^+, Na^+ and Ca^{2+}, differences in selectivity sequences obtained from measurements of conductance and permeability ratios, changes

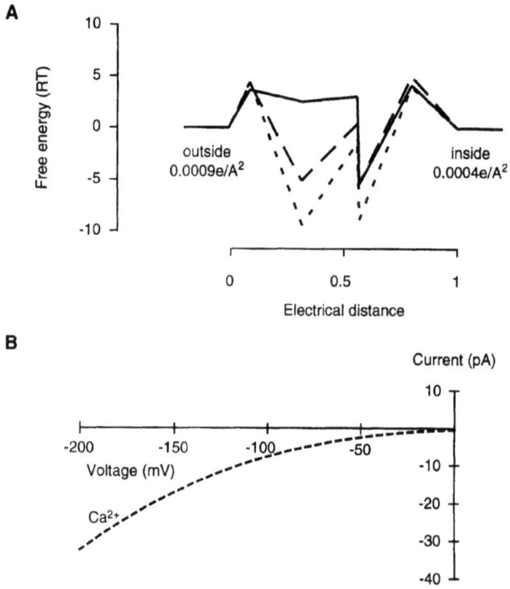

Figure 2. (*A*) Free energy profiles, defined at 20 °C with a solution reference state of 55.5 *M*, for the permeation of K^+ (——), Na^+ (– – –) and Ca^{2+} (- - - - -) through the pore of the maxi cation channel in the plasma membrane of rye roots. Significant repulsion occurs between cations within the pore, which increases the free energy of the unoccupied binding site in an occupied pore by 1.16 to 4.64 RT units, depending upon the valency of the interacting cations. Significant surface charge is present in both the external (0.0009 e/Å^2) and cytoplasmic (0.0004 e/Å^2) vestibules. (*B*) The predicted voltage-dependence of the Ca^{2+} current through a single, open maxi cation channel under physiological ionic conditions at 20 °C. The extracellular K^+, Na^+ and Ca^{2+} activities were 0.90, 0.93 and 0.70 m*M*, respectively. The cytoplasmic K^+, Na^+ and Ca^{2+} activities were 71.5 m*M*, 3.5 m*M*, and 100 n*M*, respectively. For further details see White and Ridout (1999).

in relative cation permeabilities with solution ionic composition and the complex effects of Ca^{2+} on K^+ currents through the channel (White and Ridout, 1999).

Using these models for channel kinetics and cation permeation the unitary currents and net ion fluxes through the maxi cation channel were predicted under physiological ionic conditions during the SAP (White and Ridout, 1999). Calcium influx through an active (open) maxi cation channel was predicted assuming activities of 0.90 m*M* K^+, 0.93 m*M* Na^+ and 0.70 m*M* Ca^{2+} in the extracellular medium and cytoplasmic activities of 71.5 m*M* K^+, 3.5 m*M* Na^+ and 100 n*M* Ca^{2+} (Figure 2B). Under these conditions Ca^{2+} influx dominated the current through this channel (White and Ridout, 1999) and E_{rev} varied from −42.5 mV at 18 °C to −43.6 mV at 23 °C. The Ca^{2+} influx

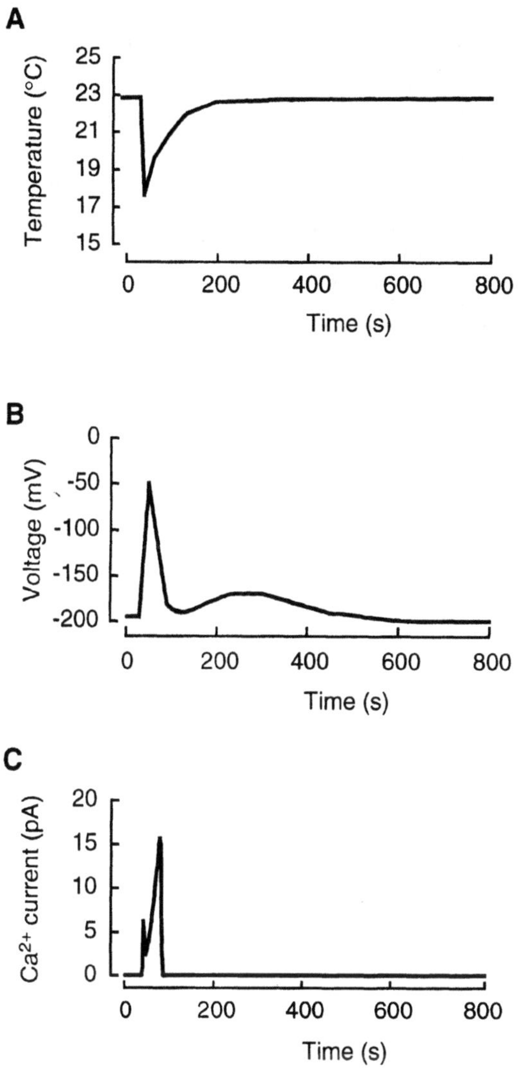

Figure 3. (*A*) Temperature perturbation imposed by Minorsky and Spanswick (1989) to elicit slow action potentials (SAP) in cucumber roots. (*B*) A typical SAP in cucumber roots induced after 30 s by rapid cooling (20 °C min^{-1}) from 23 °C to 18 °C followed by re-warming. Data taken from Minorsky and Spanswick (1989). (*C*) The mean Ca^{2+} influx predicted to occur through a single maxi cation channel during the SAP shown in panel *B*. The Ca^{2+} influx was calculated as the probability of the channel being active multiplied by the Ca^{2+} current through the open channel.

through the open channel was multiplied by P_a to predict the instantaneous Ca^{2+} influx through the maxi cation channel during the SAP (Figure 3C). This was in the form of a single peak, which lagged slightly behind the depolarising stimulus. This parallels the transient rise in $[Ca^{2+}]_{cyt}$ which is observed, for example, when *Arabidopsis* seedlings are cooled (Knight et al., 1996; Lewis et al., 1997). Thus, it is possible for Ca^{2+} influx through the maxi cation channel to transduce a transient voltage-mediated signal by increasing $[Ca^{2+}]_{cyt}$.

Acknowledgements

This work was supported by the Biotechnology and Biological Sciences Research Council (UK). The Horticulture Research Association provided funds through the Vegetable Research Trust for PJW to attend the Fifth International Symposium on Structure and Function of Roots. We thank Malcolm Bennett and Sean May (Warwick University, UK), Frans Maathuis and Dale Sanders (York University, UK), Miguel Piñeros (Cornell University, USA) and Helen Moore (HRI) for their contributions to the work presented here. We thank Julia Davies and Ed Kiegle (Cambridge University, UK) for productive discussions and Richard Napier (HRI) for reading the manuscript.

References

Alvarez O, Villarroel A and Eisenman G 1992 Calculation of ion currents from energy profiles and energy profiles from ion currents in multibarrier, multisite, multioccupancy channel models. Methods Enzymol. 207, 816–854.

Blumwald E, Aharon G S and Lam B C-H 1998 Early signal transduction pathways in plant-pathogen interactions. TIPS 3, 342–346.

Clarkson D T 1993 Roots and the delivery of solutes to the xylem. Phil. Trans. Roy. Soc. (London) B 341, 5–17.

Gaymard F, Pilot G, Lacombe B, Bouchez D, Bruneau D, Boucherez J, Michaux-Ferrière, Thibaud J-B and Sentenac H 1998 Identification and disruption of a plant shaker-like outward channel involved in K^+ release into the xylem sap. Cell 94, 647–655.

Huang J W, Grunes D L and Kochian L V 1994 Voltage-dependent Ca^{2+} influx into right-side-out plasma membrane vesicles isolated from wheat roots: Characterization of a putative Ca^{2+} channel. Proc. Natl. Acad. Sci. USA 91, 3473–2477.

Kiegle E, Haseloff J and Tester M 1998 Calcium currents in GFP-expressing endodermal protoplasts from *Arabidopsis. In* Abstracts from the 11th International Workshop on Plant Membrane Biology, Cambridge. Abstract 198.

Knight H, Trewavas A J and Knight M R 1996 Cold calcium signalling in *Arabidopsis* involves two cellular pools and a change in calcium signature after acclimation. Plant Cell 8, 489–503.

Lewis B D, Karlin-Neumann G, Davis R W and Spalding E P 1997 Ca^{2+}-activated anion channels and membrane depolarizations induced by blue light and cold in *Arabidopsis* seedlings. Plant Physiol. 114, 1327–1334.

Maathuis F J M, May S T, Graham N S, Bowen H C, Jelitto T C, Trimmer P, Bennett M J, Sanders D and White P J 1998 Cell marking in *Arabidopsis thaliana* and its application to patch-clamp studies. Plant J. 15, 843–851.

Marschner H 1995 Mineral Nutrition of Higher Plants. Second edition. Academic Press, London.

Marshall J, Corzo A, Leigh R A and Sanders D 1994 Membrane potential-dependent calcium transport in right-side-out plasma membrane vesicles from *Zea mays* L. roots. Plant J. 5, 683–694.

Minorsky P V 1989 Temperature sensing by plants: a review and hypothesis. Plant Cell Environ. 12, 119–135.

Minorsky P V and Spanswick R M 1989 Electrophysiological evidence for a role for calcium in temperature sensing by roots of cucumber seedlings. Plant Cell Environ. 12, 137–143.

Piñeros M and Tester M 1995 Characterization of a voltage-dependent Ca^{2+}-selective channel from wheat roots. Planta 195, 478–488.

Piñeros M and Tester M 1997 Calcium channels in plant cells: selectivity, regulation and pharmacology. J. Exp. Bot. 48, 551–557.

Thion L, Mazars C, Nacry P, Bouchez D, Moreau M, Ranjeva R and Thuleau P 1998 Plasma membrane depolarization-activated calcium channels, stimulated by microtubule-depolymerizing drugs in wild-type *Arabidopsis thaliana* protoplasts, display constitutively large activities and a longer half-life in *ton* 2 mutant cells affected in the organization of cortical microtubules. Plant J. 13, 603–610.

White P J 1993 Characterization of a high-conductance, voltage-dependent cation channel from the plasma membrane of rye roots in planar lipid bilayers. Planta 191, 541–551.

White P J 1994 Characterization of a voltage-dependent cation-channel from the plasma membrane of rye (*Secale cereale* L.) roots in planar lipid bilayers. Planta 193, 186–193.

White P J 1996 Specificity of ion channel inhibitors for the maxi cation channel in rye root plasma membranes. J. Exp. Bot. 47, 713–716.

White P J 1998 Calcium channels in the plasma membrane of root cells. Ann. Bot. 81, 173–183.

White P J and Ridout M S 1999 An energy-barrier model for the permeation of monovalent and divalent cations through the maxi cation channel in the plasma membrane of rye roots. J. Membr. Biol. 168, 63–75.

Wymer C L, Wymer S A, Cosgrove D J and Cyr R J 1996 Plant cell growth responds to external forces and the response requires intact microtubules. Plant Physiol. 110, 425–430.

Recent Advances of Plant Root Structure and Function, 103–109
Edited by O. Gašparíková, M. Čiamporová, I. Mistrík and F. Baluška
© 2001 *Kluwer Academic Publishers.*

Nitrogen and base cation uptake in seedlings of *Acer pseudoplatanus* and *Calamagrostis villosa* exposed to an acidified environment

Vít Gloser* and Jan Gloser
Department of Plant Physiology, Faculty of Science, Masaryk University, Kotlarska 2, 611 37 Brno, Czech Republic

Key words: acid soil, ammonium, calcium, magnesium, nitrate

Abstract

The effects of solution acidity and form of nitrogen on net nutrient uptake rates in *Acer pseudoplatanus* and *Calamagrostis villosa* seedlings were examined as part of a complex ecological study. Uptake rates were measured by the depletion method under controlled conditions (temperature 20 °C, irradiance 400 μmol m^{-2} s^{-1} PAR) from a nutrient solution containing 1.5 m*M* nitrogen in the form of nitrate or ammonium or an equimolar mixture of both. The solution acidity was kept constant at pH 5.5 (control treatment), 4.5 or 3.5 (low pH treatments). Strongly acid pH decreased or stopped the uptake rates of NO_3^-, Mg^{2+} and Ca^{2+}, but the uptake of NH_4^+ was not changed in both species. Ammonium ions reduced the uptake rate of NO_3^- in *Acer* but increased the uptake rate in *Calamagrostis.* Ammonium as the sole source of nitrogen had a strong negative impact on the uptake rates of calcium and magnesium and this effect was independent of the media acidification usually connected with NH_4^+ uptake and assimilation. However, the negative effect of ammonium ions on the base cation uptake was more pronounced at low pH values.

Introduction

Progressive forest disturbance is one of the most significant ecological problems in central Europe. The rapid expansion of perennial grasses has been observed in many disturbed forests and in open deforested areas. The new grass stands are surprisingly stable for a long period of time and prevent spontaneous reforestation by tree seedlings. In addition to the aboveground competition for light (Gloser, 1996; Gloser and Gloser, 1996), the interactions of tree seedlings and grasses below-ground should be considered.

Low soil pH values are usually found on declining forest sites. However, H^+ toxicity *per se* is rarely the factor limiting plant growth in these areas (Marschner, 1991). The effect of low soil pH on ion uptake processes, and, consequently, on plant mineral nutrition seems to be more important. The uptake rates of various ions are changed in an acidic environment but not always to the same extent. Thus, the resulting imbalance in the uptake rates of inorganic ions may have a strongly negative effect on plant growth and survival (Oren and Schulze, 1989). Generally, the uptake rates of cations decline with increasing acidity, whereas the uptake of anions is less affected (Marschner, 1995).

The uptake of ammonium ions, which dominate in acidic soils, results in a lower cation/anion ratio of plant biomass. Moreover, the utilisation of NH_4^+ ions as the main nitrogen source leads to the acidification of the root environment. Changes in the chemical composition of plants following ammonium fertilization are well known (Gijsman, 1990; Raab and Terry, 1995; Troelstra et al., 1995). However, the effects of NH_4^+nutrition alone, or in combination with low pH, on the processes of ion uptake are not clearly understood.

The aim of this work was to determine the specific rates of net uptake of selected ions by intact roots of grass and tree seedlings and also the sensitivity of the net uptake rates to low pH and the presence of ammonium ions in the root environment.

* FAX No: +420 (0) 5-412-11214.
E-mail:vitgloser@sci.muni.cz

Materials and methods

Plant cultivation

Seedlings of the tree species *Acer pseudoplatanus* and perennial grass species *Calamagrostis villosa* were used in our experiments. After germination on fine sand moistened by half-strength nutrient solution, the plants were cultivated in hydroponics in fully controlled conditions. The irradiance was 400 μmol m^{-2} s^{-1} PAR for 14 hours and the day/night air temperature was 20/15 °C. The modified Hoagland nutrient solution contained, in μM – 603 $Ca(NO_3)_2$, 795 KNO_3, 190 KH_2PO_4, 270 $MgSO_4$, 2 $MnSO_4$, 0.85 $ZnSO_4$, 0.15 $CuSO_4$, 20 H_3BO_3, 0.25 Na_2MoO_4, and 41 FeNa-EDTA. The solution was renewed each week and its pH was maintained at 5.5. One day before uptake measurements, the experimental plants were transferred to a nutrient solution with lower ionic concentrations (500 μM NH_4NO_3, 100 μM $CaSO_4$, 40 μM KH_2PO_4).

Ion uptake measurements

Measurements of ion uptake rates were made by the depletion method with 8 week old plants. The fresh mass of roots reached approx. 0.5 g and 0.2 g for *Calamagrostis* and *Acer* respectively. The sample plants were exposed individually in 80 mL containers filled with aerated nutrient solution. The basic nutrient solution for this measurement (190 μM $MgSO_4$, 200 μM $CaSO_4$, 40 μM KH_2PO_4) was mixed with either $NaNO_3$ or $(NH_4)_2SO_4$ or NH_4NO_3 stock solutions to prepare three different nitrogen treatments. The nitrogen concentration at the beginning of the measurement was 1.5 mM in all solutions. The solution pH in each nitrogen treatment was set at 5.5, 4.5 and 3.5 for control and low pH treatments, respectively. The exposure period was 8 h and the pH of the nutrient solution was stabilised by additions of 0.1 M H_2SO_4 or NaOH. Changes in the concentration of cations in the nutrient solution were determined by capillary isotachophoresis (Isotachophoretic analyser EA 100, LABECO, Spišská Nová Ves, Slovakia). The setup of the analysis was similar to Matejovič (1991). The concentration of NO_3^- in solution was determined colorimetrically as described in Cataldo et al. (1975). The roots of the experimental plants were dried (80 °C, 48 h) and uptake rates were expressed on a dry mass basis.

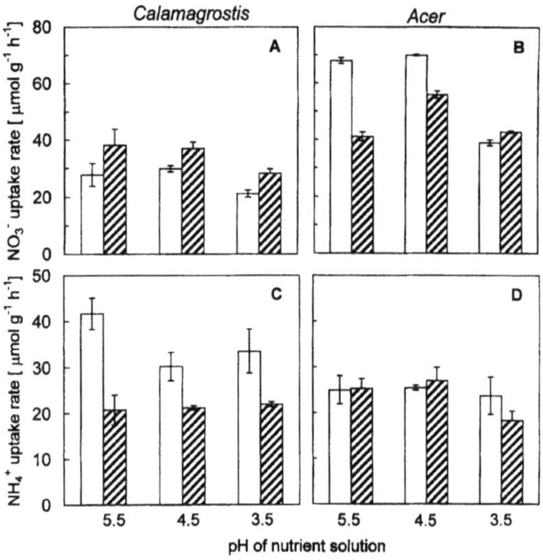

Figure 1. The effect of a sudden decrease of nutrient solution pH on specific net nitrate (A, B) and ammonium (C, D) uptake rates (uptake rate per unit of root dry matter) in *Calamagrostis villosa* and *Acer pseudoplatanus* seedlings. Measurements were made in the presence of either nitrate or ammonium (empty bars) and in the presence of an equimolar mixture of both N forms (dashed bars). The mean values from four measurements ± standard error are shown.

Results

Uptake rates of inorganic nitrogen

In *Calamagrostis* the specific nitrate uptake rate was only slightly decreased by the low pH of the nutrient solution (Figure 1A, Table 1). The presence of ammonium ions stimulated net uptake of NO_3^-. The specific nitrate uptake rate in *Acer* was about two times higher than in *Calamagrostis* (Figure 1B), but it was more sensitive to the low pH of the medium, and significantly inhibited by the presence of ammonium ions in the medium (Table 1, Figure 1B). However, the concomitant presence of ammonium and nitrate in the nutrient solution reduced the negative effect of low pH on nitrate uptake in comparison with the exclusively NO_3^- nutrition.

The specific ammonium uptake rate in *Calamagrostis* plants was not affected by the solution acidity (Table 1), but there was a significant decrease in NH_4^+ uptake in the presence of nitrate. In contrast, *Acer* plants showed no change in ammonium uptake rate either in high H^+ concentrations or in the presence of NO_3^- (Table 1, Figure 1 D).

Table 1. The results of ANOVA analysis which tested the effects of solution acidity (pH), form of nitrogen in solution (N- form) and interaction of these factors (pH × N-form) on specific uptake rates (uptake rate per unit of root dry matter) of inorganic nutrients in *Calamagrostis villosa* and *Acer pseudoplatanus*. The significance is indicated by *** $P< 0.001$, ** $P< 0.01$, * $P< 0.05$, n.s. – nonsignificant

Species	*Calamagrostis*				*Acer*			
Variance source	NO_3^-	NH_4^+	Mg^{2+}	Ca^{2+}	NO_3^-	NH_4^+	Mg^{2+}	Ca^{2+}
pH	*	n.s.	***	**	***	n.s.	**	***
N- form	**	***	***	**	***	n.s.	***	***
pH × N- form	n.s.	n.s.	**	*	***	n.s.	**	**

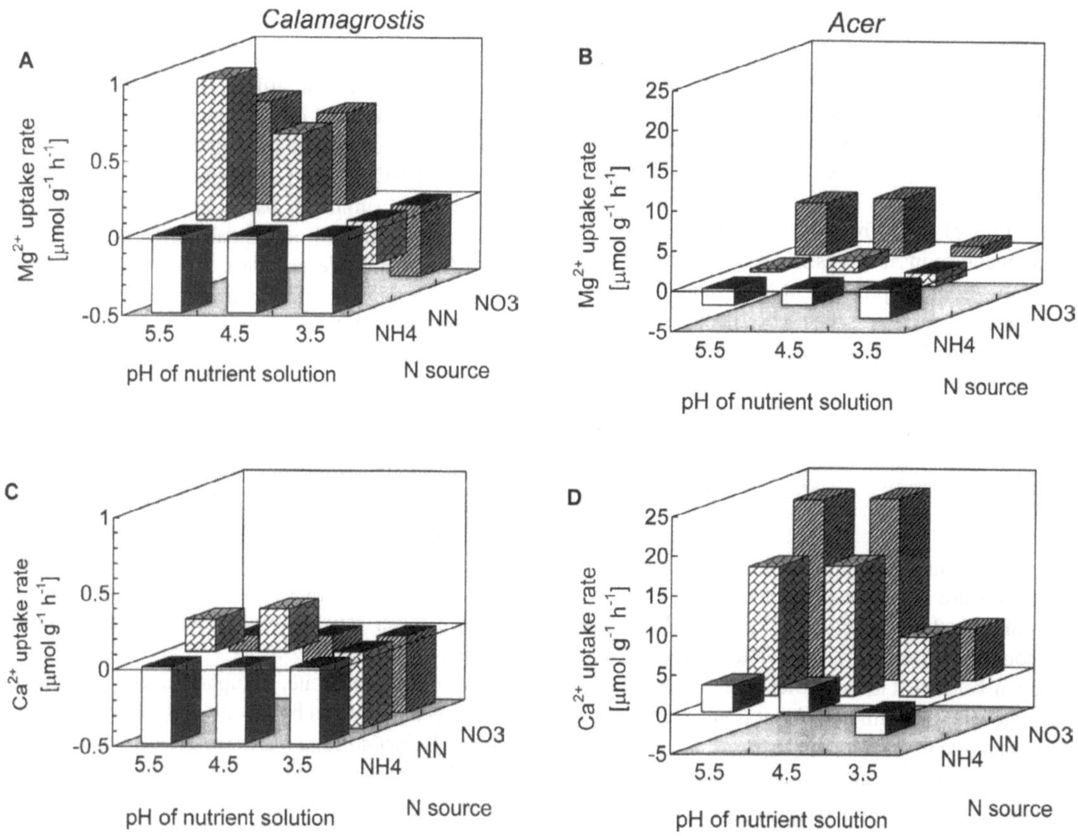

Figure 2. The specific net uptake rates of magnesium (A, B) and calcium (C, D) (uptake rate per unit of root dry matter) in *Calamagrostis villosa* and *Acer pseudoplatanus* seedlings. Nitrogen was present in the nutrient solution in the form of nitrate (NO3) or ammonium (NH4) or an equimolar mixture of both forms (NN) and the given pH of the solution was kept constant during the measurement. Note, that the scales for the cation uptake rates are different for the two species. The mean values from four measurements are shown. The statistical evaluation of this data is given in Table 1.

Uptake rates of base cations

The specific uptake rate of magnesium in *Calamagrostis* plants exposed to a solution with pH 4.5 decreased to 80% of the control rate, but at pH 3.5 only a net efflux was recorded (Figure 2A). The concurrent presence of nitrate and ammonium in solution stimulated the Mg^{2+} uptake rate in comparison with the nitrate treatment when measured at pH 5.5, whereas with ammonium as a sole nitrogen source the net uptake of Mg^{2+} in all pH treatments stopped. *Acer* seedlings showed almost ten times higher specific uptake rates of Mg^{2+} than *Calamagrostis*. These interspecific differences were especially large in the nitrate treatment (Figure 2B). *Acer* plants with nitrate as the sole source of N maintained a net uptake of Mg^{2+} in low pH 3.5, which was remarkably different from the *Calamagrostis* plants. There was, however, no net magnesium uptake in *Acer* plants in solutions with ammonium as the sole source of nitrogen.

The uptake rates of calcium in *Calamagrostis* seedlings were influenced by experimental treatments in a similar way to the uptake of magnesium. Both low pH and NH_4^+ ions as the sole nitrogen source led to a net efflux of Ca^{2+} from the roots (Figure 2C). The calcium uptake in *Acer* was much faster than in grass seedlings. Both the presence of NH_4^+ and a high H^+ concentration decreased the Ca^{2+} uptake rate considerably (Figure 2D). However, the net calcium uptake in *Acer* plants was completely stopped only under the combination of ammonium nutrition and pH 3.5.

Discussion

The experimental technique used in our work allowed us to measure simultaneously changes in the uptake rates of several ions (Ca^{2+}, Mg^{2+}, NH_4^+, NO_3^-) under controlled conditions. Much attention was paid to nitrogen, because it is the nutrient with the highest uptake rates and the form in which it is taken up plays an important role in plant ion balance.

The fast uptake and assimilation of nitrate ions is of special importance for plants growing in acidified areas, because these processes are coupled with substantial increases of pH values around the roots. Consequently, the roots may be less threatened by nutrient disharmony and by toxic aluminium ions. The concentration of nitrate ions is usually very low in acidic soils, and, therefore, knowledge of interspecific differences in potential uptake characteristics could be helpful for the prediction of growth responses and the competitive ability of various species.

Changes in nitrogen uptake rates

It is generally accepted that the pH of the uptake medium affects the net uptake of NO_3^- rate by plant roots, but variable responses have been reported. In some plant species, nitrate uptake was highest under acidic conditions (pH 3 to 4) and declined rapidly as the pH was increased (Deane-Drummond, 1984; McClure et al., 1990; Rao and Rains, 1976; Zsoldos and Haunold, 1982). Our results support findings of some other authors that the optimum pH for NO_3^- uptake may be in some cases much higher (pH 5.5 to 8) (Aslam et al., 1995; Doddema and Telkamp, 1979; Lyclama, 1963). Recently Aslam et al. (1995) used a NO_2^- incubation technique with barley for separation of the effect of acidic pH on NO_3^- influx from its effect on efflux. However, Kronzucker et al. (1999) showed on the same plant material that NO_3^- influx cannot be modelled quantitatively with the use of NO_2^- as an analogue of NO_3^-. A study with $^{13}NO_3^-$ influx in barley at various pH (Glass et al., 1990) showed only small differences between pH 4.5 and 6.0. Nevertheless, these results indicate that the response of nitrate transporters to pH depends on ambient NO_3^- concentration (cf. pH effect on ammonium uptake).

The specific uptake rate of nitrate in *Acer* was two to three times higher than in *Calamagrostis* plants. Concomitant with higher NO_3^- uptake rates, nitrate reductase (NR) activity in *Acer* roots was also > 3-fold higher than in *Calamagrostis* (Seidlová 1998). Higher nitrate uptake rates and NR activity per unit of root biomass, however, did not translate into higher growth rate of *Acer* plants (data not shown) primarily due to a much lower ratio of root to whole plant biomass compare to *Calamagrostis*.

In contrast to nitrate, the uptake rate of ammonium ions was not affected by low pH in both species. Results with *Pseudotsuga menziesii* indicate that the rate of NH_4^+ uptake is independent of pH in a range from 3.0 to 6.5 (Rygiewicz et al. 1986). Conversely, the short-term (1 h) $^{15}NH_4^+$ uptake rate in rice plants was more than 30% reduced under pH 3 compared to pH 6 (Zsoldos and Erdei, 1982) and similar results were obtained during 24 days long cultivation of soybean plants (Tolley-Henry and Raper, 1986). Detailed studies of NH_4^+ uptake in rice found different pH optima for the two ammonium transporters (Wang et al.,

1993). The high affinity transport system mediated the highest uptake rates at about neutral pH, whereas low-affinity transporters attained optimum rates between pH 4.5 and 6.0. Therefore, the response of the net uptake rate to solution pH probably is influenced by the NH_4^+ concentration in solution, and comparison of data from different experiments should be done with caution.

In the presence of ammonium in nutrient solution, nitrate uptake was decreased in *Acer* seedlings, but it was stimulated in *Calamagrostis* plants. Inhibition of NO_3^- uptake in presence of ammonium ions often has been observed (Aslam et al., 1994; Jackson and Volk, 1995; Rao and Rains, 1986) but also stimulation has been reported (Bloom and Sukrapanna, 1990). Short-term NH_4^+ effects on $^{13}NO_3^-$ uptake were recently investigated in barley (Kronzucker et al., 1999). The observed inhibition of net NO_3^- uptake in nitrate-induced plants was transient, and mediated primarily by inhibiting NO_3^- influx. They proposed that the quick inhibition of NO_3^- uptake is connected with a direct effect of NH_4^+ on the nitrate transport system in plasma membranes and is reversible after withdrawal of ammonium from solution. It seems, thus, unlikely that this direct inhibitory mechanism would cause significant underestimation of NO_3^- uptake from NH_4^+ free solution in our plants which were pre-conditioned in solution containing NH_4^+ions. However, it is not possible to exclude completely an after-effect of NH_4^+ on NO_3^- uptake in the plants exposed to NH_4NO_3 during 1 day pre-conditioning, e.g. by products of NH_4^+ assimilation. Nevertheless, our experiments showed clear interspecific differences in sensitivity of NO_3^- uptake to NH_4^+.

The uptake rate of both NO_3^- and NH_4^+ ions measured in a mixed N treatment were less affected by low pH than the uptake rates of these ions measured in separate NO_3^- and NH_4^+ treatments. There was a higher total N uptake rate recorded in plants grown in a mixed N treatment in comparison with nitrate or ammonium alone. Our results suggest that the beneficial effects of concurrent ammonium and nitrate nutrition, well known from long-term experiments (Vessey et al., 1990; Volk et al., 1992), could be even more pronounced in acidified environments.

Changes in magnesium and calcium uptake

There is very limited information concerning the uptake rates of base cations obtained by direct measurements of uptake by roots. The lack of experimental work on this topic is partly due to methodological problems and partly because of the complicated interpretation of results.

Decreasing pH of the solution from 5.5 to 4.5 did not dramatically change the uptake rates of base cations in both species. However, after a change of pH from 5.5 to 3.5 there was no net Mg^{2+} and Ca^{2+} uptake in *Calamagrostis* and a considerably reduced net uptake in *Acer* seedlings. This result is in line with other empirical findings, that the critical pH for plant growth is about 4 (Kennedy, 1992). One possible explanation for this effect is increased competition between base cations and H^+ ions for binding sites on roots.

A striking difference was found in the specific rates of base cation uptake between *Acer* and *Calamagrostis* seedlings. The *Acer* root system is rather short and thick, whereas *Calamagrostis* has long fine roots with about twice as large absorption area per root dry mass unit than *Acer*. But the specific requirements for base cation supply seems to be much higher in *Acer* plants than in *Calamagrostis*, since the ratio between nitrogen and Mg^{2+} or Ca^{2+} uptake rates is much higher in grass seedlings. In order to meet the plant's demand for base cations, the less developed roots of *Acer* need to possess much more efficient uptake mechanisms per unit of root surface area than grass seedlings.

The disturbance of membrane integrity in strongly acidic environments is another process which may interfere with the net uptake rates of ions (Zsoldos and Erdei, 1981). The rapid leak of ions from root cells through plasma membranes due to changes in permeability could be an explanation for the relatively high net efflux rates of base cations, which were recorded especially with *Calamagrostis* plants.

The second investigated ecological factor influencing base cation uptake rates was the presence of ammonium ions in the nutrient solution. When ammonium was the sole source of nitrogen, the net uptake rates of base cations were changed in a similar way to exposure of roots to solutions with extremely low pH. Moreover, the adverse impact of NH_4^+ and high H^+ concentrations on the base cation uptake rate was additive, so that the combined treatment of NH_4^+ nutrition and a pH of 3.5 was completely unfavourable for ion uptake in both investigated species. The interference between ammonium ions and base cation uptake often has been explained as an effect mediated by a decrease of rhizosphere pH, which follows NH_4^+ uptake and assimilation (Gijsman, 1990; Rayar and Hai, 1977). Our experiments showed an inhibition of base cation uptake when NH_4^+ was the sole N source even though the solution pH was stabilised at 5.5. This indicates that the influence of ammonium ions on uptake rates of other cations is independent of media acidification. However, the negative impact of ammonium ions on the base cation uptake can be more pronounced at strongly acidic pH.

Ecological implications

The reforestation of extensive polluted sites in the Czech Republic with strongly acidic soils is generally considered to be the best landscape management strategy for the future. Broad-leaf tree species including *Acer pseudoplatanus* are more desirable for planting than conifers, because they have a significant ameliorative effect on the soil chemistry (Ulrich, 1987). Any reforestation trial should be based on detailed knowledge of the physiological traits of both the tree seedlings and the competing grasses dominated by *Calamagrostis villosa.* The results of our laboratory experiments can not be, of course, directly applied to field conditions. Nevertheless, the possible ecological implications of our data may show future research prospects.

It is well known that young seedlings of *A. pseudoplatanus* are readily suppressed by grasses, in spite of their extremely high shade tolerance (Jones, 1944). Therefore, our interest was focused on the comparative measurements of nutrient uptake rates under changing environmental conditions, which might be helpful in finding an explanation for interspecific competitive relations within the rhizosphere. At least two physiological traits that can be of potential advantage to the grass *C. villosa* growing on acidic soil can be concluded from our results: (1) its nitrogen uptake rate (both in nitrate and ammonium form) is not markedly depressed even at very low pH, and (2) its demand for base cations is very low, which is also a characteristic of some other grass species (Kinzel, 1982). The second feature may be of special importance, because base cations are very scarce in acidic soils (Schulze, 1989). In such adverse conditions the species with the least requirement will be the superior competitor, as predicted by Tilman's theory of competition (Tilman, 1988).

The relatively high specific uptake rates of the roots of *Acer* are probably not very advantageous in nutrient poor acidic soils, because the relative absorption area of the roots (per DM unit of the whole plant) is rather low (Hájek, 1998). This is probably why the spontaneous regeneration of *A. pseudoplatanus* is always most successful on nutrient rich soils with a high amount of base cations (Jones, 1944).

References

Aslam M, Travis R L and Huffaker R C 1995 Effect of pH and calcium on short-term NO_3^- fluxes in roots of barley seedlings. Plant Physiol. 108, 727–734.

Aslam M, Travis R L and Huffaker R C 1994 Stimulation of nitrate and nitrite efflux by ammonium in barley (*Hordeum vulgare* L.) seedlings. Plant Physiol. 106, 1293–1301.

Bloom A J and Sukrapanna S S 1990 Effects of exposure to ammonium and transplant shock upon the induction of nitrate absorption. Plant Physiol. 94, 85–90.

Cataldo D A, Haroon M, Schrader L E and Youngs V L 1975 Rapid colorimetric determination of nitrate in plant tissue by nitration of salycilic acid. Commun. Soil Sci. Plant Anal. 6, 71–80.

Deane-Drummond C E 1984 Mechanisms of nitrate uptake into *Chara coralina* cells: lack of evidence for obligatory coupling to proton pump and new NO_3^-/NO_3^- exchange model. Plant Cell Environ. 7, 317–323.

Doddema H and Telkamp G P 1979 Uptake of nitrate by mutants of *Arabidopsis thaliana* disturbed in uptake or reduction of nitrate. II. Kinetics. Physiol. Plant. 45, 332–338.

Gijsman A J 1990 Nitrogen nutrition of Douglas-fir (*Pseudotsuga menziesii*) I. Growth, nutrient uptake and ioinic balance. Plant Soil 126, 53–61.

Glass A D M, Siddiqi M Y, Ruth T J and Rufty T W Jr. 1990 Studies of the uptake of nitrate in barley. II. Energetics. Plant. Physiol. 93, 1585–1589.

Gloser V and Gloser J 1996 Acclimation capability of *Calamagrostis epigejos* and *C. arundinacea* to changes in radiation environment. Photosynthetica 32, 203–212.

Gloser J 1996 Impact of elevated CO_2 concentration on interactions between seedlings of Norway spruce (*Picea abies*) and perennial grass (*Calamagrostis epigejos*). *In* Carbon Dioxide, Populations and Communities. Eds. C Körner and F Bazzaz. pp. 319–331. Academic Press, San Diego.

Hájek J 1998. The sensitivity of *Acer pseudoplatanus* L. seedlings to photoinhibition. M.Sc. thesis, Masaryk University, Brno.

Jackson W A and Volk R J 1995 Attributes of the nitrogen uptake systems of maize (*Zea mays* L.): Maximal suppression by exposure to both nitrate and ammonium. New Phytol. 130, 327–335.

Jones E W 1944 Biological flora of the British isles – Acer L. J. of Ecol. 32, 215–252.

Kennedy I R 1992 Acid Soil and Acid Rain. Research Studies Press, Taunton.

Kinzel H 1982 Pflanzenökologie und Mineralstoffwechsel. E. Ulmer, Stuttgart.

Kronzucker H J, Glass A D M, Siddiqi, M Y 1999 Inhibition of nitrate uptake by ammonium in barley. Analysis of component fluxes. Plant Physiol. 120, 283–291.

Lycklama J C 1963 The absorption of ammonium and nitrate by perennial rye-grass. Acta Bot. Neerl. 12, 361–423.

Marschner H 1991 Mechanisms of adaptations of plants to acid soils. Plant Soil 134, 1–20.

Marschner H 1995 Mineral Nutrition of Higher Plants. Academic Press, Cambridge.

Matejovič I 1991 Simultaneous study of NH_4^+, Na^+, Mg^{2+}, Ca^{2+}, NO_3^-, SO_4^{2-}, (NO^{2-}), $H_2PO_4^-$ and Cl^- uptake by intact winter wheat seedlings in a single depletion experiment. Biol. Plant. 33, 366–376.

McClure P, Kochian L V, Spanswick R M and Shaff J E 1990 Evidence for cotransport of nitrate and protons in maize roots. II. Measurements of NO_3^- and H^+ fluxes with ion-selective microelectrodes. Plant Physiol. 93, 290–294.

Oren R and Schulze E-D 1989 Nutritional disharmony and forest decline: A conceptual model. *In* Air Pollution and Forest Decline. Eds. E-D Schulze et al. pp 425–443. Springer Verlag, Berlin.

Raab T K and Terry N 1995 Carbon, nitrogen, and nutrient interactions in *Beta vulgaris* L. as influenced by nitrogen source, NO_3^- versus NH_4^+. Plant Physiol. 107, 575–584.

Rao K P and Rains D W 1976 Nitrate absorption by barley. Plant Physiol. 57, 55–58.

Rayar A J and Hai T V 1977 Effect of ammonium on uptake of phosphorus, potassium, calcium and magnesium by intact soybean plants. Plant Soil 48, 81–87.

Rygiewicz P T and Bledsoe C S 1986 Effect of pretreatment conditions on ammonium and nitrate uptake by Douglas-fir seedlings. Tree Physiol. 1, 145–150.

Schulze E-D 1989 Air pollution and forest decline in a spruce (*Picea abies*) forest. Science 244, 776–783.

Seidlová L 1998. Assimilation of NO_3^- in roots and leaves of *Acer pseudoplatanus*, *Calamagrostis epigejos* and *Lolium perenne*. M.Sc. thesis, Masaryk University, Brno.

Tilman G D 1988 Plant Strategies and the Dynamics and Structure of Plant Communities. Princeton University Press, Princeton.

Tolley-Henry L and Raper C D 1986 Utilization of ammonium as a nitrogen source. Plant Physiol. 82, 54–60.

Troelstra S R, Wagenaar R and Smant W 1995 Nitrogen utilization by plant species from acid heathland soils: I. Comparison between nitrate and ammonium nutrition at constant low pH. J. Exp. Bot. 46, 1103–1112.

Ulrich B. 1987 Stability, elasticity and resilience of terrestrial ecosystems with respect to matter balance. *In* Potentials and Limitations of Ecosystem Analysis. Eds. E-D Schulze and H Zwölfer. pp 11–49. Springer Verlag, Berlin.

Volk R, Chaillou S, Mariotti A and Morot-Gaudry J F 1992 Beneficial effects of concurrent ammonium and nitrate nutrition on the growth of *Phaseolus vulgaris*: A ^{15}N study. Plant Physiol. Biochem. 30, 487–493.

Wang M Y, Siddiqi M Y, Ruth T J and Glass A D M. 1993. Ammonium uptake by rice roots. II. Kinetics of $^{13}NH_4^+$ influx across the plasmalemma. Plant Physiol. 103, 1259–1267.

Zsoldos F and Erdei L 1981 Membrane and ion transport properties in cereals under acidic and alkaline stress. Physiol. Plant. 53, 468–474.

Zsoldos F and Haunold E 1982 Influence of 2,4-D and low pH on potassium, ammonium and nitrate uptake by rice roots. Physiol. Plant. 54, 63–68.

Section editor: A C Borstlap

Recent Advances of Plant Root Structure and Function, 111–117
Edited by O. Gašparíková, M. Čiamporová, I. Mistrík and F. Baluška
© 2001 *Kluwer Academic Publishers.*

Nitrate uptake by bean (*Phaseolus vulgaris* L.) roots under phosphate deficiency

A. Gniazdowska[1] and A. M. Rychter
Institute of Experimental Plant Biology, University of Warsaw, Pawińskiego 5A, 02-106 Warsaw, Poland.
[1]*Corresponding author**

Key words: nitrate reductase, *Phaseolus vulgaris*, phosphate deficiency, plasma membrane ATPase, roots

Abstract

Bean (*Phaseolus vulgaris* L.) plants were cultured for 19 d on complete or on phosphate deficient culture media. Low inorganic phosphate concentration in the roots decreased ATP level and nitrate uptake rate. The mechanisms which may control nitrate uptake rate during phosphate deficiency were examined. Plasma membrane enriched fractions from phosphate sufficient and phosphate deficient plants were isolated and compared. The decrease in total phospholipid content was observed in plasma membranes from phosphate deficient roots, but phospholipid composition was similar. No changes in ATPase and proton pumping activities measured in isolated plasma membrane of phosphate sufficient and phosphate deficient bean roots were noted. The electron microscope observations carried out on cortical meristematic cells of the roots showed that active ATPases were found in plasma membrane of both phosphate sufficient and phosphate deficient plants. The decrease in inorganic phosphate concentration in roots led to increased nitrate accumulation in roots, accompanied by a corresponding alterations in NO_3 distribution between shoots and roots. Nitrate reductase activity in roots of phosphate deficient plants estimated *in vivo* and *in vitro* was reduced to 50–60% of the control. The increased NO_3 concentration in root tissue may be explained by decreased NR activity and lower transport of nitrate from roots to shoots. Therefore, the reduction of nitrate uptake during phosphate starvation is mainly a consequence of nitrate accumulation in the roots.

Abbreviations: NEDHC – naphthyl etylenediamine dihydrochloride BTP – bis-tris-propane PEG – polyethylene glycol +P – control plants –P – phosphate deficient plants LPC – lisophosphatidylcholine PA – phosphatidic acid PC – phosphatidylcholine PE – phosphatidylethanolamine PG – phosphatidylglycerol PI – phosphatidylinositol PS – phosphatidylserine

Introduction

Phosphate starvation stress affects plant growth and metabolism. The observed changes result from the decreased Pi concentration in the tissue; in bean plants after 2 weeks of starvation the inorganic phosphate concentration in the roots was only 6% of that of the control plants, grown on complete (+P) nutrient medium (Gniazdowska et al., 1998). An increase in root mass:shoot mass ratio and carbohydrate concentration in the roots is one of the first observed effects of phosphate starvation (Ciereszko et al., 1996; Rychter and Randall, 1994). The relative growth rate of bean roots was higher until 12 d of culture and later dropped to the rate similar to the control, whereas shoot growth was reduced from the beginning of starvation (Gniazdowska et al., 1998). It was also observed that the uptake and assimilation of nitrogen is altered when plants are deprived of phosphate (Rufty et al., 1990, 1993). During phosphate starvation of bean plants, the net rate of nitrate uptake (NNUR) at the beginning of culture was similar to that of the control, but later was 8 times lower (Gniazdowska et al., 1998). The NO_3 uptake measured as the decrease in NO_3 concentration in the nutrient medium was about two times lower than for the control (Gniazdowska et al., 1999a). It was suggested by Rufty that the decrease in NO_3 uptake

* FAX No: 22658404.
E-mail: gniazdowska@ibbrain.ibb.waw.pl

could result from feedback control factors and from limited ATP availability (Rufty et al., 1993).

Low phosphate concentration in the tissue influenced root respiration rate (Rychter and Mikulska, 1990; Rychter et al., 1992) and increased the reduced:oxidized pyridine nucleotides concentration ratio (Juszczuk and Rychter, 1997). The lower respiration rate and increased participation of alternative, non phosphorylating pathway resulted in the decline of ATP concentration in the roots of plants cultured on Pi deficient media to only 25% of the control (Gniazdowska et al., 1998).

Low Pi concentration, as well as a decreased ATP level, may have an influence on phospholipid concentration of plasma membrane which in turn may alter its uptake properties. The energy driving nitrate uptake comes from the proton gradient maintained across the plasma membrane by the H^+–ATPase (Crawford and Glass, 1998). H^+–ATPase in plasma membrane pumps protons out of the cell, producing both a pH and an electrical gradient. Nitrate transporters cotransport two protons per nitrate into the cell (Crawford and Glass, 1998). In the cell, nitrate can be reduced to NH_4 or transported across the tonoplast and stored in the vacuole or moved within the symplast, loaded into xylem and transported to the shoot (Crawford, 1995; Crawford and Glass, 1998). The assimilation of nitrate is an energy-consuming process and its first step involves the reduction of nitrate to nitrite catalysed by nitrate reductase (Kaiser and Huber, 1994).

Therefore, the low ATP concentration in phosphate deficient bean roots may influence nitrate uptake rate by:

1. changing membrane uptake properties (e.g. phospholipid composition and ATPases activity),
2. affecting in the first step of nitrate assimilation pathway (nitrate reductase activity).

Nitrate transport from the roots to the shoot may be lower, and in turn, its high concentration in the roots may limit nitrate uptake.

The purpose of our study was to examine which of the above changes may be associated with mechanisms controlling nitrate uptake rate during phosphate deficiency.

Material and methods

Bean (*Phaseolus vulgaris* L. cv Złota Saxa) seeds were germinated in the darkness for 4 d and then transferred to 5 l containers (20 plants in each) filled with complete or phosphate deficient Knop medium (Rychter and Mikulska, 1990). The culture medium was continuously aerated and changed every 3 d to avoid nutrient depletion. The cotyledons were removed when the seedlings were 9 d old. Plants were grown in a growth chamber under a 16 h photoperiod, photon flux density of 150 μmol m^{-2} s^{-1}, day/night temperature 24/20 °C and air humidity of about 90%. Plants at the age of 19 d (after 15 d of culture on Pi deficient medium) were taken for the analyses described below.

Nitrate determination

Fresh tissue (1 g of root, leaves or steam) was ground in a cool mortar and homogenised with 4 ml redistilled water, boiled and centrifuged at 10 000 *g* for 10 min. Nitrate in the supernatant was determined after reduction to nitrite (Cataldo et al., 1975).

Nitrate reductase assay

NR activity was measured *in vivo* and *in vitro* in freshly harvested material. NR *in vivo* was measured in root fragments as described by Stulen et al. (1981). The reaction was stopped by adding 0.75 ml 1 g kg^{-1} sulfanilamide in 3N HCl. The nitrite formed was measured colorimetrically by adding 0.02 g kg^{-1} naphthyl etylenediaminedihydrochloride (NEDHC) according to Sanderson and Cocking (1964). NR *in vitro*: NR was extracted from root tissue as described by Lewis et al. (1982). The reaction medium contained: 0.05 M phosphate buffer pH 7.5, 0.07 mM NADH, 0.01 M KNO_3. The reaction was stopped after 15 min by addition of 1 ml 1 g kg^{-1} sulfanilamide in 3 N HCl. The nitrite formed was determined according to Sanderson and Cocking (1964).

H^+–ATPase assays

ATPase activity and H^+ pumping were detected simultaneously as described by Widell and Larsson (1990) in the presence of 2 mM ATP–BTP (bis-tris-propane) in the reaction medium.

Preparation of the membrane fractions

Microsomal fraction of bean roots was prepared according to Widell and Larsson (1981). Plasma membranes were isolated by partitioning in an 18 g aqueous polymer two-phase system (Widell and Larsson, 1981) containing 6.1 g kg^{-1} Dextran T (Pharmacia Fine Chemicals, Uppsala, Sweden) and 6.1 g kg^{-1} PEG

3350 (Sigma) in 5 mM potassium phosphate buffer (pH 7.8) with 250 mM sucrose and 4 mM KCl.

Cytochemical assay

Active ATPases in cortical cells of meristematic root section were localised using the lead precipitation method of Wachstein and Maisel (1957) with some modifications according to Podbielkowska and Wałęza (1986). After embedding in epon, the material was sectioned into ultra thin sections and contrasted according to Reynolds (1963). Observations were made using a Tesla MB 500 microscope.

Lipid analysis

Total lipids were extracted from microsomal and plasma membrane fractions using chloroform:methanol (2:1, v/v) as described by Allen et al. (1966), with modifications according to Sikorska and Kacperska-Palacz (1980). Total lipids content in microsomal and plasma membrane fractions were determined as described by Sikorska and Kacperska-Palacz (1980). Individual phospholipids of the microsomal and plasma membrane fractions were separated by two - dimensional TLC using acetone:benzene:water (91:30:8) and with chloroform:methanol:water (65:25:4). Phospholipid-containing spots were detected with iodine vapour and scraped off for phosphorus analysis. Each phospholipid component was identified by TLco-chromatography with the authentic standard phospholipid from Sigma. Quantitative analysis of phospholipids was performed according to Ames (1966).

Protein determination

Protein in the tissue extracts was determined using Bradford reagent (Bradford, 1968). Proteins in plasmamembrane and microsomal fraction of the roots were determined according to Bearden's (1978) method in which Triton X-100 was included to solubilize the membrane protein.

Results and discussion

To examine the plasma membrane properties, we isolated plasma membrane enriched fraction from plants grown under phosphate deficient (−P) and phosphate sufficient (+P) conditions using an aqueous two phase partitioning system Dextran/PEG. Estimations of marker enzymes activities in plasma membrane preparations confirmed their purity (Gniazdowska et al., 1999b). Membrane protein concentration was similar in +P and −P both microsomal fraction and plasma membrane (Gniazdowska et al., 1999b). A decrease in total phospholipid content was observed in plasma membranes from phosphate deficient roots (Table 1), while in microsomal fraction, total phospholipid content was not altered by phosphate deprivation. Content of individual phospholipids in microsomal fraction was different in +P and −P plants; the main phospholipids in control roots were PC and PE (80% of total phospholipids), whereas in phosphate deficient roots, PC and PE content decreased to 50% (Table 1). The two-fold decline in the concentration of total phospholipids in plasma membrane does not alter significantly the mol % of main individual phospholipids (PC, PE, PA). Although no changes in the relative ratio of the different phospholipids in plasma membrane were detected, the other changes in plasma membrane composition may affect its function. In wheat plants deprived of inorganic phosphate increase in the double-bond index of fatty acids (Carvajal et al., 1996b) and fluidity of plasma membrane isolated from roots was observed (Carvajal et al., 1996a). The fluidity of the plasma membrane may influence root hydraulic conductivity (Carvajal et al., 1996a), which decreased in phosphate deficiency (Carvajal et al., 1996a; Radin and Matthews, 1989). This in turn may cause lower NO_3 transport from root to shoot and result in an increased NO_3 accumulation in the roots (Gniazdowska et al., 1999a; Rufty et al., 1993).

Since the activity of H^+–ATPase depends on the phospholipid environment in the membrane (Kasamo and Sakakibara, 1995), H^+–ATPase activity was measured *in vitro* in optimal conditions in the presence of 2 mM ATP and 4 mM $mgCl_2$. No differences in the specific activity of the ATPase and the proton pump were observed in isolated plasma membrane vesicles of +P and −P roots of 19 d old plants (Table 2). It was observed that plasma membrane H^+–ATPase is activated by LPC (Palmgren, 1991). Changes in total phospholipid content in pasma membrane accompanied by increased LPC content probably do not influence ATPase activity *in vivo* in phosphate stress conditions. Additionally, ATPase activity measured in plasma membrane vesicles in presence of LPC 0.02% was stimulated similarly by about 50% in phosphate sufficient and phosphate deficient bean roots (result not shown).

Table 1. Total lipid and phospholipid composition of microsomal and plasma membrane fractions from roots of bean plants cultured for 19 d on phosphate sufficient (+P) or phosphate deficient (−P) medium

	Microsomal		Plasma membrane	
	+P	−P	+P	−P
	($g\ g^{-1}$ protein)			
Total lipids	1.2 ± 0.1	1.0 ± 0.2	2.3 ± 0.2	2.1 ± 0.4
	(n mol P mg^{-1} protein)			
Total phospholipids	322.9 ± 67.7	318.0 ± 38.0	220.0 ± 32.7	101.6* ± 53.0
LPC	6.3 ± 2.3	30.5* ± 8.8	5.7± 1.1	3.8* ± 0.4
PS +PI	29.6 ± 1.9†	42.8* ± 4.2	8.8 ± 2.2	2.2* ± 0.2
PG	8.7 ± 6.2	30.5* ± 5.4	ND	ND
PA	15.7 ± 6.4	46.2* ± 3.6	14.7 ± 5.2	7.1± 1.6
PC	142.4 ± 25.5	88.4* ± 22.4	93.1 ± 5.8	42.0 ± 7.1
PE	120.2 ± 23.2	69.6* ± 7.7	97.7 ± 22.8	46.5 ± 3.7

Data are mean values ±s.d. from three independent experiments. ND, Not detectable. † PS and PI were not separated by 2D TLC in all experiments and data for these lipids were combined. *Indicates significance at *P*=0.05 confidence limits (*t*-test), as compared with +P.

Table 2. Activity of ATPase and H^+ pump in plasma membrane isolated from root cells of 18 d old bean plants cultured on phosphate sufficient (+P) or phosphate deficient (−P) medium

ATPase [μmol ADP mg^{-1} protein min^{-1}]		H^+ pump [Δ A_{495} mg^{-1} protein min^{-1}]	
+P	−P	+P	−P
0.32 ± 0.03	0.27 ± 0.06	0.86 ± 0.11	0.71 ± 0.15

Data are mean values ±s.d. from five independent experiments.

It is possible that *in vivo* ATP concentration, being significantly lower in −P plants (Gniazdowska et al., 1998; Rychter et al., 1992), may act as a limiting factor for ATPase activity. On the other hand, no growth inhibition in −P root was observed (Gniazdowska et al., 1998). Since elongation growth depends on the activity of H^+–ATPases induced by auxins (Rayle and Cleland 1992), this observation could indicate the H^+–ATPases are functioning at the control level. Unchanged plasma membrane ATPase activity in −P plants (as compared with +P) may result from high affinity of the enzyme for the substrate, allowing its functioning in lower ATP concentration or the increased enzyme affinity for ATP in phosphate deficient conditions. No differences in kinetics properties of plasma membrane ATPase in 19 d old +P and −P plants were observed (Km of plasma membrane ATPase were: 0.06 and 0.08, Vmax: 0.263 and 0.280 in +P and −P roots, respectively), thus indicating that ATP concentration in the tissue may not limit ATPase activity. Additionally, culture of *Plantago major* spp. pleiosperma on two-fold diluted nutrient solution caused a decrease in specific ATPase activity determined in the presence of ATP and Mg^{2+} by only 20% as compared to the control in Kuiper et al. (1991).

Examining the effect of Pi deficiency on cell ultrastructure, we confirmed earlier observations: greater vacuolisation and plasma membrane invaginations in root cortical cells of −P plants (Wanke et al., 1998). Phosphate deficiency seems not to influence the localisation of ATPases stimulated by Mg^{2+}in cortical meristematic cells of 18 d old bean root. In both +P and −P roots active ATPases were localised mainly in plasmalemma (Figure 1). No significant difference in the intensity of the lead precipitation was observed.

All the above data indicate that the alterations in plasma membrane compositions and properties may not directly influence the observed decrease in nitrate uptake during phosphate deficiency.

Next, we examined the possibility that nitrate uptake rate may be affected by its assimilation in the roots. Nitrate reductase activity was determined*in vivo* and *in vitro* in 18 d old +P and −P roots. In −P roots, NR determined *in vivo* assay was reduced to 60% of the control (Figure 2). The effect of Pi deficiency on NR activity in *in vitro* assay was even more pronounced, whereas in +P roots it remained constant during the culture, after 18 d of culture in −P roots it decreased to 50% of the control (Figure 2). Reducing power is considered as one of the major factors regulating NR activity *in vivo* (Oaks, 1985). Since the concentration of NADH was not affected while NADPH increased during phosphate starvation (Juszczuk and Rychter, 1997), it seems that NR activity is not limited by reducing power. These results are complementary to the data obtained by Jeschke et al. (1997), who did not measure NR activity in Pi deficient castor bean plants, but found the great depression of NO_3 reduction in roots calculated from the net nitrate reduction in xylem and phloem sap.

Nitrate concentration in the roots of phosphate deficient plants was higher as compared with the control (Figure 3 A), similarly as that observed in soybean (Rufty et al., 1990). Phosphate starvation influenced distribution of NO_3 ions between roots and shoot. In control plants, 70% of nitrate ions were located in the shoot and only 30% in the root, whereas in −P plants,

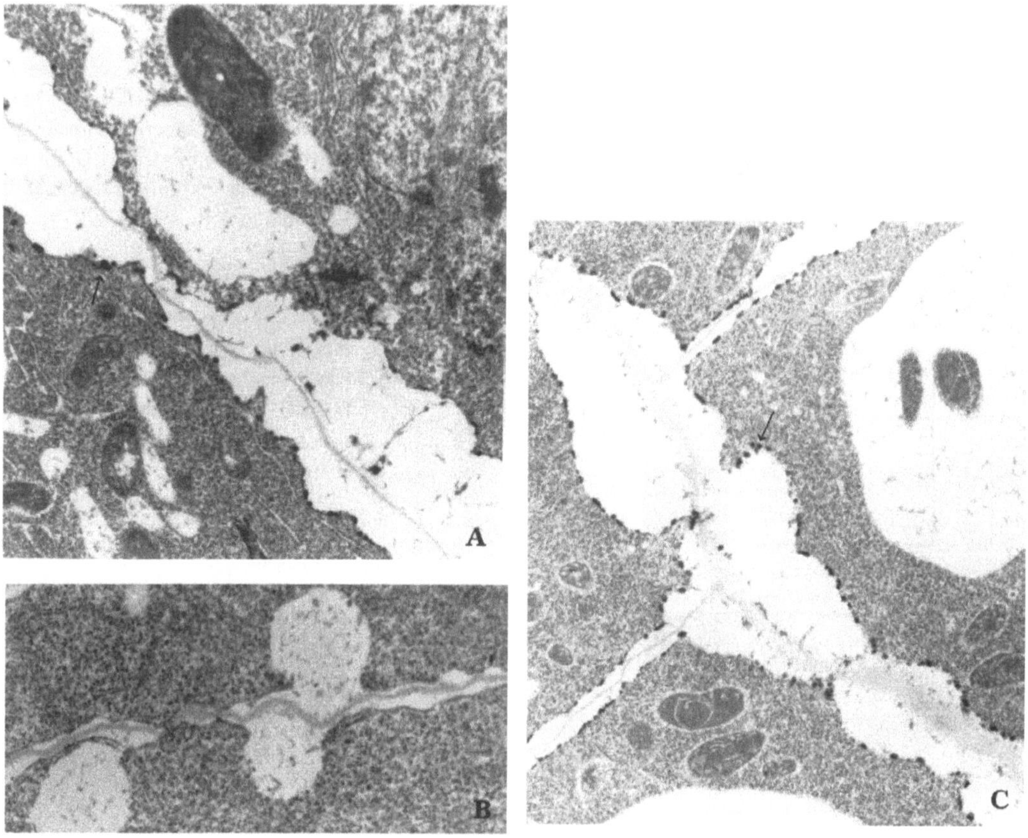

Figure 1. Localisation of active ATPases in cortical cells from the meristematic zone of roots of 18 d old bean plants, fixed in 1% glutaraldehyde + 4% formaldehyde. (A) plants grown on phosphate sufficient (+P) medium, ATPases present in plasma membrane (arrow). Magnification 20 000 ×. (B) plants grown on phosphate sufficient (+P) medium, control sample, negative reaction. Magnification 12 000 ×. (C) plants grown on phosphate deficient (−P) medium, ATPases present in plasma membrane (arrow). Magnification 18 000 ×.

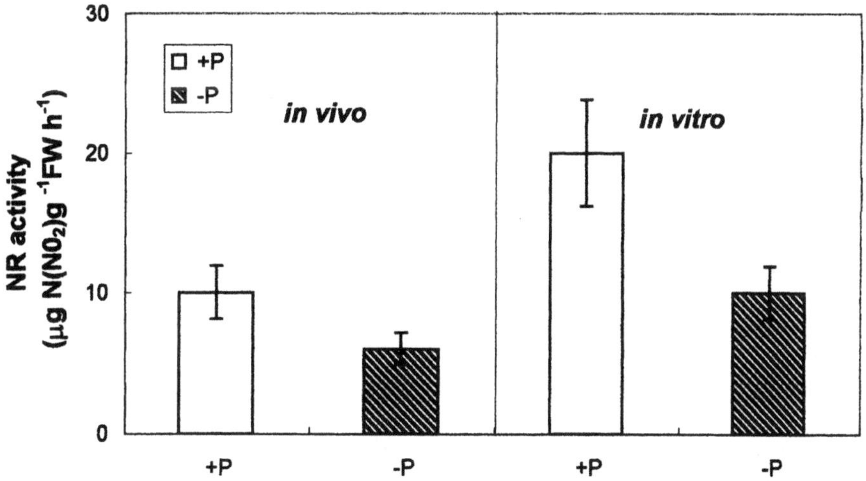

Figure 2. Nitrate reductase activity (*in vivo* and *in vitro*) in roots of 18 d old bean plants cultured on phosphate sufficent (+P) or phosphate deficient (−P) medium (after Gniazdowska et al., 1999a).

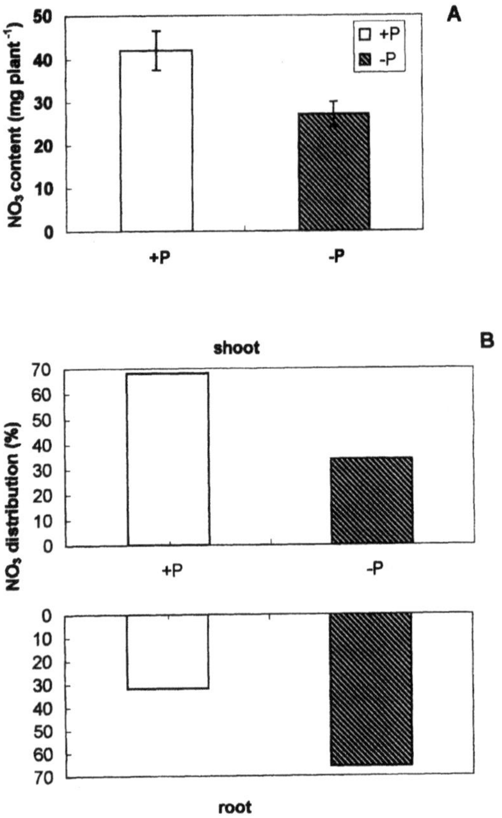

Figure 3. NO_3 content (A) and NO_3distribution (B) between shoot and root in 18 d old bean plants cultured on phosphate sufficient (+P) or phosphate deficient medium (after Gniazdowska et al., 1999a).

70% of NO_3 ions were found in the roots and only 30% in the shoot (Figure 3 B). The different distribution of NO_3 between shoot and root in Pi deficient plants indicated that NO_3 transport to the shoot may be also lower as was found in *Ricinus communis* (Jeschke et al., 1996, 1997). Thus, the increased nitrate concentration in roots may be explained by lower NO_3 transport and lower NR activity. In NR deficient barley mutants, nitrate accumulation lowered nitrate uptake rate (King et al., 1993).

Our data support the notion (Jeschke et al., 1997; Rufty et al., 1993) that during phosphate starvation, the decrease in nitrate uptake mainly results from feedback inhibition as a consequence of NO_3 accumulation in the roots. However, the changes in plasma membrane composition may also directly or indirectly affect these processes.

Acknowledgements

We wish to thank Dr Maria Podbielkowska for help in cytochemical localisation of active ATPases. Thanks are given to Prof. E. Kirkby for critical reading of manuscript. This work was supported by Grant no 6 PO4C 03413 from the State Committee for Scientific Research (Poland) given to A.R.

References

Allen C F, Good P, Davis H F, Hisum P and Fowler S D 1966 Methodology for the separation of plant lipids and application to spinach leaf and chloroplast lamellae. J. Am. Oil Chem. Soc. 43, 223-230.

Ames B N 1966 Assay of inorganic phosphate, total phosphate and phosphatases. *In* Methods in Enzymology. Eds. S P Colowick and N Kaplan. Vol 8. pp 115-118. Acad. Press, New York.

Bearden J C Jr 1978 Quantitation of submicrogram quantities of protein by an improved protein-dye binding assay. Biochim. Biophys. Acta 533, 525-529.

Bradford M M 1976 A rapid and sensitive method for the quantitation of microgram quantities of protein utilizing the principle of protein-dye binding. Anal. Biochem. 72, 248-254.

Carvajal M, Cooke D T and Clarkson D T 1996a Plasma membrane fluidity and hydraulic conductance in wheat roots: interaction between root temperature and nitrate or phosphate deprivation. Plant Cell Env. 19, 1110-1114.

Carvajal M, Cooke D T and Clarkson D T 1996b Responses of wheat plants to nutrient deprivation may involve the regulation of water-channel function. Planta 199, 372-381.

Cataldo D A, Haroon M, Schrader L E and Youngs V E 1975 Rapid colorimetric determination of nitrate in plant tissue by nitration of salicylic acid. Comm. Soil. Sci. Plant Anal. 6, 71-80.

Ciereszko I, Gniazdowska A, Mikulska M and Rychter A M 1996 Assimilate translocation in bean plants (*Phaseolus vulgaris* L.) during phosphate deficiency. J. Plant Physiol. 149, 343-348.

Crawford N M 1995 Nitrate: Nutrient and Signal for Plant Growth. The Plant Cell 7, 859-868.

Crawford N M and Glass A D M 1998 Molecular and physiological aspects of nitrate uptake in plants. Trends Plant Sci. 3, 381-395.

Gniazdowska A, Mikulska M and Rychter A M 1998 Growth, nitrate uptake and respiration rate in bean roots under phosphate deficiency. Biologia Plantarum 41, 217-226.

Gniazdowska A, Krawczak A, Mikulska M and Rychter A M 1999a Low phosphate nutrition alters bean plants ability to assimilate and translocate nitrate. J. Plant Nutr. 22, 551-563.

Gniazdowska A, Szal B and Rychter A M 1999b The effect of phosphate deficiency on membrane phospholipid composition of bean (*Phaseolus vulgaris* L.) roots. Acta Physiol. Plant. 21 (in press).

Jeschke W D, Peuke A, Kirkby E A, Pate J S and Hartung W 1996 Effects of P deficiency on the uptake, flows and utilization of C, N and H_2O within intact plants of *Ricinus communis* L. J. Exp. Bot. 47, 1737-1754.

Jeschke W D, Kirkby E A, Peuke A D, Pate J S and Hartung W 1997 Effect of P deficiency on assimilation and transport of nitrate and phosphate in intact plants of castor bean (*Ricinus communis* L.). J. Exp. Bot. 48, 75-91.

Juszczuk I M and Rychter A M 1997 Changes in pyridine nucleotide levels in leaves and roots of bean plants (*Phaseolus vulgaris* L.) during phosphate deficiency. J. Plant Physiol. 151, 399-404.

Kaiser W M and Huber S C 1994 Posttranslational regulation of nitrate reductase in higher plants. Plant Physiol. 106, 817-821.

Kasamo K and Sakakibara Y 1995 The plasma membrane H^+–ATPase from higher plants: functional reconstitution into liposomes and its regulation by phospholipids. Plant Sci. 11, 117-131.

King B J, Siddiqi M Y, Ruth T J, Warner R H and Glass A D M 1993 Feedback regulation of nitrate influx in barley roots by nitrate, nitrite and ammonium. Plant Physiol. 102, 1279-1286.

Kuiper D, Sommarin M and Kylin A 1991 The effect of mineral nutrition and benzyladenine on the plasmlemma ATPase activity from roots of wheat and *Plantago major* spp. pleiosperma, Physiol. Plant. 81, 169-174.

Lewis O A M, Watson E F and Hewitt E J 1982 Determination of nitrate reductase activity in barley leaves and roots. Ann. Bot. 49, 31-38.

Oaks A 1985 Nitrogen metabolism in roots. Ann. Rev. Plant Physiol. Plant Mol. Biol. 36, 407-414.

Palmgren M G 1991 Regulation of plant plasma membrane H^+–ATPase. Physiol. Plant. 83, 314-323.

Podbielkowska M and Wałęza M 1987 The suitability of the lead precipitation method for the localization of active ATPases in the meristematic cells of *Allium cepa* L. roots. Acta Soc. Bot. Pol. 56, 191-200.

Radin J W and Matthew M A 1989 Water transport properties of cortical cells in roots of nitrogen- and phosphorus-deficient cotton seedlings. Plant Physiol.89, 264-268.

Rayle D L and Cleland R E 1992 The acid growth theory of auxin-induced cell elongation is alive and well. Plant Physiol. 99, 1271-1274.

Reynolds E S 1963 The use of lead citrate at high pH as an electron opaque in electron microscopy. J. Cell Biol. 17, 208-213.

Rufty T W Jr, MacKown C T and Israel D W 1990 Phosphorus stress effects on assimilation of nitrate. Plant Physiol. 94, 328-333.

Rufty T W Jr, Israel D W, Volk R J, Qui J and Sa T 1993 Phosphate regulation of nitrate assimilation in soybean. J. Exp. Bot. 44, 879-891.

Rychter A M and Mikulska M 1990 The relationship between phosphate status and cyanide-resistant respiration in bean roots. Physiol. Plant. 79, 663-667.

Rychter A M and Randall D D 1994 The effect of phosphate deficiency on carbohydrate metabolism in bean roots. Physiol. Plant. 91, 383-388.

Rychter A M, Cheauveau M, Bomsel J-L and Lance C 1992 The effect of phosphate deficiency on mitochondrial activity and adenylate levels in bean roots. Physiol. Plant. 84, 80-86.

Sanderson G W and Cocking E C 1964 Enzymatic assimilation of nitrite in tomato plants. I. Reduction of nitrate to nitrite. Plant Physiol. 39, 416-422.

Sikorska E and Kacperska-Palacz A 1980 Frost-induced phospholipid changes in cold-acclimated and non-acclimated rape leaves. Physiol. Plant. 48, 201-206.

Stulen I, Lanting L, Lambers H, Posthumus F, van de Dijk S J and Hofstra R 1981 Nitrogen metabolism of *Plantago major* spp. major as dependent on the supply of mineral nutrient. Physiol. Plant. 52, 108-114.

Wachstein M and Meisel E 1957 Histochemistry of hepatic phosphatases at physiologic pH with special reference to the demonstration of bile canaliculi. Ann.J. Clin. Path. 27, 12-23.

Wanke M, Ciereszko I, Podbielkowska M and Rychter A M 1998 Response to phosphate deficiency in bean (*Phaseolus vulgaris* L.) roots. Respiratory metabolism, sugar localization and changes in ultrastructure of bean root cells. Ann. Bot. 82, 809-819.

Widell S and Larsson C 1981 Separation of presumptive plasma membranes from mitochondria by partition in an aqueous polymer two-phase system. Physiol. Plant. 51, 368-374.

Widell S and Larsson C 1990 A critical evaluation of markers used in plasma membrane purification. *In* The Plant Plasma Membrane. Eds. C Larsson and I M Möller. pp 17-39. Springer-Verlag Berlin Heidelberg.

Section editor: A C Borstlap

Recent Advances of Plant Root Structure and Function, 119–122
Edited by O. Gašparíková, M. Čiamporová, I. Mistrík and F. Baluška
© 2001 Kluwer Academic Publishers.

Nitrate reductase in roots: Succinate- and NADH-dependent plasma membrane-bound forms

Christine Stöhr, Stefanie Wienkoop & Wolfram R. Ullrich[1]
Institut für Botanik, Technische Universität, Schnittspahnstrasse 10, D-64287 Darmstadt, Germany. [1]*Corresponding author**

Key words: nitrate reductase, plasma membrane-bound nitrate reductase, root, succinate, tobacco

Abstract

This short review summarizes recent experimental data on root plasma-membrane-bound nitrate reductase (PM-NR) mainly from tobacco roots. Although root PM-NR apparently comprises the same domains for partial reactions as leaf PM-NR and the soluble NR of roots and leaves, only this enzyme uses succinate as electron donor, in addition to NADH. Various properties and the regulation of root PM-NR by the preceding nitrate nutrition of the plants differ considerably, not only from soluble NR in the roots, but also from soluble NR and PM-NR in the leaves of the same plants.

Introduction

Nitrate assimilation is a complex process implying nitrate uptake, nitrate and nitrite reduction, as well as ammonium assimilation. All of them are regulated reactions and located in different compartments of the plant cells or even different tissues. Nitrate transport and mainly nitrate reduction are strongly regulated: Genetically, by induction and repression of nitrate reductase transcription, as well as by post-translational modifications of the protein.

Nitrate reductase has been detected in roots as well as in shoots. Former generalisations assuming that in trees, the bulk of nitrate reduction is located in the roots, whereas in herbaceous plants it is mainly located in the energetically well-supplied leaves, have now been much relativated because external factors can strongly influence the allocation of the enzymes. Hence, under-ground and above-ground nitrate reduction have to be regarded as of similar importance.

Intracellular location of nitrate reductase has been discussed even for a longer time. Nowadays there is agreement that the bulk enzyme is soluble and occurs in the cytosolic compartment (cytosolic NR). A variable part of the enzyme has been found to be associated with various membranes. More recently, a plasma membrane-bound nitrate reductase (PM-NR) has been identified with only a few percent of total NR activity (Tischner et al., 1989; Ward et al., 1988), isolated (Stöhr et al., 1993) and is shown to be fixed to the outer surface of the plasma membrane with a glycosyl-phosphatidyl-inositol anchor (Kunze et al., 1997; Stöhr et al., 1995). Both the intracellular and PM-bound forms use NADH (in some cases also NADPH) as the electron donor. Thus, NADH has to become available in the extracellular space for PM-NR. Recently, an activity of the root PM-enzyme with succinate as the electron donor has been discovered (Stöhr and Ullrich, 1997). Our present work is concentrated on this succinate-dependent enzyme activity and its regulation in comparison with the NADH-dependent activity of PM-NR in roots and leaves (Stöhr, 1999a, b; Wienkoop et al., 1999). The present article gives a short and comprehensive review on this work.

Materials and methods

In most of the experiments reported in this paper, the material was 6-week-old plants of tobacco (*Nicotiana tabacum* cv. samsun) grown in a sand culture. Some

* FAX No: +49-6151-164808. TEL No: +49-6151-163706.
E-mail: ullrich@bio.tu-darmstadt.de

other plant species (*Hordeum vulgare*, *Zea mays*, *Dactylis glomerata* and *Phleum arenarium*) at similar age and grown in the same conditions were used for comparison (for details see Stöhr, 1999a; Stöhr and Ullrich, 1997).

Cytosolic NR was extracted from the tissue after homogenising in liquid N_2 (Stöhr and Ullrich, 1997). For the preparation of PM-NR, the plant material was gently homogenised in a preparation medium containing 5% PVP and 1.5% PVPP, 100 mM ascorbate and 5 mM dithiothreitol as stabilising agents for the succinate-dependent activity. The two-phase system with polyethyleneglycol and dextran was used to separate plasma membrane vesicles (Larsson, 1985; Stöhr and Ullrich, 1997). The hydrophobic PM-NR was then separated from possible contaminations of membrane-associated soluble NR by the temperature-dependent phase separation technique with Triton X-114 (Werck-Reichhart et al., 1991), performed with a temperature shift from the ice-bath to 15 °C and low-speed centrifugation. The lower (detergent) phase contained the hydrophobic PM-NR. The activity could be clearly separated from mitochondrial succinate dehydrogenase (Stöhr and Ullrich, 1997). Separate activities of the PM-NR from cytosolic and membrane-associated forms could also be shown by hydrophobic interaction (Stöhr et al., 1993) or anion exchange chromatography (Stöhr and Ullrich, 1997). The enzyme assay had to be performed in the absence of ascorbate and other protectants.

Table 1. Summary of the properties of PM-NR in tobacco roots (from own experiments). n.d., not determined

Effector			NADH	Succinate
Malonate			No effect	Inhibition
Temperature optimum			30 °C	50 °C
pH optima			n.d.	30 °C : 5.6/8.0
				50 °C : 7.0
K_m (μM)		Nitrate	NADH	Succinate
30 °C	pH 5.6 / pH 8.0	35 / 117	n.d.	60 / 220
50 °C	" / "	13 / 125	"	36 / 250

Results and discussion

In roots of all plants investigated, PM-NR has always been found together with soluble and membrane-associated soluble NR, though with varying activities. A survey of the different NR forms and nitrate transport in root cells is shown in Figure 1.

Properties of root PM-NR

NAD(P)H was the electron donor for the enzyme of leaves and roots as well as for cytosolic NR, while succinate only worked for root PM-NR, which usually exhibited higher activity with succinate than with NADH. This root enzyme showed a slight inhibition of the nitrate reduction rate when both electron donors, NADH and succinate, were supplied together. Thus, the activities were not additive, which suggests that both reactions use the same protein. Malonate was a strong inhibitor of the succinate reaction, but not of the NADH-dependent reaction, which indicates different binding sites for the two electron donors. The temperature optimum for the NADH reaction was 30 °C, while for the succinate reaction it was 50 °C (Table 1). The pH curve of the succinate activity showed two peaks at 30 °C, at pH 5.6 and 8.0, but only one high peak at 50 °C at pH 7.0. Michaelis constants were found to be lowest at pH 5.6 and 50 °C, K_m for nitrate reaching values below 15 μM under these conditions. At pH 8, K_m for nitrate was about 120 μM at both temperatures and thus more similar to those reported for cytosolic NR in the literature. Also, the K_m values for succinate were highly pH-dependent and the lowest values were found at pH 5.6 and 50 °C (Table 1) (Wienkoop et al., 1999). A direct comparison of the K_m values for nitrate and NADH with the same enzyme preparations is not available yet.

Partial reactions

Cytosolic NR is known to be composed of three domains, the diaphorase domain with FAD, the heme domain with cytochrome b_{556} and the terminal reductase containing the molybdo-pterin complex (MoCo) (Campbell, 1996). While PM-NR of roots could reduce ferricyanide with NADH, succinate could not serve as an electron donor in this reaction, thus indicating that structural differences of the FAD domain might be responsible for the succinate-dependent nitrate reduction. Among other partial reactions tested, nitrate reduction with $FADH_2$ (via heme and molybdo-pterin) was the slowest and resembled in its temperature dependence the overall reaction from succinate to nitrate. The partial reaction from methylviologen to nitrate was much more rapid and showed a very different temperature profile, which, in

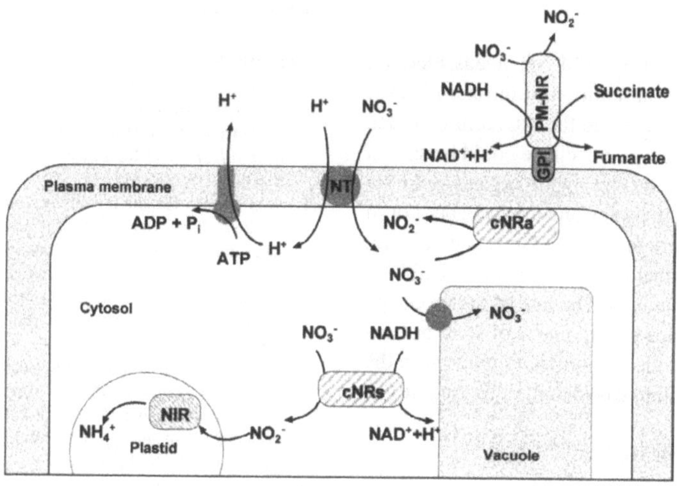

Figure 1. Nitrate assimilation proteins in a root cell. NT: nitrate transporter for NO_3^-/H^+ cotransport; PM-NR: plasma membrane-bound nitrate reductase; cNRs: cytosolic nitrate reductase, soluble form; cNRa: cytosolic nitrate reductase, membrane-associated form; NIR: nitrite reductase.

Table 2. Relative NR activities in roots of tobacco plants cultured with different nitrate supply (% of maximum activity). Activities calculated on a protein basis, hence cytosolic NR and PM-NR not quantitatively comparable. For PM-NR: 100% in μmol NO_2^- mg^{-1} protein h^{-1}: 1.13 with succinate, 0.65 with NADH. No activity with succinate was found for cytosolic NR

NO_3^- supply (mM)	2	5	10	15	20	25
Cytosolic NR	52	100	70	20	13	11
PM-NR Succinate	2	14	13	42	68	100
" NADH	10	27	27	53	52	100

contrast to other data, suggests that it proceeds without involvement of the heme domain. Concluding from our experiments, the heme domain seems to account for the limitation of the overall NR reaction.

Regulation of PM-NR

Cytosolic NR activity is known to be greatly dependent on genetic and post-translational regulation, so PM-NR had to be tested if it is a regulated enzyme as well and if the regulation is different from that of cytosolic NR. This was tested by comparing the activities of the NR forms after different nitrate treatment of the plants. Growth of plants of tobacco and of several other species, such as *Dactylis glomerata* and *Phleum arenarium*, achieved its optimum when a 10 mM nitrate solution was supplied daily to the sand cultures (Stöhr, 1999a). Soluble NR of the roots had highest activities at 5 mM nitrate and was apparently repressed at higher concentrations. NADH- and succinate-dependent activities of PM-NR in the roots were very low below 5 mM nitrate, increased very much with higher nitrate supply up to 25 mM and showed a decrease beyond that concentration (Table 2). In the roots, succinate-dependent activities exceeded those with NADH, especially at high nitrate. In the leaves, all NR activities reached a high level at 10 mM and then remained in the same range at higher nitrate concentrations (data not shown). Irregularly high nitrate accumulation by high nitrate supply, when cytosolic NR was repressed or suppressed, could not be found in the plant material up to 25 mM external nitrate. Instead, tissue nitrate was even less at 25 than at 10 or 15 mM. The same was shown for total nitrogen in the tissues of roots, leaves and the whole plants, so the tissue nitrate and total nitrogen contents were apparently in parallel with plant growth. C/N ratios were high only at nitrate supply below 5 mM, otherwise they remained constant over the whole range, even at very high external nitrogen supply (Stöhr, 1999a). Like the response to nitrate supply, diurnal regulation of cytosolic NR and PM-NR was found to be completely different: While cytosolic NR had the optimum activity during the light period, PM-NR was low at daytime and showed its optimum during the night.

Availability of substrates

Obviously, for its function, PM-NR needs electron donors in the apoplast in addition to nitrate. NADH, like ATP, seems to be present in the extracellular space of plant and animal tissues. Succinate could be shown in our experiments to be the major organic acid released from the roots in barley and tobacco, but not in the leaf apoplast, where malate release was dominant. From this, it appears that the reactions described here can easily take place *in vivo*. The fate of the produced nitrite is not well known yet, much of it is not assimilated within the cells, but probably reduced in the extracellular space and released to the rhizosphere.

Distribution over plant species

So far, all plant species tested (tobacco, barley, maize, wild grasses from nitrogen rich and nitrogen poor stands) showed PM-NR activity. All of them also contained succinate-dependent PM-NR in the roots, not in their leaves (Stöhr, 1999b).

Possible role of PM-NR

This wide distribution together with the low K_m values at the low pH reported for the extracellular space, the correlation between high activity and over-optimum nitrate supply and the regulation separate from that of cytosolic NR suggest a regulatory or signal role of PM-NR for the nitrate uptake systems, for cytosolic NR activities and probably also for the remarkably stable C/N ratio in the plant at high nitrate supply (Stöhr, 1999a).

Acknowledgements

Sincere thanks are due to Gisela Marx for careful technical assistance. The work was financially supported by the Deutsche Forschungsgemeinschaft (Sonderforschungsbereich 199).

References

Campbell W 1996 Update on biochemistry. Nitrate reductase biochemistry comes of age. Plant Physiol. 111, 355–361.

Kunze M, Riedel J, Lange U, Hurwitz R and Tischner R 1997 Evidence for the presence of GPI-anchored PM-NR in leaves of *Beta vulgaris* and for PM-NR in barley leaves. Plant Physiol. Biochem. 35, 507–512.

Larsson C 1985 Plasma membranes. *In* Modern Methods in Plant Analysis, New Series, Vol. 1, Cell Components. Eds HF Linskens and JF Jackson. pp 85–104. Springer, Berlin.

Stöhr C 1999a Relationship of nitrate supply with growth rate, plasma membrane-bound and cytosolic nitrate reductase and tissue nitrate content in tobacco plants. Plant Cell Environ. 22, 169–177.

Stöhr C 1999b Plasma membrane-bound nitrate reductase in algae and higher plants. *In* Plasma Membrane Redox Systems and Their Role in Biological Stress and Disease. Eds H Asard, A Bérczi and RJ Caubergs. pp 103–119. Kluwer Academic Publishers, Dordrecht, The Netherlands.

Stöhr C, Tischner R and Ward M R 1993 Characterization of the plasma-membrane-bound nitrate reductase in *Chlorella saccharophila* (Krüger) Nadson. Planta 191, 79–85.

Stöhr C, Schuler F and Tischner R 1995 Glycosyl-phosphatidylinositol-anchored proteins exist in the plasma membrane of *Chlorella saccharophila* (Krüger) Nadson: Plasma-membrane-bound nitrate reductase as an example. Planta 196, 284–287.

Stöhr C and Ullrich W R 1997 A succinate-oxidising nitrate reductase is located at the plasma membrane of plant roots. Planta 203, 129–132.

Tischner R, Ward M R and Huffaker R C 1989 Evidence for a plasma-membrane-bound nitrate reductase involved in nitrate uptake of *Chlorella sorokiniana*. Planta 178, 19–24.

Ward M R, Tischner R and Huffaker R C 1988 Inhibition of nitrate transport by anti-nitrate reductase IgG fragments and the identification of plasma membrane associated nitrate reductase in roots of barley seedlings. Plant Physiol. 88, 1141–1145.

Werck-Reichhart D, Benveniste I, Teutsch H, Durst F and Gabriac B 1991 Glycerol allows low-temperature phase separation of membrane proteins solubilized in Triton X-114: Application to the purification of plant cytochromes P-450 and b_5. Anal. Biochem. 197, 125–131.

Wienkoop S, Ullrich W R and Stöhr C 1999 Kinetic characteristics of succinate-dependent plasma membrane-bound nitrate reductase in tobacco roots. Physiol. Plant. 105, 609–614.

Recent Advances of Plant Root Structure and Function, 123–134
Edited by O. Gašparíková, M. Čiamporová, I. Mistrík and F. Baluška
© 2001 Kluwer Academic Publishers.

Impact of *Agrobacterium tumefaciens*-induced stem tumors on NO_3^- uptake in *Ricinus communis*§

Igor Mistrik[1], Jan Pavlovkin[2], Rebecca Wächter[3], Katja S. Pradel[3], Katja Schwalm[3], Wolfram Hartung[4], Ulrike Mathesius[5], Christine Stöhr[3] and Cornelia I. Ullrich[3,*]
[1]*Institute of Botany, Slovak Academy of Sciences, Dúbravská cesta 14, SK-84223 Bratislava, Slovakia;* [2]*Institute of Experimental Phytopathology, Slovak Academy of Sciences, SK-90028-Ivanka pri Dunaji, Slovakia;* [3]*Institut für Botanik, Technische Universität, Schnittspahnstr. 3, D-64287-Darmstadt, Germany;* [4]*Julius-von-Sachs-Institut für Biowissenschaften, Universität Würzburg, Julius-von-Sachs-Platz 2, D-97082 Würzburg, Germany;* [5]*Plant-Microbe Interaction Group, Research School of Biological Sciences, Australian National University, PO Box 475, Canberra City, ACT 2601, Australia*

Key words: abscisic acid, *Agrobacterium tumefaciens*-induced tumors, amino acids, ethylene, indole-3-acetic acid and GH3:GUS, NO_3^- and $H_2PO_4^-$ uptake

Abstract

Developing tumors induced by *Agrobacterium tumefaciens*, strain C58, on stems of *Ricinus communis* L. var. *gibsonii* cv. Carmencita were shown to be strong metabolic sinks for sucrose and amino acids, thus causing higher nutrient demand in the host plant. However, NO_3^- uptake and, to a lesser extent, also $H_2PO_4^-$ uptake were strongly inhibited. Correspondingly, NO_3^- concentration was lower in tumorised than in the control plants. NO_3^-reductase activity was the same in both plant types, but it was completely suppressed in the tumors. The electrical membrane potential difference of root cells was unaffected in tumorised plants when soil-grown, but significantly lowered when grown hydroponically. Consistent with the low NO_3^- uptake rate, NO_3^--dependent membrane depolarisation at the onset of $NO_3^-/2H^+$-cotransport was nearly zero. In the phloem sap, sucrose and amino acid concentrations were considerably lower in tumorised than in control plants, and lower below than above the tumor. The qualitative pattern of amino acids of the phloem sap of stems was almost the same in tumorised and control plants. It is concluded that neither the overall amino acid concentration nor special amino acids nor ammonium in the transport phloem suppress NO_3^- uptake in the roots. Aminocyclopropane-carboxylate, the precursor of ethylene, which is produced in the tumors in high amounts, was low in the stems and the same in both plant types. Thus, ACC and ethylene were ruled out as directly interfering with nutrient uptake in the roots. Root morphology was strongly affected during tumor development. Root fresh weight decreased to 50% of the controls and lateral root development was almost completely prevented. This suggests that the high tumor ethylene production, together with an increasing concentration of phenolic compounds, severely inhibits the basipetal auxin flow to the roots. Auxin accumulation and retention was confirmed by specifically enhanced expression of the auxin-responsive promoter of the soybean gene GH3:GUS in tumors induced in transgenic *Trifolium repens* L. Hence, root development is poorer and anion uptake inhibited in tumorised plants. This may be aggravated by abscisic acid accumulation in the tumor and its basipetal export into the roots. Moreover, sucrose depletion of the sieve tubes leads to energy shortage at the root level for maintaining energy-dependent anion uptake.

Introduction

Integration and expression of the oncogenes encoded in the T-DNA of the Ti-plasmid of *Agrobacterium tumefaciens* leads to rapid tumor (crown gall) cell proliferation supported by differentiation of well-

§ Dedicated to Professor Wilhelm Simonis on the occasion of his 90th birthday.

* FAX No: +49 6151 164630. E-mail: uleb@bio.tu-darmstadt.de

organised vascular bundles. Overexpression of the phytohormone genes, i.e., the synthesis of high concentrations of auxin and cytokinins (Kado, 1984), promotes the production of considerable amounts of ethylene (Aloni et al., 1998; Wächter et al., 1999) and abscisic acid. Ethylene causes structural changes, as summarised in the 'gall constriction hypothesis', explaining the mechanism that gives priority in water and nutrient supply to the growing tumor over the host shoot: the xylem below the tumor is considerably increased in extent, while adjacent to the tumor it shows tracheas of much smaller diameter surrounded by unusual multiseriate unlignified rays and xylem parenchyma cells of high metabolic activity. Disruption of the epidermis and cuticle on the tumor surface causes high transpiration rates (Aloni et al., 1995; Schurr et al., 1996).

The strong accumulation of assimilates, solutes and phytohormones together with these structural changes infer hormonal signals from the tumor into the root and shoot. Such signals are assumed to enhance ion absorption and NO_3^- reductase (NR) activity supporting the synthesis of opines, which serve for the ecological explanation of this phenomenon of natural genetic engineering. The aim of the present study was to investigate the relationship between NO_3^- absorption and the metabolite concentration of the tumor in comparison with NO_3^- and amino acid concentration and NRA of the shoot, roots and phloem sap of both tumorised and non-infected *Ricinus communis*. Measurements of the transmembrane electrical potential were designed to determine the energisation state of root tissue under tumor stress.

Materials and methods

Tumor induction

Ricinus communis L. var. *gibsonii* cv. Carmencita (Walz Samen, Stuttgart, Germany) was grown in standard potting soil (LD 80) under universal white fluorescent lamps (L 58 W/25; Osram, München, Germany) in growth chambers. They were exposed to 14 h light (200 μmol m^{-2} s^{-1}) at 27 °C and 10 h darkness at 21 °C with 60–80% relative humidity.

Ten days after sowing, after appearance of the first leaves, the plant hypocotyls were wounded with a razor blade 10 mm below the cotyledons and were inoculated with a bacterial pellet. The wild-type strain C58 of *Agrobacterium tumefaciens*, obtained from the Max-Planck-Institut für Züchtungsforschung (Köln, Germany), was grown for 24 h in YEB medium (1 g yeast extract, 5 g beef extract, 5 g peptone, 5 g sucrose per liter and 2 m*M* $MgSO_4$, pH 7.8). For inoculation the bacteria were harvested by centrifugation for 10 min at 3600 *g*. Except for ion uptake experiments and some electrophysiological measurements, soil grown plants have been used throughout all analyses.

Sucrose and amino acids

Sucrose concentration of tissues was determined in methanol extracts (80%) after Cardini (in Büttner et al., 1985), total amino acid concentration with ninhydrin. For determination of transport sucrose and amino acids, xylem sap was collected from stems detopped 1 cm above the tumor and phloem sap from exudation after gentle incision of the stem. Contamination of exuding phloem sap by xylem sap does not occur during strong transpiration in the light. In the case of simultaneous wounding by mistake of the xylem no sap exuded at all. The amino acid pattern of xylem and phloem sap was obtained by HPLC analysis (Biotronik, Maintal, Germany), using the ninhydrin reaction.

Uptake of NO_3^- and $H_2PO_4^-$

Plants were grown for 3 weeks in a complete and well aerated nutrient solution, containing 3.96 m*M* KNO_3, 5.47 m*M* $CaCl_2$, 1.22 m*M* $MgSO_4$, 1.47 m*M* KH_2PO_4, 18 μM Fe-EDTA, 1.5 μM $MnCl_2$, 8.1 μM H_3BO_3, 0.5 μM Na_2MoO_4, pH 4.8. Prior to the experiments the plants were transferred into a solution without NO_3^- or $H_2PO_4^-$ for up to further 3 weeks in order to stimulate ion uptake. Twelve hours before the measurement of NO_3^- uptake, 100 μM KNO_3 was resupplied to a 1 m*M* $CaSO_4$ solution for the induction of NO_3^- uptake. The plants were then transferred into a fresh $CaSO_4$ solution containing 100 μM KNO_3 or 100 μM KH_2PO_4, pH 5.6. Ion uptake experiments were performed with intact plants in the light. Samples of the incubation solution were withdrawn at intervals and NO_3^- or $H_2PO_4^-$ concentrations were determined colorimetrically (Braun-SysteMatic, Methodenblatt N60; Strickland and Parsons, 1965, respectively). The uptake rates were calculated on the basis of NO_3^- and $H_2PO_4^-$ depletion in the uptake solution and related to the root fresh weight.

NO_3^- reductase activity (NRA)

After homogenisation with a microdismembrator (Braun, Melsungen, Germany) of plant material that had been frozen in liquid nitrogen, the homogenate was extracted, centrifuged and the supernatant was assayed according to Stöhr and Ullrich (1997). Tissue NO_3^- concentration was determined in aqueous extracts by ion chomatography (Spectraphysics, Fremont, CA, USA), i.e. separation on a Partisil SAX column in 3 m*M* NaH_2PO_4+H_3PO_4 pH 2.4 at 40 °C at a flow rate of 1.5 mL min^{-1}, at 210 nm with a UV detector.

Electrophysiology

Roots were mounted in a vertical Plexiglass chamber that was perfused with 1X solution (containing 1 m*M* KCl, 1 m*M* NaH_2PO_4, 1 m*M* $Ca(NO_3)_2$ and 0.25 m*M* $MgSO_4$ pH 5.6) at a rate of 170 μl s^{-1}. Voltage changes were recorded by means of glass microelectrodes, filled with 3 *M* KCl as electrolyte and connected to a high-impedance electrometer amplifier (L/M-1, List, Darmstadt, Germany). The diffusion potential (E_D) was determined by addition of 1 m*M* NaCN plus 1 m*M* salicylhydroxamic acid (pH 5.6) to the perfusion solution. To test whether the P-ATPase activity of tumorised and control root cells could be differently enhanced either directly at the membrane level or indirectly by energy supply, the P-ATPase stimulating toxin fusicoccin (FC) or sucrose were added to the perfusion solution. Further details have been described earlier (Marx and Ullrich-Eberius, 1988).

ACC in stem tissue

1-Aminocyclopropane-1-carboxylic acid was determined with a gas chromatograph according to Lizada and Yang (1979) as described in detail recently (Wächter et al., 1999).

IAA and ABA

To determine auxin (IAA) and abscisic acid (ABA), lyophilised and homogenised plant material was extracted with 80% methanol containing 1 mg L^{-1} butylated hydroxytoluene as antioxidant. The supernatant was purified with a SEP-PAK C_{18}-cartridge, extracted with ethyl acetate at pH 3 and dried. Samples for ABA determination were re-dissolved in Tris buffer, pH 7.5 (50 m*M*, 1 m*M* $MgCl_2$, 10 m*M* NaCl). For IAA determination, samples were methylated with

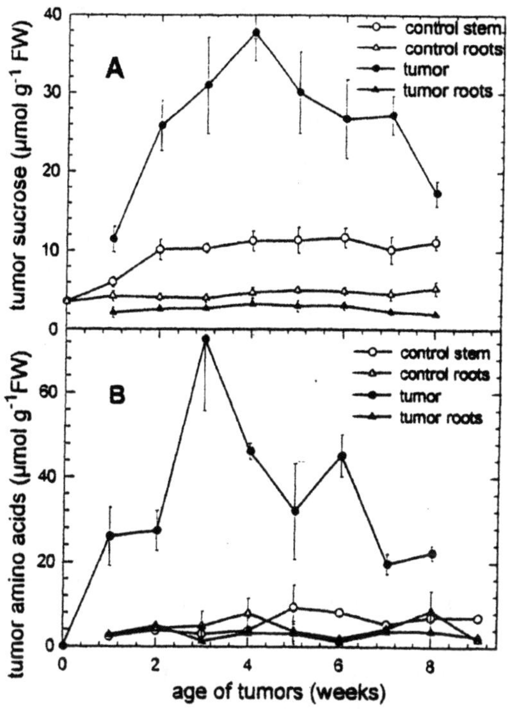

Figure 1. Sucrose and amino acid accumulation in tumors, uninfected stems and roots of *Ricinus*, in dependence of tumor age. (A) Sucrose concentration (n=3–20), (B) Total amino acid concentration (n=3–6). Mean values ±SD.

ether-dissolved diazomethane, and after ether evaporation, dissolved in distilled water and diluted in Tris buffer (100 m*M* pH 7.5). The concentrations of both hormones were determined with the enzyme immunoassay (ELISA) Phytodetek (Idetek, San Bruno, CA, USA). ABA concentration of the phloem sap was determined as described recently (Jeschke et al., 1997b).

Auxin-responsive promoter GH3:GUS assay

In order to detect and visualise auxin accumulation and retention in crown galls, transformed white clover was infected with the wild-type strain C58 of *Agrobacterium tumefaciens*. Transgenic plants of *Trifolium repens* L. cv. Haifa contained the 749bp auxin-responsive promoter of the soybean gene GH3 (Hagen et al., 1991), fused to the GUS reporter gene (Jefferson et al., 1987). White clover transformation, vector plasmids and *Agrobacterium* strains have recently been described in detail (Larkin et al., 1996).

GUS enzyme activity was localised in tissues with the histochemical assay using 5-bromo-4-chloro-3-

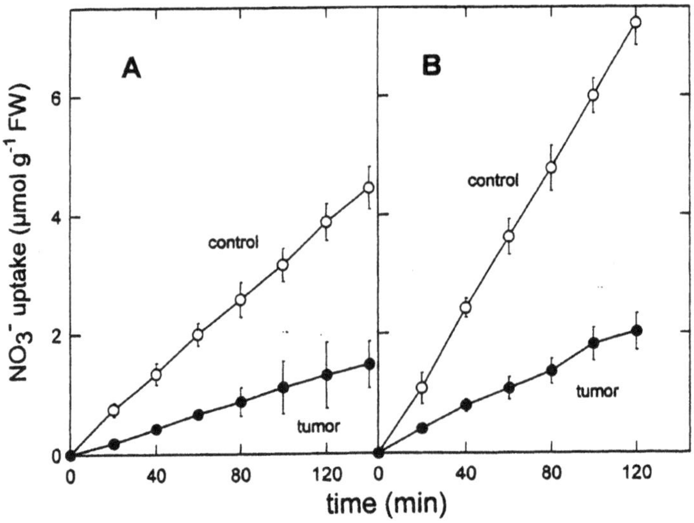

Figure 2. Time course of NO_3^- uptake. (A) NO_3^- uptake into tumorised plants with a higher biomass than the control plants of the same age and (B) tumorised plants with lower biomass than the control plants of the same age (4-week-old tumors). External KNO_3 concentration was 100 μM in 1 mM $CaSO_4$ solution. Mean values ±SD (n=5).

indolyl-glucuronide (X-gluc) as substrate (Jefferson et al., 1987). Tissue sections of 100 μm thickness were prepared with a Lancer Series 1000 Vibratome. For staining, tissues were incubated for one h at 37 °C in 50 mM phosphate buffer pH 7, containing 10 mM EDTA, pH 7, 0.5 mM $K_4Fe(CN)_6$, 0.5 mM $K_3Fe(CN)_6$ and 1 mM X-gluc.

Phenolic compounds

Phenolic compounds were extracted from frozen, homogenised tissue with methanol and assayed with the Folin-Ciocalteu reagent in the presence of Na_2CO_3 according to Rapp and Ziegler (1973).

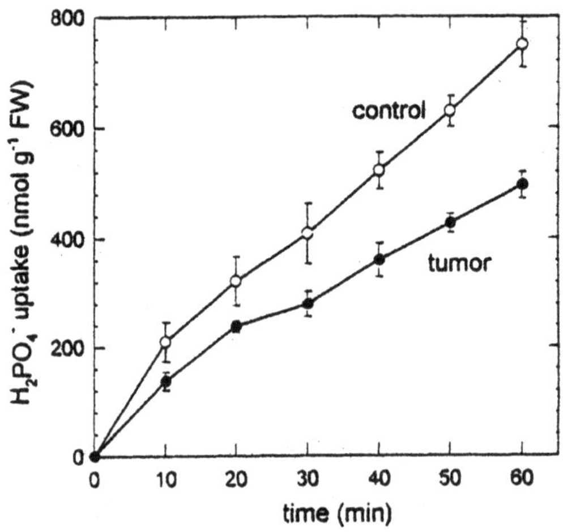

Figure 3. $H_2PO_4^-$ uptake in roots of tumorised (4-week-old tumors) and control plants. External KH_2PO_4 concentration was 100 μM at pH 5.6. Mean values ±SD (n=5).

Results

Sucrose accumulation in stem tumors was strongest between 3 and 5 weeks after bacterial infection. With 40 μmol g^{-1} FW it was up to 4 times that of the control stems (Figure 1A). In contrast, sucrose concentration in the roots of tumorised plants was about one half that of roots of control plants with 5 μmol g^{-1} FW on average (Figure 1A).

Amino acid accumulation was strongest in 3-week-old tumors, up to 40 times higher than in the control stem tissue (Figure 1B). Similar to sucrose concentration, amino acid concentration was nearly always higher in the control roots than in the roots of tumorised plants (Figure 1B).

Surprisingly, NO_3^- absorption rates of tumorised plants, measured with hydroponically-grown plants, were less than 50% of the control plants (Figure 2). There was no correlation in these plants between biomass production (Table 1) and the NO_3^- uptake rate (Figure 2A, B). NO_3^- uptake was inhibited in both

Table 1. Biomass of tumorised and control plants of the same age, used in NO_3^- uptake experiments (Figure 2A and B), 4-week-old tumors each

	g FW			
	Tumor plant A	Control plant A	Tumor plant B	Control plant B
Shoot	16.6	15.6	8.0	12.5
Tumor	3.0		2.0	
Root	6.5	4.0	1.3	3.0
Total biomass	26.1	19.6	11.3	15.5

Table 2. NO_3^- concentration and nitrate reductase activity (NRA) of tumorised and control *Ricinus* plants. 4-week-old tumors. Values of leaves are mean values of samples taken at 8, 12 and 16 h. Values of tumorised stems are mean values of tissue samples taken from above, at the level of or below the tumor. Mean values ± SD (*n*=8–12). nd=not determined

Plants	NO_3^- (μmol g^{-1} FW)	NRA (μmol g^{-1} FW h^{-1})	NRA (μmol g^{-1} protein h^{-1})
Control plant			
Leaf	55±15	4.7±1.3	161±26
Stem	114±12	0.3±0.1	219±96
Roots	nd	0.03±0.02	20±18
Tumor plant			
$Leaf_{next\ to\ tumor}$	12±4	5.4±1.4	215±56
Stem	104±8	0.7±0.4	236±92
Tumor	14±2	0	0
Roots	nd	0.07±0.05	35±39

types of tumorised plants, those with higher and with smaller overall biomass.

The NO_3^- concentration was lower in leaves of tumorised than in the control plants, that of stems was not significantly different. The overall NO_3^- concentration of the tumors was in the same range as that of the leaves of infected plants (Table 2).

NO_3^- reductase activity (NRA) in leaves and roots of tumorised plants was not significantly different from that of the control tissue, independent of whether related to fresh weight or protein concentration of the respective organs (Table 2). Surprisingly, no activity could be discovered in the tumors (Table 2).

To find out whether tumorised plants suffer from lower energisation of their roots, the electrical membrane potential (E_m) was determined in root cells about 3 cm from the tip (epidermis, cortex and stele). E_m was dependent on growth conditions. When plants had been grown in soil, the root E_m of about -170 mV was the same in control and tumorised plants (Table 3). The same applied to the energy-independent diffusion potential (E_D). Exposing plants to additional stress, e.g. by growing them hydroponically, resulted in a drastic decrease of E_m and a slightly lower diffusion potential in both root types (Table 3). Such plants had roots with a membrane potential close to the diffusion potential, relatively lower under tumor stress. Membranes could be hyperpolarised by sucrose or fusicoccin in both root types but only control cells could be permanently stimulated to values as high as in roots of soil-grown plants. In soil-grown roots of the control plants, the NO_3^--dependent membrane depolarisation at the onset of NO_3^- uptake was 20 mV on average, indicating active NO_3^- uptake. In contrast, depolarisation in root cells of tumorised plants did not exceed about 5 mV and was not enhanced by pretreatment with sugars as a signal or energy source (Table 3).

Table 3. Electrical membrane potential difference (E_m) and diffusion potential (E_D) of cells of the early root hair zone of tumorised (tumor) *Ricinus* plants. Additions to the perfusion solution: For E_D determination, 1 m*M* NaCN plus 1 m*M* salicylhydroxamic acid, pH 5.6; for proton pump activation, fusicoccin (FC), applied in 0.1% ethanol, or sucrose. Maximum membrane depolarisation upon addition of 1 m*M* NO_3^- to a nitrate-free perfusion solution (ΔE_m). Plants grown in soil or hydroponically, 4-week-old tumors. Mean values ±SD (*n*)

Plants	(mV)				
	E_m	E_D	FC (10 μM)	Sucrose (10 mM)	ΔE_m (1 mM NO_3^-)
Soil					
Control	−172±6 (58)	−97±3 (15)	−195±4 (3)	−194±11 (4)	21±6 (8)
Tumor	−174±6 (62)	−94±3 (10)	−193±5 (3)	−197±2 (3)	6±4 (12)
Hydroponics					
Control	−112±11 (11)	−88±9 (14)	−173±19 (13)	−192±19 (10)	11±1 (4)
Tumor	−102±11 (45)	−89±8 (4)	−140±13 (4)	−130±13 (11)	3±3 (11)

Table 4. Aminocyclopropane-carboxylate (ACC) concentration in tumorised and control plants. Mean values ±SD (*n*). For comparison: 2-week-old tumors contain 8000 pmol ACC g^{-1} FW

Age of tumors	ACC (pmol g^{-1} FW)		
(weeks)	Control stem	Stem below tumor	Stem above tumor
1	47±11 (4)	220±133 (11)	209±46 (10)
2	219±184 (3)	415±171 (5)	260±103 (6)
3	95±20 (4)	443±181 (7)	196±77 (8)
4	186±36 (3)	298±192 (4)	323±210 (5)
5	165±68 (3)	231±87 (5)	219±177 (5)
6	219±107 (3)	156±35 (4)	483±162 (4)

To test the specificity of inhibition of ion uptake, $H_2PO_4^-$ uptake was investigated as well (Figure 3). In tumorised plants, $H_2PO_4^-$ uptake was also inhibited, but only by about 25%.

Since tumors were found to emit large amounts of ethylene which could be the cause for the inhibition of ion absorption, stem tissue was analysed for its concentration in the ethylene precursor, aminocyclopropane-carboxylate (ACC). Table 4 shows that the ACC concentration in control stems was not significantly different from that of tumorised stems. Also in the stem tissue above and below the tumor it was not significantly different at any time (Student's *t*-test with a significance level of P=5%).

Amino acids like glutamine, glutamic acid, asparagine and aspartic acid were suspected to specifically inhibit NO_3^- uptake. Total amino acid concentration of the phloem sap was consistently higher in control plants than in tumorised ones (Figure 4B). This difference increased with the length of the light period and with tumor age. In addition, amino acid concentration of the phloem sap collected above the tumor was higher than below the tumor. These differences were most conspicuous 5 weeks after tumor induction. Also sucrose concentration of the phloem sap decreased with tumor age and was lower in tumorised than in control plants (Figure 4A).

Concentrations of amines or glutamic and aspartic acid, determined by HPLC, were slightly higher in methanolic extracts of roots of tumorised plants. In stems almost all amino acids were up to twice as concentrated in the xylem exudate of tumorised plants in comparison with control plants (Table 5). However, xylem volume flow was about five times smaller in tumorised than in control stems, caused by flow resistance due to the smaller diameter of the vessels (Aloni et al., 1995, 1998; Wächter et al., 1999). Decrease in flow rate generally results in increase of xylem sap solute concentration (Else and Jackson, 1998). On the other hand, leaf transpiration of tumorised plants was half that of control plants (Schurr et al., 1996). Hence,

Table 5. Pattern of amino acid concentration in root extracts, xylem and phloem sap of tumorised (tumor) *Ricinus* and uninfected tissue (control). Phloem sap was collected from stems above or below tumors (four- to five-week-old). Total age of soil-grown plants 6 weeks. nd, not detectable at the concentrations of the samples used. Mean values ±SD (n=2-3)

	Roots (μmol g^{-1} FW)		Xylem sap (μM)		Phloem sap (mM)		
	Control	Tumor	Control	Tumor	Control	Above	Below
Pser	0.06	nd			0.15±0.03	0.11±0.02	0.12±0.02
Asp	0.68	1.18	61	110±51	12.0±2.9	12.2±3.22	14.9±0.9
Thr	0.11	0.08	59	125±11	2.7±0.6	2.2±0.2	1.9±0.1
Ser	0.29	0.44	nd	20±16	5.7±1.3	7.5±2.7	6.1±0.4
Asn	nd	0.63	331	516±120	0.7±0.1	0.6±0.3	0.7±0.2
Glu	0.89	1.05	40	35±7	10.8±0.6	12.0±4.8	12.9±0.6
Gln	2.99	2.20	4550	4790±260	53.4±5.8	34.6±1.8	32.1±9.0
Pro	nd	nd	nd	18±10	0.8±0.2	0.6±0.1	0.6±0.1
Gly	0.06	0.08			0.29±0.04	0.49±0.14	0.37±0.01
Ala	0.18	0.29	nd	8±2	2.3±0.4	1.9±0.9	1.9±0.1
Val	0.13	0.12	151	277±36	3.0±1.1	2.2±0.2	2.0±0.1
Cys					0.13±0.004	0.07±0.001	0.07±0.03
Met			15	33±11	0.26±0.03	0.17±0.03	0.13±0.01
Ile	0.05	nd	112	209±38	1.4±0.2	1.0±0.2	0.9±0.2
Leu	0	0	38	80±18	0.7±0.1	0.53±0.02	0.4±0.1
Phe	nd	nd	nd	16±1	0.8±0.2	0.52±0.01	0.4±0.1
GABA	nd	0.38	23	68±54	0.4±0.1	0.33±0.07	0.6±0.1
Orn					0.4±0.1	0.14±0.02	0.15±0.02
Lys			84	172±84	0.8±0.2	0.56±0.004	0.5±0.1
His			150	164±68	1.0±0.3	0.70±0.2	0.72±0.001
Try	nd	0.26			0.22±0.04	0.28±0.02	0.2±0.1
Arg	0.09	0.13	66	127±89	0.4±0.1	0.20±0.01	0.2±0.1
NH_4	0.45	1.92	560	318±13	0.63±0.02	0.67±0.02	0.7±0.3
Total	5.98	8.76	6240	7086±296	99.0±9.0	79.6±1.0	78.6±3.3

the amino acid transport rate of the xylem sap in intact plants can be assumed to be similar in both plant types.

Only the amino acid concentration measured in the phloem sap is supposed to reflect the actual pattern of that in intact plants (Table 5). Sap was collected from plants with four- to five-week-old tumors, when NO_3^- uptake was measured. Total amino acid concentration in the phloem sap as determined by HPLC was higher in uninfected than in tumorised stems, as found already in the overall assay (Figure 4B). There was no difference in asp, glu, asn and NH_4^+ concentrations between both plant types, that of gln was higher in the control phloem sap.

On average, in tumorised plants root development was inhibited, fresh weight being only 50% of the root fresh weight of control plants (Figure 5A). Most strikingly, the development of lateral roots was very poor in tumorised plants (Figure 5B).

In three- to four-week-old tumors the IAA concentration was three times that in control stems (Figure 6). Interestingly, in comparison with the uninfected stem the IAA concentration of tumorised stems was slightly higher above the tumor, below the tumor it was only one third of that above. IAA accumulation and retention in tumors could be visualised in *Trifolium repens* (Figure 7), transformed with the soybean auxin-responsive promoter, GH3, fused to the GUS gene (β-glucuronidase). Upon induction of crown gall growth at stems of such transformed white clover, GH3:GUS expression was specifically enhanced in developing seven-week-old tumors, as indicated by the intensity of the developing indigo-blue colour (Figure 7B, arrowheads). It was not visible in the transformed control stems of the same age (Figure 8A) and in the stem tissue opposite the tumor (Figure 7B).

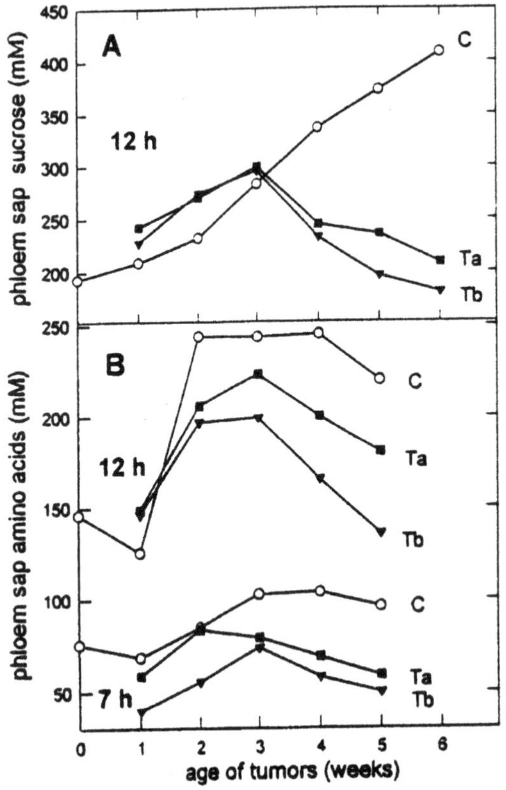

Figure 4. (A) Sucrose and (B) amino acid concentration of phloem sap collected from stems of tumorised (Ta, above tumors and Tb, below tumors) and control (C) plants, in dependence of tumor age, at the onset of the light period (7 h) and 5 h later (12 h). Sap from 5 different plants was pooled together at each time (3 repetitions).

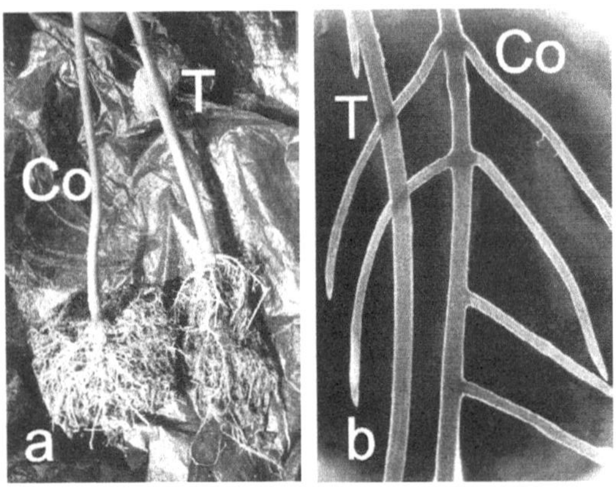

Figure 5. Root development under tumor stress. (a) Tumorised *Ricinus* plant (T) and uninfected control (Co); (b) Lack of lateral roots in tumorised (T) in comparison with uninfected control plants (Co) grown under identical conditions.

The ABA concentration was found to be more than 45 times higher in tumors than in uninfected stems (Figure 6). In the phloem sap it was on average ten-fold above and four-fold below the tumor in comparison with control stems (Figure 6).

For comparison with phytohormone accumulation in the tumors, accumulation of phenolic compounds was followed as well. Phenols rapidly increased with tumor age to considerable concentrations of more than 100 μmol g^{-1} FW (Figure 8).

Discussion

Rapid growth of *Agrobacterium tumefaciens*-induced plant tumors creates strong metabolic sinks on the host plants (Malsy et al., 1992; Pradel et al., 1996, 1999). Such tumors world-wide cause damage and losses in yield of sugar-beet, apple and pear trees and grapevine (Bien et al., 1990; Goodman, 1986; Lawson, 1991). Not only growth, but also considerable metabolite accumulation contributes to the sink properties. Sucrose accumulated up to 4 times, amino acids up to 40 times (Figure 1A, B). In some plant/parasite interactions such as *Cuscuta*/*Ricinus* or *Coleus*, photosynthetic CO_2 assimilation, transpiration and NO_3^- uptake were stimulated upon infection (Jeschke et al., 1997a). Similar effects were expected for tumorised *Ricinus* plants.

However, the present results did not confirm this assumption. Instead, NO_3^- uptake rates decreased by more than 50%. Even when the total biomass of tumorised plants exceeded that of control plants of the same age, NO_3^- uptake was consistently lower than in the controls (Figure 2 and Table 1). This indicates that in this system growth dynamics does not simply regulate NO_3^- uptake, as known from non-tumorised *Ricinus* grown under various environmental conditions (Schurr, 1998).

NO_3^- reductase activity (NRA) could not be detected in tumors (Table 2). Since it is known that ethylene inhibits NRA (Stacewicz-Sapuncakis et al., 1973), it is assumed that the high ethylene production of up to 400 pmol g^{-1} FW of such tumors (Wächter et al., 1999) suppresses NRA. Thus, NO_3^- taken up by the roots will predominantly be reduced and converted to amino acids in the leaves and accumulated and metabolised to opines within the tumors. Amino acid consumption seems to exceed NO_3^- uptake, since the amino acid concentration in the phloem sap considerably decreases with increasing tumor age. The sink property

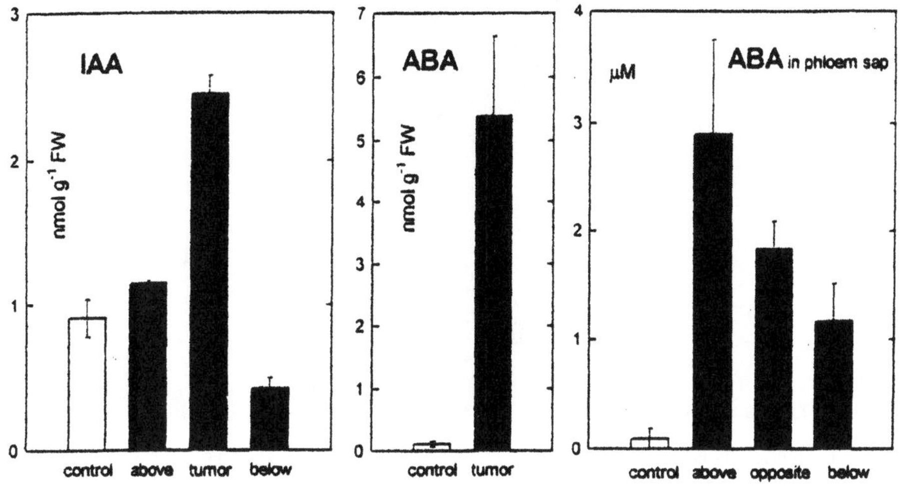

Figure 6. Auxin (IAA) and abscisic acid (ABA) accumulation in three- to four-week-old *Ricinus* tumors. Immunological assay of tissue of uninfected stems (control), of tumor tissue and of tumorised stems above or below the tumor. Phloem sap was collected from uninfected stems (control) and from tumorised ones, above, opposite of and below the tumor. Mean values ±SD (n=2–8).

of the tumor is clearly revealed by the fact that in the phloem sap collected above and below the tumors, the sucrose and amino acid concentrations below the tumor were lower than those above and in tumorised considerably lower than in uninfected plants (Figure 4).

Inhibition of NO_3^- uptake could be due to a specific inhibition by higher concentrations of glutamine, asparagine or NH_4^+ (Muller and Tourraine, 1992; Zhuo et al., 1999). But HPLC analysis of the roots and more pronounced of phloem sap revealed that the concentration of the critical amino acids and NH_4^+ in tumorised and in control plants are within the same range (Table 5). Hence, specific amino acids cannot be the reason for the inhibition of NO_3^- uptake. This conclusion is supported by recent NO_3^- uptake experiments with the split root system, where an essential role of phloem transport of amino acids is ruled out in the shoot-to-root control of NO_3^- uptake in *Ricinus communis* (Tillard et al., 1998).

Not only NO_3^- but also $H_2PO_4^-$ uptake into tumorised plants was inhibited, though to a smaller extent (Figure 3). Therefore, the inhibition must have a more general cause. The most likely reason is an interaction of phytohormones. Ethylene production in the tumors is preceded by high accumulation of ACC of more than 8000 pmol g^{-1} FW (Wächter et al., 1999). Since long-distance ACC transport is well documented for plants (Else and Jackson, 1998), ACC could be exported from the tumors into the roots and converted into ethylene. This may impair anion uptake. However, ACC analysis of tumorised stems did not indicate any significant ACC export from the tumors into the roots (Table 4).

Abscisic acid (ABA), known to inhibit NO_3^- and $H_2PO_4^-$ uptake (Suleiman et al., 1990), was about 45 times more concentrated in tumors than in uninfected stems (Figure 6). Below tumors the ABA concentration of the phloem sap was about four times that of the controls. Hence, ABA may contribute to the inhibition of ion uptake.

Due to considerable sucrose depletion of the phloem sap by the tumors (Figure 4A), energisation of the roots may become the limiting factor for ion uptake. The energy-dependent electrical membrane potential (E_m) of roots of tumorised, soil-grown plants was not different from the controls. But it was close to the diffusion potential (E_D) upon the additional stress, when plants were grown hydroponically (Table 3). NO_3^--dependent membrane depolarisation at the onset of $NO_3^-/2H^+$-cotransport was nearly zero in such roots and very small in roots of soil-grown, tumorised plants, in comparison with the controls. These results with soil-grown plants confirm the inhibition of NO_3^- uptake in tumorised plants as a general phenomenon.

Phenolic compounds, necessary for induction of the *vir*-region of the Ti plasmid (Stachel et al., 1985), are reported to play an important role in nodule form-

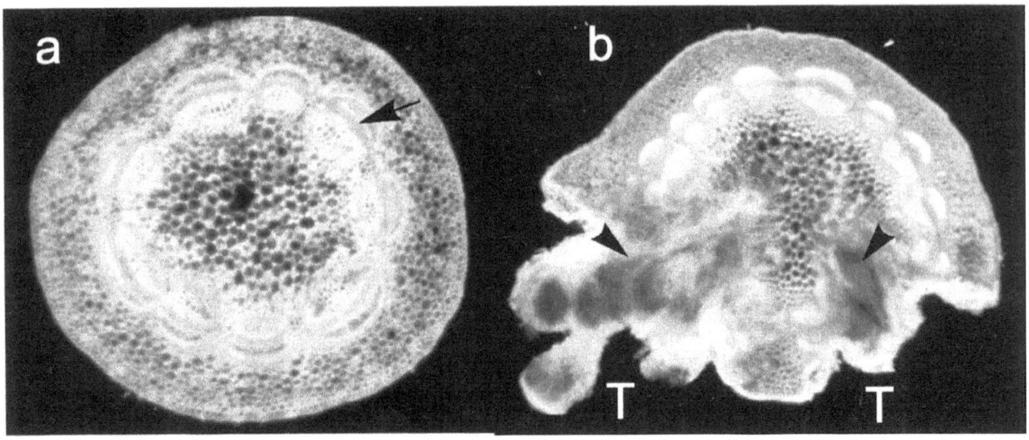

Figure 7. GUS histochemical activity in stem and tumor tissue of transgenic *Trifolium repens* containing the soybean auxin-responsive promoter GH3:GUS. (a) GUS activity is absent in the cambium of vascular bundles (arrow) of uninfected stems (×28). (b) Strong GUS activity (arrowheads) is restricted to developing 7-week-old tumors (T). Note that vascular bundles opposite the tumor do not show GUS activity (×18).

ation in legumes by rhizobia. In particular, flavonoids have been found to inhibit basipetal auxin export and to prevent IAA oxidation, thus leading to IAA accumulation and nodule proliferation (Jacobs and Rubery, 1988; Mathesius et al., 1998). In *A. tumefaciens*-induced tumors the concentration of phenolic compounds showed a 20 fold rise beyond that of the control (Figure 8). They may have a similar effect in retaining IAA in the tumor, which additionally accelerates tumor proliferation. IAA accumulation and retention indeed occur in such tumors (Figure 6). Increased IAA concentration in tumors was indicated and visualised as well by inducing tumors in transgenic white clover (Figure 7), containing the auxin-responsive promoter of the soybean gene GH3 (Hagen et al., 1991) fused to the GUS gene (Larkin et al., 1996). The GH3 promoter responds rapidly and specifically to biologically-active auxins (Guilfoyle et al., 1993), but shows little or no expression in untreated plants.

High ethylene production of tumors induces structural differentiation of vascular bundles in the contact zone of the tumor with the host stem: larger xylem with smaller diameter of vessels and multiseriate, unlignified rays developped (Aloni et al., 1995, 1998; Wächter et al., 1999). Simultaneously, root development was poor. Total fresh weight decreased by up to 50% and lateral root development was abolished (Figure 5). These structural and morphological changes indicated auxin deficiency in the roots. Ethylene is known to inhibit basipetal flow of auxin (Davies, 1995; Liu and Reid, 1992; Roberts et al., 1988).

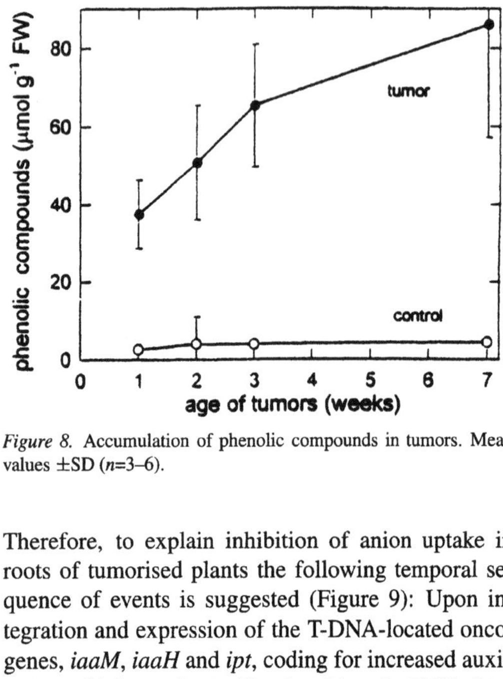

Figure 8. Accumulation of phenolic compounds in tumors. Mean values ±SD (*n*=3–6).

Therefore, to explain inhibition of anion uptake in roots of tumorised plants the following temporal sequence of events is suggested (Figure 9): Upon integration and expression of the T-DNA-located oncogenes, *iaaM*, *iaaH* and *ipt*, coding for increased auxin and cytokinin synthesis (Zambryski et al., 1989), IAA considerably accumulates in tumors in comparison with control stems. IAA promotes ethylene synthesis (Yang and Hoffman, 1984) which in turn inhibits basipetal IAA transport from the shoot to the root and triggers ABA synthesis in the tumors and its export into shoots and roots. IAA deficiency and increased ABA supply, together with a decrease in energisation of the membranes, which is a precondition for anion uptake, gradually leads to a decrease in total nitrogen

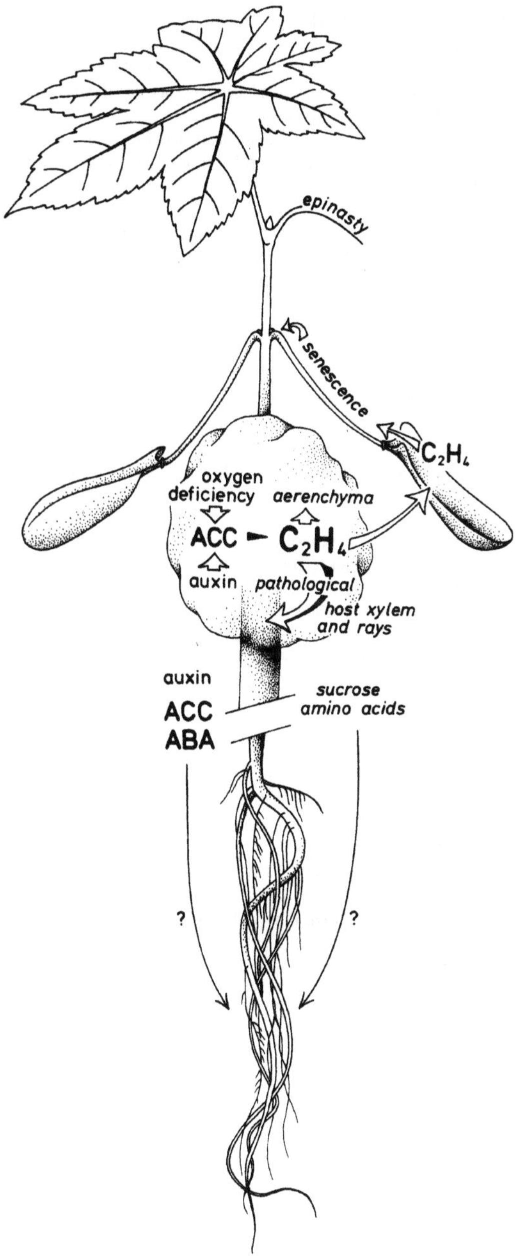

Figure 9. Summarising model, explaining inhibition of anion absorption by roots of tumorised *Ricinus* plants. T-DNA coded IAA expression and O_2 deficiency induce ACC accumulation in the tumor. ACC is converted to ethylene. Ethylene impairs basal IAA flow, leading to inhibition of lateral root development and anion uptake. This effect is assumed to be aggravated by assimilate retention by the tumor and additional basipetal release of ABA.

assimilate concentration. Additional assimilate consumption by the tumor contributes to an accelerated weakening of the whole plant.

Acknowledgements

This work was supported by grants of the Deutsche Forschungsgemeinschaft (SFB 199 and 251) to C.I.U. and W.H. We thank Sylvia Lenz and Barbara Dierich for valuable technical assistance.

References

Aloni R, Pradel K S and Ullrich C I 1995 The three-dimensional structure of vascular tissues in *Agrobacterium tumefaciens*-induced crown galls and in the host stems of *Ricinus communis*. Planta 196, 597–605.

Aloni R, Wolf A, Feigenbaum P, Avni A and Klee H J 1998 The *Never ripe* mutant provides evidence that tumor-induced ethylene controls the morphogenesis of *Agrobacterium tumefaciens*-induced crown galls on tomato stems. Plant Physiol. 117, 841–849.

Bien E, Lorenz D, Eichhorn K and Plapp R 1990 Isolation and characterization of *Agrobacterium tumefaciens* from the German vineregion Rheinpfalz. Z. Pflanzenkr. Pflanzenschutz 97, 313–322.

Büttner G, de Fekete M A R and Vieweg G H 1985 Changes in fructan content in developing barley caryopses. Angew. Bot. 59, 171–177.

Davies P J 1995 The plant hormones: their nature, occurrence, and functions. *In* Plant Hormones: Physiology, Biochemistry and Molecular Biology. Ed. P J Davies, pp 1–12. Kluwer Academic Publishers, Dordrecht.

Else M A and Jackson M B 1998 Transport of 1-aminocyclopropane-1-carboxylic acid (ACC) in the transpiration stream of tomato (*Lycopersicon esculentum*) in relation to foliar ethylene production and petiole epinasty. Aust. J. Plant Physiol. 25, 453–458.

Goodman R N 1986 Crown gall in grapes. Am. Wine Soc. J. 18, 80–81.

Guilfoyle T J, Hagen G, Li Y, Ulmasov T, Liu Z and Gee M 1993 Auxin-regulated transcription. Aust. J. Plant Physiol. 20, 489–502.

Hagen G, Martin G, Li Y and Guilfoyle T J (1991) Auxin-induced expression of the soybean GH3 promoter in transgenic tobacco plants. Plant Mol. Biol. 17, 567–579.

Jacobs M and Rubery P H 1988 Naturally occurring auxin transport regulators. Science 241, 346–349.

Jefferson R A, Kavanagh T H and Bevan M W 1987 GUS fusion: beta-glucuronidase as a sensitive and versatile gene fusion marker in higher plants. The EMBO J. 6, 3901–3907.

Jeschke W D, Baig A and Hilpert A 1997a Sink-stimulated photosynthesis, increased transpiration and increased demand-dependent stimulation of nitrate uptake: Nitrogen and carbon relations in the parasitic association *Cuscuta refexa-Coleus blumei*. J. Exp. Bot. 48, 915–925.

Jeschke W D, Peuke A D, Pate J S and Hartung W 1997b Transport, synthesis and catabolism of abscisic acid (ABA) in intact plants of castor bean (*Ricinus communis* L.) under phosphate deficiency and moderate salinity. J. Exp. Bot. 48, 1737–1747.

Kado C I 1984 Phytohormone-mediated tumorigenesis by plant pathogenic bacteria. *In* Genes Involved in Microbe-Plant Interactions. Eds. D P S Verma and Th Hohn. pp 311–336. Springer-Verlag, Wien.

Lawson M 1991 First to market. Nature 353, 687.

Larkin P J, Gibson J M, Mathesius U, Weinman J J, Gartner E, Hall E, Tanner G J, Rolfe B G and Djordjevic M A 1996 Transgenic white clover. Studies with the auxin-responsive promoter, GH3, in root gravitropism and lateral root development. Transgen. Res. 5, 325–335.

Liu J-H and Reid D M 1992 Auxin and ethylene-stimulated adventitious rooting in relation to tissue sensitivity to auxin and ethylene production in sunflower hypocotyls. J. Exp. Bot. 43, 1191–1198.

Lizada M C C and Yang S F 1979 A simple and sensitive assay for 1-aminocyclopropane-1-carboxylic acid. Analyt. Biochem. 100, 140–145.

Malsy S, van Bel A J E, Kluge M, Hartung W and Ullrich C I 1992 Induction of crown galls by *Agrobacterium tumefaciens* (strain C 58) reverses assimilate translocation and accumulation in *Kalanchoë daigremontiana*. Plant Cell Environ. 15, 519–529.

Marx S and Ullrich-Eberius C I 1988 Solute accumulation and electrical membrane potential in *Agrobacterium tumefaciens*-induced crown galls in *Kalanchoë daigremontiana* leaves. Plant Sci. 57, 27–36.

Mathesius U, Schlaman H R M, Spaink H P, Sautter C, Rolfe B G and Djordjevic M A 1998 Auxin transport inhibition precedes root nodule formation in white clover roots and is regulated by flavonoids and derivatives of chitin oligosaccharides. Plant J. 14, 23–34.

Muller B and Touraine B 1992 Inhibition of NO_3^- uptake by various phloem-translocated amino acids in soybean seedlings. J. Exp. Bot. 43, 617–623.

Pradel K S, Rezmer C, Krausgrill S, Rausch T and Ullrich C I 1996 Evidence for symplastic phloem unloading with a concomitant high level of acid cell-wall invertase in *Agrobacterium tumefaciens*-induced plant tumors. Bot. Acta 109, 397–404.

Pradel K S, Ullrich C I, Santa Cruz S and Oparka K J 1999 Symplastic continuity in *Agrobacterium tumefaciens*-induced tumours. J. Exp. Bot. 50, 183–192.

Rapp A and Ziegler A 1973 Bestimmung der Phenolcarbonsäuren in Rebblättern, Weintrauben und Wein mittels Polyamid-Dünnschichtchromatographie. Vitis 12, 226–236.

Roberts L W, Gahan P B and Aloni R 1988 Vascular Differentiation and Plant Growth Regulators. Springer-Verlag, Berlin. 154 p.

Schurr U 1998 Dynamics of nutrient transport from the root to the shoot. Progr. Bot. 60, 234–253.

Schurr U, Schuberth B, Aloni R, Pradel K S, Schmundt D, Jähne B and Ullrich C I 1996 Structural and functional evidence for xylem-mediated water transport and high transpiration in *Agrobacterium tumefaciens*-induced tumors of *Ricinus communis*. Bot. Acta 109, 405–411.

Stachel S E, Messens E, van Montagu M and Zambryski P 1985 Identification of the signal molecules produced by wounded plant cells that activate T-DNA transfer in *Agrobacterium tumefaciens*. Nature 318, 624–629.

Stacewicz-Sapuncakis M, Marsh H V, Vengris J, Jennings P H and Robinson T 1973 Participation of ethylene in common purslane response to Dicamba. Plant Physiol. 52, 466–471.

Stöhr C and Ullrich W R 1997 A succinate-oxidising nitrate reductase is located at the plasma membrane of plant roots. Planta 203, 129–132.

Strickland J D H and Parsons T R 1965 A manual of sea water analysis. Bull. Fish. Res. Board Can. 125, 203.

Suleiman S, Hourmant A and Penot M 1990 Influence de l'acide abscissique sur le transport d'ions inorganiques chez la pomme de terre (*Solanum tuberosum* cv. Bintje). Etude comparée avec quelques autres phytohormones. Biol. Plant. (Praha) 32, 128–137.

Tillard P, Passama L and Gojon A 1998 Are phloem amino acids involved in the shoot to root control of NO_3^- uptake in *Ricinus communis* plants? J. Exp. Bot. 49, 1371–1379.

Wächter R, Fischer K, Gäbler R, Kühnemann F, Urban W, Bögemann G M, Voesenek L A C J, Blom C W P M and Ullrich C I 1999 Ethylene production and ACC-accumulation in *Agrobacterium tumefaciens*-induced plant tumours and their impact on tumour and host stem structure and function. Plant Cell Environ. 22. (In press).

Yang S F and Hofman N E 1984 Ethylene biosynthesis and its regulation in higher plants. Annu. Rev. Plant Physiol. 35, 155–189.

Zambryski P, Tempé J and Schell J 1989 Transfer and function of T-DNA genes from *Agrobacterium* Ti and Ri plasmids in plants. Cell 56, 193–201.

Zhuo D, Okamoto M, Vidmar J J and Glass A D M 1999 Regulation of a putative high-affinity nitrate transporter (Nrt2;1At) in roots of *Arabidopsis thaliana*. Plant J. 17, 563–568.

Section editor: H Lambers

Recent Advances of Plant Root Structure and Function, 135–141
Edited by O. Gašparíková, M. Čiamporová, I. Mistrík and F. Baluška
© 2001 Kluwer Academic Publishers.

Fast shoot responses to root treatment. Are hormones involved?

G. Kudoyarova, D. Veselov, M.Symonyan, A. Mustafina, A. Dedov, L. Vysotzkaya & S. Veselov
Institute of Biology of Russian Academy of Scinces. Pr. Octybrya, 69, Ufa 450054, Russia

Key words: abscisic acid, cytokinin, indole-3-acetic acid, maize, stress, wheat

Abstract

Hormone concentration was measured in shoots of maize and wheat seedlings shortly after addition of PEG or ice to nutrient medium, or after dipping the roots for 0.5 min in a concentrated solution of potassium chloride. The concentration of IAA (indole-3-acetic acid), ABA (abscisic acid) and cytokinins was immunoassayed using specific antibodies. Very fast changes in hormone content were observed in shoots of plants following their root treatment. Evidence is presented which supports the involvement of shoot metabolic processes in the control of hormone content changes. Both accumulation of ABA in the shoot of PEG-treated plants and the decrease in cytokinin in the shoots of cooled plants correlated with the cessation of shoot extension in stressed plants. Possible mechanisms of fast signal transfer from roots to shoots are discussed.

Introduction

The growth and behaviour of roots is known to be coupled closely to the growth and behaviour of shoots (Davies and Jeffcoat, 1990; Jackson, 1994). Root–shoot interactions imply the exchange of signals between them. Possible signals might include mineral nutrients or hormones. Moreover root–shoot communication might be maintained by means of electric or hydraulic signals.

Of these possible means of signal transmission, hormones have been paid particularly close attention. Hormones in small amounts influence many processes in plants and are mobile in the plant. The literature contains many examples of changes in hormone export from roots to shoots in response to soil stresses (Jackson, 1994, 1997). Accordingly, hormones are strong candidates as messenger molecules that help co-ordinate development throughout the plant.

Although root–shoot interactions have been thoroughly studied, many questions still need to be answered. One of them concerns the possible rapidity of signal transfer in plants. The literature cites several examples of a very fast appearance of responses in plant organs distant from the site of stimulation. Thus, cold-shock of leaves changed ion uptake by roots (Opritov et al., 1972) reduced phloem translocation sharply more than 15 cm from the site of stimulation (Fromm and Bauer, 1994). Addition of PEG to the nutrient medium inhibited shoot growth in only a few minutes (Chazen and Neumann, 1994). Not in all cases did roots serve as the site of stimulation, but all these examples demonstrate a rapid transfer of signals from one organ to another.

What kind of signal might be sent from one organ to another in the above considered cases? Some authors concentrate on action potentials and other electrical signals, since they are characterised by high rate of spreading (Davies et al., 1991). Although hormones were not forgotten, the possibility of fast changes in their concentration has not been tested. These researchers were satisfied with some indirect evidence that seemed to exclude the involvement of hormones in fast signal transmission. Auxin is known to be transported from cell to cell at a very slow rate (Goldsmith, 1977). Fromm et al. (1995) tested the possibility of auxin involvement in quick signal transfer. They showed that it took about 1 h for auxin applied to the site of stimulation to reach the organ where changes in respiration were induced 5 min after the stimulation. In the case of root-produced hormones, the rate of their export with the transpiration flow is rather high. However, the hormone concentration in xylem sap is reported to be much lower than in shoot tissues (Wagner and Beck, 1994). So it obviously needs a considerable time for changes in hormone delivery to influence significantly their content in the shoot. Thus, the system of hormone transport probably does not operate promptly enough to enable quick changes in hormone concentration in responding tissue to occur. However, hormone concentration is controlled not only through their transport, but by means of other mechanisms too. For example, hormones are known to be released from bound forms (Bandurski, 1978). It is not clear from the literature how quickly the interconversion of free and bound forms might occur in plants. However, the rate of hormone release or conjugation could be rather high. Bearing this possibility in mind, we decided to study hormone concentration in shoots, soon after stress treatments of the root systems. It was of interest to find out if hormone content might change prompt enough to be involved in the control of fast responses of shoot–root treatment.

Materials and methods

Seedlings of durum wheat (*Triticum durum* Desf., cv. Bezenchukskaya 139) and maize (Zea mays L. cv. Krasnodarsky 303) were grown in darkness at 24 °C between the hollow glass tubes connected at the edges, tied together and floating in containers filled with 0.1 strength Hoagland-Arnon nutrient solution. Three days later, containers were transferred into the light, and plants were further grown under illumination of 90 W/m^2 PAR from ZN and DNAT-400 fluorescent lamps, at a 14-h photoperiod and a temperature of 21 °C. Four day old maize seedlings and 9-d-old wheat plants were used for the study of the effect of root treatment on leaf extension, electrical tissue potential and hormone content.

Salt-stress experiments

Roots of maize seedlings were dipped for 0.5 min in 1 M KCl. Shoots were sampled for IAA immunoassay prior to and after the treatment and stored in liquid nitrogen. Electric tissue potential was measured by means of non polarizable Ag/AgCl electrodes which were connected to the stem base and leaf by an agar block prepared on KCL solution. The electrode signal was continuously recorded on a chart recorder.

Cold-shock treatment

Nutrient medium of wheat seedlings was cooled to 4 °C with small pieces of iced nutrient solution. An analogue inductive electromechanical position sensor was used to monitor extension growth. The output signal from the sensor was tracked continuously and was registered with a chart recorder. Samples for cytokinin analysis were collected prior to and after the treatment.

PEG-treatment

PEG 2000 was added to the nutrient medium of wheat seedlings to yield 15% concentrations. Roots were also killed by dipping in water at 100 °C for 30 s (shoots were protected from heating by rolling in sheets of foil). Afterwards, half of the plants were placed in a nutrient medium with 15% PEG, while the rest were transferred into standard nutrient medium (control). Wheat seedlings were sampled for ABA immunoassay.

Experiments with detached leaves

Leaves of 1 week-old wheat seedlings were detached and placed into tubes with nutrient medium. Sampling for cytokinin immunoassay was carried out once per day.

Xylem sap collection

Wheat seedlings were cut under water and shoots reconnected to the roots with graduated silicon tubes. Comparison of transpiration rate of intact and cut plants by weighing techniques showed that the procedure did not affect it significantly. After 1 h, tubes were disconnected, root exudate was collected and sampled for determination cytokinin concentration. After the purification and immunoassay of cytokinins according to the protocol described below, delivery of cytokinins from the shoot was calculated by multiplying their concentration and transpiration rate.

Hormone purification and immunoassay

The tissue was homogenised in 80% ethanol and incubated throughout the night at +4 °C. After filtration and vacuum evaporation of extracts until all ethanol was removed, aqueous residue was further processed for hormone purification. IAA and ABA were purified by means of solvent partitioning according to modified scheme described earlier (Veselov et al., 1992). Immunoassay of IAA and ABA was carried out using corresponding antibodies (Mustafina et al., 1997, 1998; Veselov et al., 1992). Cytokinin content in aqueous residue was immunoassayed using antibodies

Table 1. The effect of different components of incubation media on the process of degradation of iPA in the presence of protein fraction from maize seedlings

Composition of incubation medium	Quantity of degraded iPA (pmol mg^{-1} protein h^{-1})
0.1 M Tris (pH 7.0)	243
0.1 M Tris (pH 7.0) + 0.16 mM $CuSO_4$	480
0.1 M imidazole (pH 7.0)	348
0.1 M imidazole (pH 7.0) + 0.16 mM $CuSO_4$	742
0.1 M imidazole (pH 7.0) + 0.16 mM $CuSO_4$ +1,3-diphenylurea	490

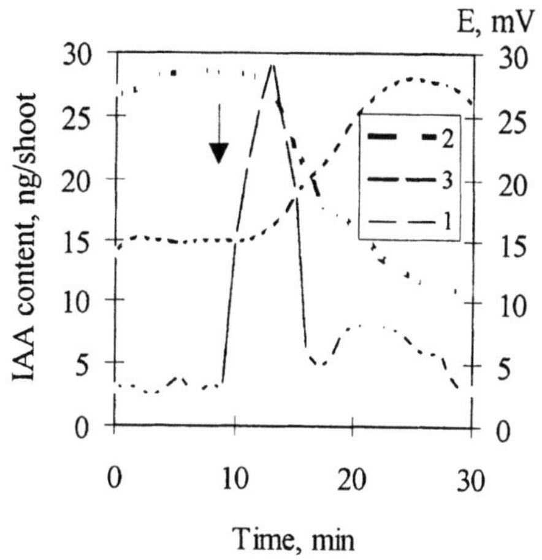

Figure 1. Effect of root salt stress on electric tissue potential (1), free (2) and bound (3) auxin content in shoot of maize seedlings.

to zeatin riboside (Kudoyarova et al, 1998). Measurements were carried out in three biological and four chemical replications and presented as mean values and their standard errors.

Estimation of the activity of cytokinin oxidase

Maize and wheat seedlings were detached from endosperm, homogenised in 0.1 M Tris-buffer (pH 7.0) and centrifuged at 12 000 *g* for 30 min. To separate the sample from endogenous hormones, equal volumes of saturated solution of ammonium sulphate was added to supernatant and the suspension was once again centrifuged. The pellet was resuspended in one of the buffers indicated in Table 1 and 30 ng of cytokinin isopentenyladenosine (iPA) added to the mixture which was then incubated for 1 h at 37 ° C. Cold ethanol was added to the samples to yield 60% concentration, the suspension was incubated at 4 °C for 40 min and then centrifugation was carried out at 5000 *g* for 10 min. The quantity of iPA lost by degradation was determined by means of immunoassay. The absence of any conversion of iPA to zeatin or dehydrozeatin derivatives was confirmed with the help of immunoassay using corresponding antibodies.

Table 1 shows that the presence of imidazole and Cu^{2+} accelerated the rate of the iPA disappearance, while 1,3-diphenylurea inhibited the process. The results confirm that the decrease in iPA content is due to its conversion to adenosine catalysed by cytokinin-oxidase which activity is known to be increased by imidazole and Cu^{2+} and inhibited by 1,3-diphenylurea (Hare and Van Staden, 1994).

Results and discussion

Figure 1 presents results of experiments carried out with maize seedlings. Their roots were dipped for 0.5 min in a concentrated solution of potassium chloride. Five minutes after the treatment, the concentration of IAA began to increase and had doubled after 10 min.

Figure 1 shows the decrease in the content of bound form in the hormone. Consequently, accumulation of IAA might be due to its release from the storage form. But what induced the changes in hormone content in shoots? Figure 1 shows alteration in electrical tissue potential registered between the stem base and a leaf which preceded hormonal response. It was also suggested that electrical signals might cause the release of auxin from its bound forms (Chisnell and Bandurski, 1988). Thus, our results are in accordance with that assumption.

Similar results have been observed in an inverse experiment where the shoots were stimulated and the response was recorded in the roots (Kudoyarova et al., 1990). In this case, the tips of coleoptiles of etiolated seedlings were illuminated for 15 s. Once again, a very fast increase in free IAA content was observed in roots which might be due to its release from bound forms since the content of the latter decreased. IAA is known to be transported from shoots to roots at a very low rate (Goldsmith, 1977). Nevertheless, the fast changes

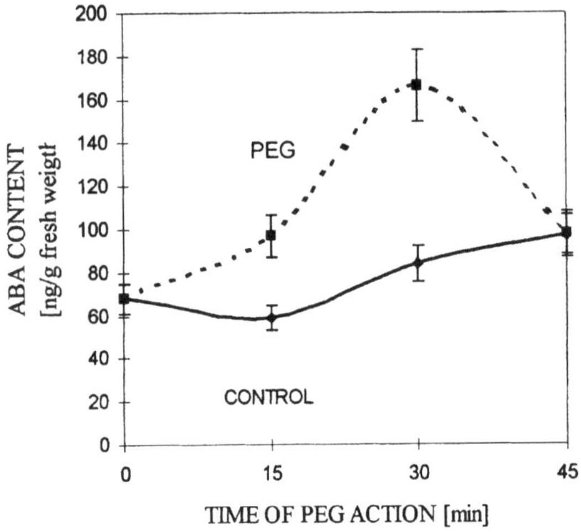

Figure 2. ABA content in shoot of control and PEG-treated wheat seedlings with live roots.

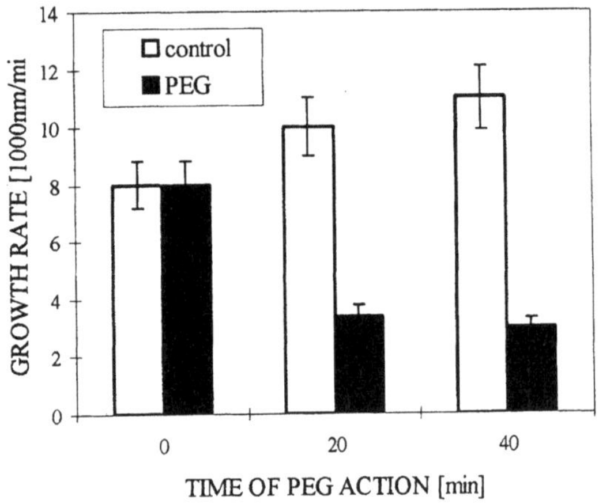

Figure 4. Growth rate of the second leaf of control and PEG-treated wheat seedlings with killed roots.

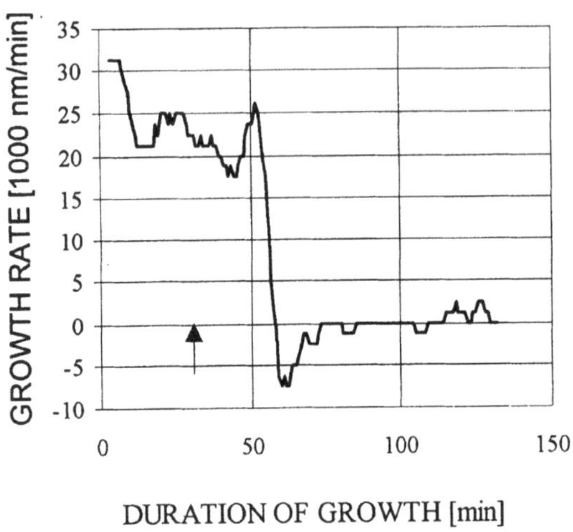

Figure 3. Growth of the second leaf of wheat seedlings with live roots. Arrow indicate the moment of the addition of PEG to the nutrient medium.

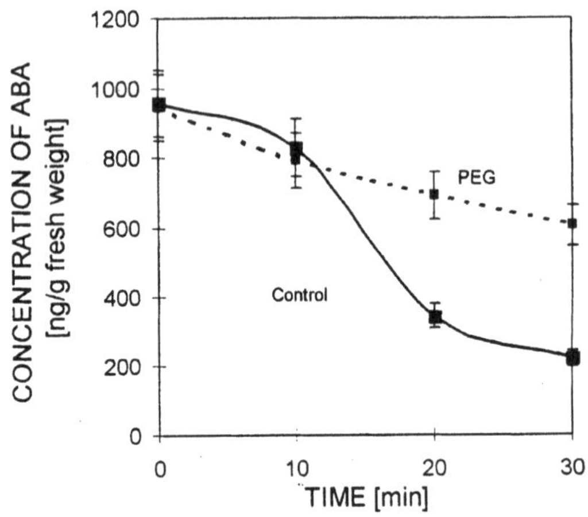

Figure 5. ABA content in shoots of control and PEG-treated wheat seedlings with killed roots.

in tissue electrical potential of root system have been observed and might stimulate the release of IAA from its bound form.

The dynamics of changes in ABA concentration were followed in shoots of wheat seedlings which experienced addition of PEG to the nutrient medium. Figure 2 shows a fast accumulation of ABA in shoots of treated plants which was accompanied by complete cessation of leaf growth 15 min after the addition of the osmolyte (Figure 3). Chazen and Neumann (1994) showed that in these circumstances, shoot growth is inhibited by a decrease in cell-wall extensibility. Some regulative signal presumably passed from root to shoot and led to the fast appearance of a reaction in the shoot. Since ABA is known to inhibit shoot growth (Evans, 1984), the results suggest that the increase in its shoot concentration is at least a part of the mechanism implicated in shoot response to osmotic stress. This assumption is in accordance with the data of Saab et al. (1990) who observed the inhibition of shoot growth by increased ABA content under long action of low water potential. Our results show that similar effects might occur much more quickly.

But what was the source of ABA accumulation in plants? Roots are known to respond to a worsening environment by an increased production of ABA which is exported from the roots and is believed to influence shoot response (Jackson, 1997; Zhang et al., 1987). However, Chazen and Neumann (1994) excluded the possibility of the implication of root-produced ABA in the fast transfer of signal from PEG-treated roots. They killed roots of maize seedlings by boiling them, but this procedure did not prevent fast inhibition of shoots growth by PEG-treatment of the killed roots. Hormone concentration changes resulting from this treatment was not measured by Chazen and Neumann (1994), but our analysis is given in Figure 4. The growth rate of shoots of wheat seedlings with killed roots was decreased by osmotic stress as compared with the corresponding control, that is plants with killed roots which did not experience any additional treatment. Thus, wheat seedlings showed the same response as reported for maize seedling. The comparison of Figures 2 and 5 shows that boiling roots resulted in a huge accumulation of ABA in the shoots which might be released from dead root cells. However, its concentration in shoot declined with time. Accumulation of ABA in shoots of plants after boiling attracts attention. However, the study of the effect of roots boiling itself was not the aim of our work. We used this procedure just to exclude possibility of generation of root-produced hormone signal, since hormones could not be produced by killed roots. Although the release of ABA from cells of killed roots might serve as a signal of roots damage by boiling, those roots were already dead when PEG was added and could not participate in the response to osmoticum by increased hormone production. Nevertheless, osmotic stress retarded the decline in hormone concentration in shoots of plants with killed roots and it became higher than in the control (shoots of plants with killed roots which did not experience PEG-treatment) (Figure 5). Thus, higher ABA content in shoots of PEG-treated plants than in shoots of corresponding control plants was a characteristic response to osmotic stress for both seedlings with killed and live roots. Since killed roots were unable to respond to additional (osmotic) stress by ABA production, its elevated content in shoots of PEG-treated plants as compared to corresponding control might be due to its production or decreased decay in shoots themselves. The time-course of ABA content seems to favour the latter assumption. The changes in metabolic processes in the shoots are likely to be triggered by some quickly moving signal which arrived from stimulated roots. The latter might be hydraulic or electrical signals. It is difficult to distinguish between these possibilities since electrical waves were shown to be linked with water currents in the vessels (Julien et al., 1991).

The rapid change in ABA content might be important in the control of the fast growth response of shoot to root treatment. Moreover, ABA is well-known to control stomatal aperture (Mansfield and McAinsch, 1995). Consequently the fast changes in ABA content might also be involved in the control of water relations under its reduced availability. In many cases, fast bulk changes in ABA concentration have not been found and redistribution of hormone between leaf tissues have been suggested to control stomatal apperture (Slovik and Hartung, 1992). However, the changes in bulk ABA concentration must also be important in the control of processes when they do take place, as in our case.

A third example of a fast change in hormone concentration involves cytokinins. Five-fold reductions in their concentration were observed in shoots of wheat seedlings 15 min after the addition of ice to the nutrient medium (Kudoyarova et al., 1998). Since shoots are believed to receive cytokinins from roots (Incoll and Jewer, 1990), the simplest explanation of the results might be their reduced production by cooled roots. However, in this case, the rate of cytokinin delivery from roots and their turnover in shoots should be much higher than has been reported. Thus, Wagner and Beck (1994) showed that *Urtica dioica* plants need a whole day to receive from roots the amount of cytokinins they have in their shoots. These results were obtained with adult plants. So there were some grounds to suggest that the young seedlings used here might have different characteristics. Since it is rather difficult to collect root exudate from wheat seedlings, an indirect approach has been used initially. The dynamics of cytokinin content was followed in detached leaves of wheat seedlings which had no roots to receive cytokinins from. As expected, cytokinin content declined gradually in isolated leaves. However, it took 2 days for the cytokinin content to be reduced 5 times (from 350 to 70 ng/g FW). If we compare it with the five-fold decrease in cytokinin content that was observed just 15 min after the start of root cooling, it is difficult to attribute the latter effect to reduced cytokinin delivery from the roots. Nevertheless, it was not excluded that isolation of leaves might change the rate of cytokinin turnover. So the indirect approach did not allow us to make any definite conclusion and it was necessary to measure cytokinin output from the roots. Xylem flow is known to reduce dramatically when shoots are cut off to collect root exudate. The task is especially difficult in the case of wheat seedlings where xylem volume is small even in intact plants. Therefore, in this experiment, xylem sap was collected cutting seedlings under water and reconnecting the shoot to the roots by means of a silicon tube. The procedure did not affect the transpiration stream significantly, and in 1 hour, xylem sap was substituted for the fluid inside the tube and was collected for cytokinin immunoassay. Cytokinin content in whole shoot of 1 week-old wheat seedlings was about 100 ng, while hormone flow to the shoot comprised 40 ng per h. Thus, wheat seedlings needed only 2.5 h to receive from roots the amount of cytokinins they have in their shoots. In this way, the rate of cytokinin delivery from roots of wheat seedlings was shown to be about 10-times higher than that reported for adult plants. However, this rate of cytokinin delivery was not high enough and even its complete cessation could not lead to such a fast decline in the hormone content in shoots of plants observed after the cooling of their roots. So, once again, it became clear that cytokinin content in this experiment might be regulated through changes in the hormone metabolism in the shoots. Estimation of the activity of cytokinin oxidase confirmed this assumption. Root cooling led to a 1.4-fold increase in the activity of this enzyme in the protein fraction recovered from shoots. Consequently, cytokinin content in the shoots of cooled wheat seedlings might be reduced due to an accelerated decay. Hormone metabolism might be altered by means of a spreading action potential, since rapid cooling is known to induce it. Hydraulic signals might also participate in signal transfer, since root cooling inevitably influences water flow in plants.

The fast decrease in cytokinin content must be important in the control of shoot behaviour and correlates with the rapid inhibition of shoot growth (Kudoyarova, 1998). The same result might be achieved in different ways. In osmotically stressed plants, inhibition of shoot growth was related to ABA accumulation, while in cooled seedlings, the same effect might be attributed to a reduced cytokinin content. This is not surprising since cytokinins are known to maintain shoot growth while, in contrast, ABA inhibits it.

Summarising the results presented, it is clear that different kinds of root treatment resulted in rapid changes in the hormone content of shoots. In each case, this was not likely to be due to changes in the hormone transport from roots, but due to metabolic processes in the shoots themselves triggered by some signals which might propagate quickly from treated roots.

Acknowledgements

The work was founded by Russian Foundation of Fundamental Research Grant No 97-04-49962 and No 96-15-98030.

References

Bandurski R S 1978 Chemistry and physiology of myo-inositol ethers of indole-3-acetic acid. *In* Cycletols and Phosphoinositides. Eds. W Wells end F Eisenberg. pp 35–54. Acad. Press, London New York.

Chazen O and Neumann P M 1994 Hydraulic signals from the roots and rapid cell-wall hardening in growing maize (*Zea mays* L.) leaves are primary responses to polyethylene glycol-induced water deficits. Plant. Physiol. 104, 1385–1392.

Chisnell J and Bandurski R S 1988 Translocation of radiolabelled indole-3-acetic acid and indole-3-acetyl-myo-inositol from kernel to shoot of *Zea mays* L. Plant Physiol. 1, 79–84.

Davies W J and Jeffcoat (eds) B 1990 Importance of Root to Shoot Communication in the Responses to Environmental Stress. British Society for Plant Growth Regulation, Bristol.

Davies E, Zavadzki T and Writters D 1991 Electrical activity and signal transmission in plants: How do the plants know? *In* Plant Signalling, Plasma Membrane and Change of State. Eds. S Penel and H Greppin. pp 119–137. University of Geneva, Geneva.

Evans M L 1984 Function of hormones at the cellular level of organization. *In* Encycl. Plant. Physiol. 10. Ed. TK Scott. pp 23–79. Springer-Verlag, Berlin Heidelberg New York Tokyo.

Fromm J and Bauer T 1994 Action potentials in maize sieve tubes change phloem translocation. J. Exp. Bot. 45, 463–469.

Fromm J, Hajirezaei M and Wilke I 1995 The biochemical response of electrical signalling in the reproductive system *Hibiscus* plants. Plant Physiol. 109, 375–384.

Goldsmith M H M 1977 The polar transport of auxin. Ann. Rev. Plant Physiol. 28, 439–478.

Hare P D and Van Staden J 1994 Cytokinin oxidase: Biochemical features and physiological significance. Physiol. Plant. 91, 128–136.

Incoll L D and Jewer P C 1987 Cytokinins and water relations of whole plants. *In* Plant Hormones in Search of a Role. Eds R Horgan and B Jeffcoat. pp 85–97. British Plant Growth Regulator Group, Bristol.

Jackson M B 1993 Are Plant Hormones Involved in Root–Shoot Communications? Advances in Botanical Research 19, 104–167.

Jackson M 1997 Hormones from roots as signals for the shoots of stressed plants. Elsevier Trends J. 2, 22–27.

Julien J L, Desbiez M O, De Jaegher G and Frachisse J M 1991 Characteristics of the wave of depolarization induced by wounding in *Bidens pilosa* L. J. Exp. Bot. 42, 131–137.

Kudoyarova G R, Usmanov I U, Guyli-Zade V Z, Fattahutdinov E G and Veselov S U 1990 Interaction of spatially distant organs. Relation of electrical and hormonal sygnals. Dokl Acad Sci USSR 310, 1511–1514.

Kudoyarova G R, Farhutdinov R G, Mitrichenko A N, Teplova I R, Dedov A V, Veselov A V and Kulaeva O N 1998 Fast changes in growth rate and cytokinin content of the shoot following rapid cooling of wheat seedlings. Plant Growth Regul. 26, 105–108.

Mansfield T A and McAinsch M R 1995 Hormones as regulators of water balance. *In* Plant Hormones. Ed. PJ Davies. pp 598–616. Kluwer Academic Publisher, Dortrecht, Berlin, London.

Mustafina A R, Veselov S Yu, Valcke R and Kudoyarova GR 1997/1998 ABA and cytokinins content in shoots of wheat seedlings during their dehydration. Biol. Plant. 40, 291–293.

Opritov V A, Kraus V O and Treushnikov V M 1972 The role of electrical reaction of excitement in functional linkage between shoot and root under action of external influences on shoot apex. Fiziol. Rast. 19, 961–967.

Saab I N, Sharp R E, Pritchard J and Voetberg G S 1990 Increased endogenous abscisic acid maintains primary root growth and inhibits shoot growth of maize seedlings at low water potentials. Plant Physiol. 93, 1329–1336.

Slovik S and Hartung W 1992. *In* Stress-induced Redistribution Kinetics of ABA in Leaves: Model Consideration. Progress in Plant Growth Regulation. Eds. CM Karssen, C Van Loon and D Vreugdenhil. pp 464–473. Kluwer Academic Publishers, Dordrecht, Berlin, London.

Veselov S Yu, Kudoyarova G R, Egutkin N L, Gyuli-Zade V Z, Mustafina A R and Kof E M 1992 Modified solvent partitioning scheme providing increased specificity and rapidity of immunoassay for indole-3-acetic acid. Physiol. Plant. 86, 93–96.

Wagner B M and Beck E 1994 Cytokinins in Urtica dioica plants: production, metabolism and fluxes. *In* Physiology and Biochemistry of Cytokinins in Plants. Eds. M Kaminek, DWS Mok and E Zazimalova. pp 53–57. Kluwer Publishers, Amsterdam, The Netherlands.

Zhang J, Schurr U and Davies W J 1987 Control of stomatal behavior by abscisic acid which apparently originates in the roots. J. Exp. Bot. 39, 1174–1181.

Recent Advances of Plant Root Structure and Function, 143–1150
Edited by O. Gašparíková, M. Čiamporová, I. Mistrík and F. Baluška
© 2001 *Kluwer Academic Publishers.*

Response of tomato plants to chilling stress in association with nutrient or phosphorus starvation

Z. Starck, B. Niemyska, J. Bogdan and R. N. Akour Tawalbeh
Department of Plant Physiology, Warsaw Agricultural University, Rakowiecka 26/30, 02-528 Warsaw, Poland

Key words: chilling, chlorophyll fluorescence, nutrient, phosphorus, photosynthesis, starvation

Abstract

The experiments were conducted on two tomato cultivars: Garbo and Robin. Mineral starvation due to plant growth in 20-fold diluted nutrient solution (DNS) combined with chilling reduced the rate of photosynthesis (P_N) and stomatal conductance (g) to a greater extent than in plants grown in full nutrient solution (FNS). In phosphate-starved tomato plants the P_N rate and stomatal conductance decreased more after chilling than in plants grown on FNS. In low-P plants even 2 days after chilling the recovery of CO_2 assimilation rate and stomatal conductance was low. A resupply of phosphorus to low-P plants (low P + P) did not improve the rate of photosynthesis in non-chilled plants (NCh) but prevented P_N inhibition in chilled (Ch) plants. The greatest effect of P resupply was expressed as a better recovery of photosynthesis and stomatal conductance, especially in non-chilled low P + P plants. The F_v/F_m (ratio of variable to maximal chlorophyll fluorescence) decreased more during P starvation than as an effect of chilling. Supplying phosphorus to low-P plants caused the slight increase in the F_v/F_mratio. In conclusion, after a short-term chilling in darkness a much more drastic inhibition of photosynthesis was observed in nutrient-starved or P-insufficient tomato plants than in plants from FNS. This inhibition was caused by the decrease in both photochemical efficiency of photosystems and the reduction of stomatal conductance. The presented results support the hypothesis that tomato plants with limited supply of mineral nutrients or phosphorus are more susceptible to chilling.

Abbreviations: Ch – chilling; Ci – intercellular CO_2 concentration; DNS – diluted nutrient solution; FNS – full nutrient solution; chlorophyll fluorescence: F_m – maximal; F_0 – initial; F_v – variable; g – stomatal conductance; NCh – not chilled; P_N – net photosynthesis; RuBP – ribulose 1,5-bisphosphate; SLW – specific leaf weight

Introduction

In tomato plants chilling reduced the rate of photosynthesis (Brüggemann et al., 1992; Byrd et al., 1995; Starck et al., 1994, 1997). The response of maize (Sowiński and Maleszewski, 1989, 1990) and tomato plants (Choluj et al., 1997; Starck et al., 1996) to chilling depends on root size and /or activity in supplying shoot with mineral nutrition and water.The physiological role of roots in plant tolerance to a low temperature is also related to a modified production of root-borne phytohormones (decrease of cytokinins and increase of ABA) and their transport to the shoot during chilling (Kudoyarova et al., 1998; Radin, 1984; Tantau and Dörffling, 1991). In the case of limited nutrient supply, content of cytokinins declined both in shoots and roots of *Plantago major* (Kuiper et al., 1989). There is evidence that low N fertilisation leads to a reduced cytokinin formation in the roots and its export to the shoot (Beck, 1996).

A deficit of nitrogen in tomato and barley plants reduced the stomatal conductance even more strongly than photosynthesis, perhaps due to the increase in ABA content (Chapin et al., 1988). In spinach, the nitrogen-deficiency greatly decreased the rate of acclimation to low temperature (Martindale and Leegood, 1997).

In tomato plants, the P-starvation caused a decrease in cytokinin content (Menary and van Staden, 1976). In several other papers the effect of P and N contents in plant tissue seems to be related to an alteration of the balance between ABA and cytokinin (for discussion, see Marschner, 1995).

A long-term supply of inadequate amount of Pi decreases the rate of photosynthesis by limiting the capacity for regeneration of RuBP (Dietz and Foyer, 1986; Plesničar et al., 1994; Qui and Israel, 1994; Rao and Terry, 1989, 1995; Rao, 1996). In severely P-deficient leaves the regeneration of ATP is slowed down, the rate of photosynthesis and the export of photoassimilates decease; these processes observed in several species are usually reversible after resupply of Pi to the plants (Jacob and Lawlor, 1993; Rao and Terry, 1995). In spinach and barley plants, the negative effect of long-term P deficiency on photosynthesis was reversed by supplying plants with Pi (Dietz and Foyer, 1986). In spinach, both the stomatal conductance and the photosynthetic quantum yield recovered after 24 h (Brooks, 1986). A similar response after Pi resupply was observed in P-deficient soybean but as late as after few days (Qui and Israel, 1994).

To our knowledge reports about relationships between chilling and mineral nutrition of plants are very scanty. Under suboptimal temperatures the rate of photosynthesis in barley plants was limited by the availability of Pi (Labata and Leegood, 1988). The deficiency of Pi occurred more readily at low temperatures due to an increased demand for ATP (Mächler et al., 1984). Low contents of P and K did not affect the rate of photosynthesis but increased the negative response of tomato plants to low temperatures and retarded their recovery after chilling (Starck et al., 1997). These results suggest that the level of P, which is sufficient under optimal temperature, will no longer be sufficient under chilling. Therefore, more attention should be paid to the way in which the plant response to chilling is modified by mineral nutrient supply.

The aim of the present experiments was to compare the strategies of coping with chilling stress in tomato plants with contrasting nutrient supply. As a measure of plant response to a low temperature, the rate of photosynthesis, stomatal conductance and chlorophyll fluorescence were examined.

Material and methods

Tomato plants were grown in a greenhouse under uncontrolled light and temperature conditions on fertilised peat and were then transferred from the medium to the Hoagland nutrient solution (1.5 dm^3 per plant), that was changed once a week at the beginning and two times a week later on.

In Experiment I the plants of two cultivars, both at comparable stage of fruit development, were grown during 7 weeks (cv. Garbo) and 5 weeks (cv. Robin) before chilling. The plants were grown on: (1) full nutrient solution (FNS), as a control; or (2) 20-fold diluted full nutrient solution (DNS).

In Experiment II, 6 weeks before chilling, plants of cv. Robin were grown on: (1) full nutrient solution, (FNS), as a control; or (2) low level of P; the concentration of P was reduced to 20-fold compared with full Hoagland solution (estimated as low-P); one day before chilling the plants were transferred to a solution without P; or (3) low-P nutrient solution which one day before chilling was resupplied with P by the transfer of plants to FNS (estimated as low P + P).

In both experiments the chilling treatment was performed in refrigerators in water saturated atmosphere under the following conditions: in Experiments I and II the chilled plants (Ch) were kept at 1–5°C respectively for 16 or 20 h of prolonged night. At the same time non-chilled plants (NCh) in both experiments were kept at 18–20°C in darkness, also in the water saturated atmosphere.

The net photosynthetic rate (P_N), the leaf stomatal conductance for H_2O (g) and intercellular CO_2 concentration (Ci) were estimated as follows: in Experiment I the gas exchange was measured in a closed system under combined natural and artificial light (about 400 μmol m^{-2} s^{-1}, at the level of measured blades) with the Portable Photosynthetic System LI-6200 (LI-COR, Lincoln, NE, USA); in Experiment II the model LI-6400 with open system under artificial red light (665 nm) and blue light (470 nm) and total photon flux density 600 μmol m^{-2} $s^{-1}$was used.

The gas exchange parameters were estimated in both experiments on the leaflets situated near the apical part of the first, fully expanded leaf below the cluster with small fruitlets. In Experiment II the P-content was determined in the same leaflets. Discs cut from the same leaf were taken for estimation of the specific leaf weight (SLW). The chlorophyll fluorescence was measured at room temperature on intact plants (the same leaves as used for the gas exchange) by a modulated fluorescence measuring system, after about 30 min leaf dark adaptation, using a Bio Monitor AB Plant Stress Meter (Öquist and Wass, 1988). After 2-day recovery, the plants were harvested to measure dry matter of individual organs.

Table 1. Analysis of variance for exp. I; chilling treatments compared NCh with Ch (1) and nutrient solution treatments compared FNS with DNS (2)

Source of variations	cv. Robin After chilling			cv. Garbo After chilling			cv. Garbo One day later		
	P_N	g	Ci	P_N	g	Ci	P_N	g	Ci
Chilling (1)	**	**	**	**	**	n.s.	**	**	n.s.
Nutrient solution (2)	**	n.s.	n.s.	**	**	**	**	**	**
Interaction (1 × 2)	**	n.s.	n.s.	n.s.	*	n.s.	*	**	**

NCh, non chilled plants; Ch, chilled plants. P_N, net photosynthetic rate; g, stomatal conductance; Ci, intercellular CO_2 concentration. FNS, full nutrient solution; DNS, 20-times diluted nutrient solution.
n.s., not significant; * significant at p=0.05; ** significant at p=0.01.

The data were statistically analysed using the ANOVA procedure of Statistical Analysis System with α level for significance of 0.05 as (x) and 0.01 as (xx).

Results and discussion

Experiment I (cvs. Robin and Garbo)

The growth of tomato plants of both Garbo and Robin cultivars not sufficiently supplied with nutrients (DNS) was significantly reduced. The dry matter of blades was about one half of that in FNS plants (AkourTawalbeh, 1996, unpublished). The growth of fruits was inhibited even more. The root size of cv. Garbo in DNS-plants, estimated as their volume, was about four times lower than in the respective control plants from FNS (Bogdan, 1997, unpublished).

The gas exchange was measured 2 h after the end of chilling (plant acclimation to light after prolonged darkness). In Robin plants the nutrient starvation decreased the rate of photosynthesis (Fig. 1A, Table 1). The rate of P_N in control plants (FNS) did not respond to chilling in spite of a drastic decrease in stomatal conductance. On the contrary, in DNS-plants, a strong inhibition of both CO_2 fixation and stomatal conductance was observed. The plant nutrition did not significantly change Ci in both temperature treatments. Nevertheless, the chilling decreased the Ci in plants grown in either FNS or DNS. This suggests that when the rate of photosynthesis is limited by CO_2 diffusion through the stomata, some intercellular CO_2 seems to be assimilated.

In cv. Garbo the low nutrient supply (DNS-plants) dramatically reduced the rate of photosynthesis (Fig.1B, Table 1). The chilling decreased P_N in a

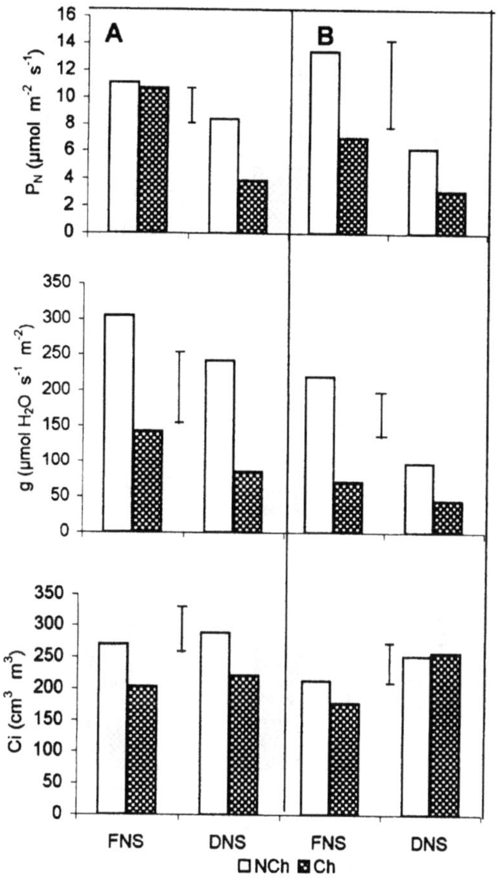

Figure 1. Response to chilling of tomato plants grown on FNS (full nutrient solution), DNS (diluted nutrient solutions), NCh (not chilled, white bars); Ch (chilled, dotted bars); (A) cv. Robin, (B) cv. Garbo. For statistical analysis: see Table 1. The bars indicate $LSD_{0.05}$ values. P_N, net photosynthetic rate; g, stomatal conductance; Ci, intercellular CO_2 concentration.

similar proportion in both the FNS and DNS-grown plants. The stomatal conductance was also strongly decreased both by the nutrient deficiency and the low temperature treatment. The values of Ci were significantly higher in DNS plants. Increased Ci values in both NCh and Ch of DNS-plants, as compared to FNS associated with low stomatal conductance, may indicate some damage of PSII activity and/or decreased activity of Calvin cycle enzymes. The low temperature treatment did not affect Ci in DNS plants. On the contrary Ci values slightly decreased in FNS-chilled plants, perhaps as a response to low stomatal conductance.

To check the post-effect of temperature treatment during darkness (16 h) of cv. Garbo all experimental

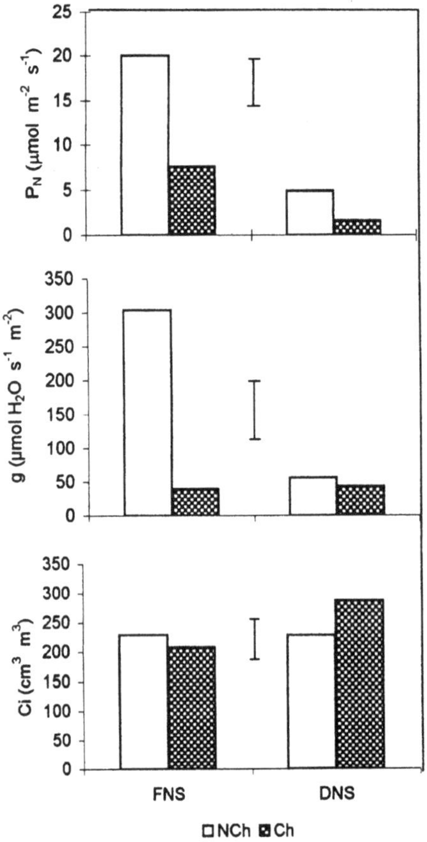

Figure 2. Post-chilling effects on tomato plants cv. Garbo grown on full (FNS) or diluted (DNS) nutrient solutions (measurements one day after chilling in darkness). All other abbreviations as in Fig. 1. For statistical analysis: see Table 1. The bars indicate LSD $_{0.05}$ values.

plants were transferred to a greenhouse with uncontrolled conditions where the temperature during daytime was above 20°C, but only 12°C at night. In FNS, non-chilled plants, the rate of P_N and stomatal conductance increased as compared to P_N rate measured one day before (compare Fig.1B with Fig. 2), perhaps due to recovery after prolonged darkness. Nevertheless, in FNS-plants, and even more in DNS-plants, 1 day after chilling the photosynthesis decreased instead of recovering, perhaps due to a low night temperature – a second chilling stress. The dramatic inhibition of photosynthesis of DNS-plants coincided with a low stomatal conductance. The changes in Ci were small, especially in the DNS-plants after chilling, where P_N rate was the lowest (significant interaction, Table 1), perhaps as result of photoinhibition as a post-chilling effect, observed also in maize (Janda et al., 1994), suggesting that the repair processes needed a temperature close to the optimal one. On the contrary, in other experiments (Kingston-Smith et al., 1997) with maize, chilling decreased the rate of CO_2 assimilation but without photoinhibition.

Table 2. Effect of P supply on dry matter distribution (in g per plant) and the shoot/root dry matter ratio in tomato plants cv. Robin (average of eight plants, Expt. II)

P supply	Blades	Petioles + stem	Cluster with fruits	Roots	Total plants	Shoot/root
Control (FNS)	12.4	13.0	12.6	8.3	46.3	4.58
Low P	8.4	11.9	2.6	6.7	29.6	3.42
Low P+P	9.2	11.5	4.6	7.6	32.9	3.33
LSD, p=0.05	1.5	2.7	3.7	1.6	5.4	-

Experiment II, cv. Robin

During 6 weeks of low-P supply the growth rate of plants decreased especially in clusters with fruits (to 21%) and blades (to 68%). The lowest inhibition was observed in the root growth. Partitioning of the biomass was also modified leading to a lower shoot/root ratio as a response to the P-starvation. The content of P in the blades decreased in low-P blades from 200 (FNS blades) to 23 μmol g^{-1}of dry matter. Transferring the low-P plants to full nutrient solution caused about a 5-fold increase in P content but had insignificant effect on the plant growth (Table 2).

The leaves of low-P plants had a slightly higher SLW (Fig. 3) possibly as an effect of reduced photosynthate export from the blades (see Rao, 1996) and as a consequence of the accumulation of starch (Starck et al., 1997). Chilling enhanced SLW of FNS and even more so in the low-P plants (Fig. 3).

Plants from FNS did not respond to chilling neither in the rate of photosynthesis nor in stomatal conductance, but both parameters were low (Table 3), probably due to the post-effect of prolonged dark period (20 h) preceding the measurement of gas exchange. Therefore, after one day recovery the P_N rate in NCh plants significantly increased and even more so did the stomatal conductance. On the contrary in chilled FNS-plants at the same time P_N did not significantly increase and the reopening of stomata was also small (Table 3 and 4).Thus one day after stress the chilled FNS plants had a P_N rate significantly lower than the respective control (FNS, NCh-plants), possibly as a post-chilling effect still observed 2 days after chilling.

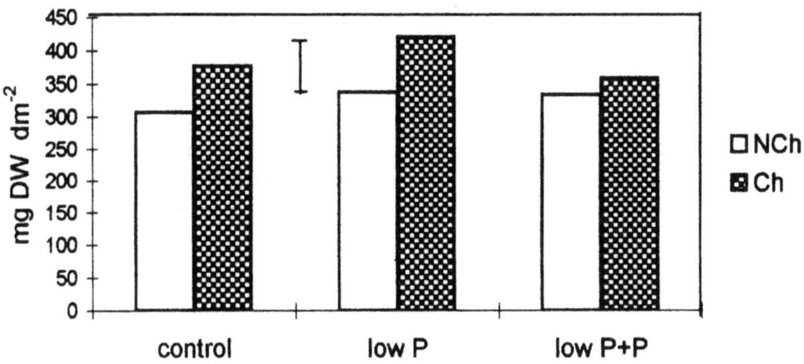

Figure 3. Specific leaf weight (SLW) of tomato plants cv. Robin affected by chilling stress and phosphorus nutrition. Control plants grown in full nutrient solution, low P plants grown for 6 weeks on 1/20 P in nutrient solution, low P+P as in low P, but one day before chilling treatment resupplied with P (transferred to FNS). Measurements: two days after the end of chilling treatment (this means 3 days after P-resupply). Analysis of variances: chilling is significant at p=0.01; the nutrition effect and the interaction were not significant. The bars indicate LSD $_{0.05}$ values. NCh, non chilled; Ch, chilled.

Table 3. Effect of mineral nutrition supply and chilling on gas exchange parameters, (Expt. II)

Measurements	Mineral nutrition	After chilling NCh	After chilling Ch	1 day recovery NCh	1 day recovery Ch	2 days recovery NCh	2 days recovery Ch
Photosynthesis P_N (μmol m^{-2} s^{-1})	FNS	9.89	9.83	14.13	10.83	12.40	10.25
	low P	4.82	2.68	5.87	3.56	4.26	4.30
	low P+P	3.42	4.67	8.03	5.81	9.59	6.58
Stomatal conductance (g) (mol H_2O m^{-2} s^{-1})	FNS	0.136	0.138	0.294	0.187	0.235	0.178
	low P	0.073	0.030	0.096	0.048	0.057	0.050
	low P+P	0.032	0.050	0.127	0.089	0.158	0.070
Intercelular CO_2 concentration Ci (μmol CO_2mol air^{-1})	FNS	214	207	247	235	239	236
	low P	233	208	245	233	227	215
	low P+P	179	179	234	231	251	206

FNS, full nutrient solution; low P, nutrient solution with 1/20 phosphorus concentration; low P+P, nutrient solution with 1/20 phosphorus, plants were resupplied with P 1 day before chilling; NCh, non-chilled plants; Ch, chilled plants.

The rates of P_N and stomatal conductance were inhibited in low-P plants but the effect was more pronounced after chilling (it decreased to 56% of the respective control, Table 3) with stronger diminution of stomatal conductance (down to 41%).

Similar observations were reported by Thakur (1990) in experiments with tomato plants, but during water deficit. As late as 2 days after rehydration, stomata reopening was observed.

Stomatal behaviour may be related to an alternation of balance between ABA and cytokinins in low-P plants as suggested by Radin (1984). In *Ricinus,* leaf conductance and reduction of transpiration rate were also greatly decreased by P starvation (Jeschke et al.,1997).

In low P + P plants, the rate of P_N and stomatal conductance were reduced similarly as in P-starved plants (differences are insignificant, Table 4). Nevertheless, their photosynthesis improved during the recovery. In both NCh and Ch plants, the reopening of stomata was observed but still incomplete even after 2 days. It suggests that the insufficient Pi level in the cells is not the only factor limiting the photosynthesis. Some damage of photosystems and/or changes in enzyme activity could perhaps be the other limiting factors, especially after chilling (see introduction).

An analysis of relationships between P_N and stomatal conductance in all the experimental plants with differentiated P supply showed a positive, highly significant correlation (r=0.93). Therefore, the rate of photosynthesis of all plants in the experiment was per-

Table 4. Analysis of variance for expt. II; chilling treatments compared NCh with Ch (1) and nutrient solution treatments compared FNS with low P and low P+P (2)

Source of variations	After chilling			After 1 day recovery			Two days recovery		
	P_N	g	Ci	P_N	g	Ci	P_N	g	Ci
Chilling (1)	n.s.	n.s.	n.s.	**	**	n.s.	**	*	**
Nutrient solution (2)	**	**	*	**	**	n.s.	**	**	n.s.
Interaction (1 × 2)	n.s.	n.s.	n.s.	n.s.	n.s.	n.s.	*	n.s.	n.s.
LSD for (1)	2.73	0.06	50	2.30	0.08	35	2.06	0.09	31
for (2)	2.24	0.05	41	1.89	0.07	29	1.70	0.07	26

P_N, net photosynthetic rate; *g*, stomatal conductance; Ci, intercellular CO_2 concentration.
n.s., not significant; * significant at p=0.05; ** significant at p=0.01.

Table 5. Effect of P nutrition and chilling stress on chlorophyll fluorescence parameters

Fluorescence	Treatments	FNS	low P	low P+P	Statistical analysis	
					chilling (1)	**
F_v/F_m	NCh	0.813	0.743	0.751	nutrition (2)	**
	Ch	0.778	0.703	0.708	interaction (1 × 2)	n.s.
					chilling (1)	**
F_v/F_0	NCh	4.45	2.82	3.03	nutrition (2)	**
	Ch	3.50	2.42	2.42	interaction (1 × 2)	n.s.

NCh, non chilled; Ch, chilled; FNS, full nutrient solution; low P, nutrient solution with 1/20 phosphorus concentration; low P+P, nutrient solution with 1/20 phosphorus, plants were resupplied with P 1 day before chilling.

haps closely determined by stomatal conductance. The fact that the stomatal closing was still an important factor limiting CO_2 assimilation in P-resupplied plants as well is related to the significant decrease of Ci values in both NCh and Ch-plants, measured immediately after the temperature treatment.

Nowadays, some contradictory points of view concerning the role of stomata in controlling photosynthesis exist. In the opinion of Jones (1998), the stomata play a relatively small part (about 20% of all limiting factors) in determining the rate of P_N.Nevertheless, he mentioned that under some conditions the stomatal resistance may play a major role in the determination of the assimilation rate. That is exactly the point presented in our experiments. In plants with low-P supply or starved of all macro- and microelements (Experiment I) the stomatal conductance may determine P_N to a much higher degree than in FNS plants. In tomato plants, Choluj et al., (1997) also observed that stomatal conductance less affected the CO_2 assimilation under optimal temperature conditions than after chilling.

Besides of stomatal limitation of photosynthesis, photoinhibition caused by P-starvation and chilling treatments the chlorophyll fluorescence was estimated as an indicator of photochemical efficiency of PSII. The measurements of chlorophyll fluorescence were caried out in vivo at room temperature in plants exposed to light for 2–3 h after the end of temperature stress and then after adaptation to dark. The correlation coefficient between P_N and F_v/F_m calculated for all experimental plants was low but significant (r=0.76). In all treatments, the chlorophyll fluorescence was affected by chilling, which indicates a reduced efficiency of PSII photochemical reaction (Table 5). Nevertheless, the lowest decrease of the F_v/F_m ratio was observed in FNS-plants. In plants transferred from low-P to full nutrient solution in both NCh and Ch treatments, the F_v/F_m ratio slightly but significantly increased, suggesting the induction of some repairing processes in chloroplasts.

As an indicator of water splitting the F_v/F_o ratio was analysed, where F_o estimates initial chlorophyll fluorescence. In the opinion of Kamps et al. (1987) based on the data of experiments carried out on various genotypes of *Lycopersicon,* the chlorophyll fluorescence, expressed as F_v/F_o ratio, is the most precise assay to quantify the chilling injury. This ratio decreased in low-P plants and those resupplied with P, even more than in plants treated with chilling (Table 5).

The presented data indicate that the reason for photosynthesis inhibition after chilling and P-starvation is both stomatal limitation and the disturbances in the chloroplast photosystems. Such a result may explain the reason for slow recovery after plant treatment with low temperature.

Various environmental factors affecting plants before or during chilling may modify the plant reaction to low temperature. The post-chilling effect reported by Szalai et al. (1996) as photoinhibition was presented in maize on the basis of F_v/F_m. In their experiments P deficit also affected the chlorophyll fluorescence similarly as in our experiments. In cucumber leaves, the chilling effects expressed as a decrease of F_v/F_m measured after dark adaptation depended very precisely on the level of low temperature (Oxborough and Ort, 1995). Damaging of PSII by chilling takes place below the threshold of temperature treatment. Jung and Steffen (1997) reported that *L. hirsutum* acclimated differently to various intensities of photosynthetic photon flux density (PPFD) before and during the chilling stress. The changes in light conditions dramatically affected the plant response to a low temperature, expressed also as F_v/F_m. A decreased efficiency of the PSII reactive center in P-deficient sunflower and maize plants was observed by Jacob and Lawlor (1993). Correlations between P deficiency, expressed as P-content and F_m in strawberry, reported also by Keutgen et al. (1997).

In conclusion, the response of plants to chilling is a syndrome of very complex processes affected by genotypic characteristics that vary under different environmental conditions. The degree of disfunction observed after chilling of tomato plant during the reproductive stage depends on temperature pretreatment, mineral nutrition, ABA content and/or activity (Starck et al., 1994, 1997, 1998). The changes in photosynthesis after chilling are greatly determined both by CO_2 diffusion and activity of photosystems. These processes are strongly associated with nutrients supply.

The presented data, together with scarce information in literature, indicate a much more dramatic decrease in P_N after the chilling period in nutrient starved or P-deficit tomato plants. It is concomitant with a decrease in photochemical efficiency of PSII and diminished stomatal conductance, perhaps as the consequence of unbalanced level of phytohormones or their activity. Therefore, the results support the hypothesis that, at least in fruiting tomato plants, there is a close relation between mineral nutrition and the sensitivity to chilling in association with other environmental factors.

Acknowledgements

The authors are grateful to Dr Waclawczyk-Lach for the estimation of chlorophyll fluorescence and Mrs M. Dzięcioł and G. Wiśniewski, M.Sc., for excellent technical assistance during experiments.

References

Beck E H 1996 Regulation of shoot/root ratio by cytokinins from roots in *Urtica dioica*. Opinion. Plant Soil 185, 3–12.

Biddinger E J, Ch Liu, R J Joly, K G Raghothama 1998 Physiological and molecular responses of aeroponically grown tomato plants to phosphorus deficiency. J. Am. Soc. Hort. Sci. 123, 330–333.

Brooks A 1986 Effects of phosphorus nutrition on ribulose-1,5-bisphosphate carboxylase activation, photosynthetic quantum yield and amounts of some Calvin-cycle metabolites in spinach leaves. Aust. J. Plant Physiol. 13, 221–237.

Brüggemann W, van der Kooij T A W and van Hasselt P R 1992 Long-term chilling of young tomato plants under low light and subsequent recovery. Planta 186, 172–178.

Chapin F S and Wardlaw I F III 1988 Effect of phosphorus deficiency on source–sink relations between the flag leaf and developing organ in barley. J. Exp. Bot. 39, 165–177.

Choluj D, Kalaji H M and Niemyska B 1997 Analysis of the gas exchange components in chilled tomato plants. Photosynthetica 34, 583–589.

Dietz K-J and Foyer C 1986 The relationship between phosphate status and photosynthesis in leaves. Reversibility of the effects of phosphate deficiency on photosynthesis. Planta 167, 376–381.

Jacob J and Lawlor D W 1993 In vivo photosynthetic electron transport does not limit photosynthetic capacity in phosphate-deficient sunflower and maize leaves. Plant Cell Environ. 16, 785–795.

Janda T, Szalai G, Kissimon J, Páldi E, Marton C and Szigeti Z 1994 Role of irradiance in the chilling injury of young maize plants studies by chlorophyll fluorescence induction measurements. Photosynthetica 30 (2) 293–299.

Jung S and Steffen K L 1997 Influence of photosynthetic flux densities before and during long-term chilling on xanthophyll cycle and chlorophyll fluorescence quenching in leaves of tomato (*Lycopersicon hirsutum*). Physiol. Plant. 100, 958–966.

Jeschke W D, Peuke A D, Pate J S and Hartung W 1997 Transport, synthesis and catabolism of abscisic acid (ABA) in intact plants of castor bean (*Ricinus communis* L.) under phosphate deficiency and moderate salinity. J. Exp. Bot. 48, 1737–1747.

Jones H G 1998 Stomatal control of photosynthesis and transpiration. J. Exp. Bot. 49, 387–398.

Kamps T L, Isleib T G, Herner R C and Sink K C 1987 Evaluation of techniques to measure chilling injury in tomato. HortSci. 22, 1309–1312.

Keutgen N, Chen K and Lenz F 1997 Responses of strawberry leaf photosynthesis, chlorophyll fluorescence and macronutrient contents to elevated CO_2. J. Plant Physiol. 150, 395–400.

Kingston-Smith A H, Harbinson J, Williams J and Foyer C H 1997 Effect of chilling on carbon assimilation, enzyme activation, and photosynthetic electron transport in the absence of photoinhibition in maize leaves. Plant Physiol. 114, 1039–1046.

Kuiper D, Kuiper P J C, Lambers H, Suit J and Staal M 1989 Cytokinin concentration in relation to mineral nutrition and benzyladenine treatment in *Plantago major* ssp. *pleiosperma*. Physiol. Plant. 75, 511–517.

Kudoyarova G R, Farhutdinov R G, Mitrichenko A N, Teplova I R, Dedov A V, Veselov S U and Kulaeva O N 1998 Fast changes in growth rate and cytokinin content of the shoot following rapid cooling of roots of wheat seedling. J. Plant Growth Regul. 26, 105–108.

Labate C A and Leegood R C 1988 Limitation of photosynthesis by changes in temperature. Planta 173, 519–527.

Marschner H 1995 Mineral Nutrition of Higher Plants, 2nd ed., Academic Press, London.

Martindale W and Leegood R C 1997 Acclimation of photosynthesis to low temperature in *Spinacia oleracea* L. II. Effects of nitrogen supply. J. Exp. Bot. 48, 1873–1880.

Mächler F, Schnyder H and Nösberger J 1984 Influence on inorganic phosphate on photosynthesis of wheat chloroplasts. J. Exp. Bot. 35, 481–487.

Oxborough K and Ort D R 1995 In situ evidence that chilling in the light does not cause uncoupling of photophosphorylation or detachment of coupling factor in chilling-sensitive plants. Photosynthesis Res. 43, 93–105.

Öquist G and Wass R 1988 A portable, microprocessor operated instrument for measuring chlorophyll fluorescence kinetics in stress physiology. Physiol. Plant. 73, 211–217.

Plesničar M, Kastori R, Petrović N and Panković D 1994 Photosynthesis and chlorophyll fluorescence in sunflower (*Helianthus annuus* L.) leaves as affected by phosphorus nutrition. J. Exp. Bot. 45, 919–924.

Qiu J and Israel D W 1994 Carbohydrate accumulation and utilisation in soybean plants in response to altered phosphorus nutrition. Physiol. Plant. 90, 722–728.

Radin J W 1984 Stomatal responses to water stress and to abscisic acid in phosphorus-deficient cotton plants. Plant Physiol. 76, 392–394.

Rao I M 1996 The role of phosphorus in photosynthesis. *In* Handbook of Photosynthesis. Ed. M Pessarakli, pp 173-194. Marcel Dekker, Inc. New York, USA.

Rao I M and Terry N 1989 Leaf phosphate status, photosynthesis, and carbon partitioning in sugar beet. I. Changes in growth, gas exchange, and Calvin cycle enzymes. Plant Physiol. 90, 814–819.

Rao I M and Terry N 1995 Leaf phosphate status, photosynyhesis, and carbon partitioning in sugar beet. IV. Changes with time following increased supply of phosphate to low-phosphate plants. Plant Physiol. 107, 1313–1321.

Sowiñski P and Maleszewski S 1989 Chilling sensitivity in maize seedlings. I. Growth and functioning of shoots and roots. Acta Physiol. Plant. 11, 165–171.

Sowiñski P and Maleszewski S 1990 Chilling sensitivity in maize seedlings. II. Effect of low temperature on transport of ^{14}C-assimilates from leaves to roots. Acta Physiol. Plant. 12, 35–40.

Starck Z, Choluj D and Niemyska B 1994 Effect of preceding temperature and subsequent condition on response of tomato plants to chilling. Acta Physiol. Plant. 16, 329–336.

Starck Z, Niemyska B and Choluj D 1996 Post-effect of chilling stress on competitive ability of sinks in tomato plants. Acta Physiol. Plant. 18, 75–83.

Starck Z, Bogdan J and Choluj D 1997 Effect of phosphorus and potassium supply on tomato plant response to chilling stress. *In* Crop Development for the Cool and Wet Regions of Europe, COST 814, pp. 49-56, EUR 17915 EN.

Starck Z, Choluj D and Gawrońska H 1998 The effect of drought hardening and chilling on ABA content in xylem sap and ABA – delivery rate from root tomato plant. Acta Physiol. Plant. 20, 41–48.

Szalai G, Janda T, Páldi E and Szigeti Z 1996 Role of light in the development of post-chilling symptoms in maize. J. Plant Physiol. 148, 378–383.

Tantau H and Dörffling K 1991 Effects of chilling on physiological responses and changes in hormone levels in two *Euphorbia pulcherrima* varieties with different chilling tolerance. J. Plant Physiol. 138, 734–740.

Thakur P S 1990 Different physiological responses of tomato (*Lycopersicon esculentum* Mill.) cultivars to drought. Acta Physiol. Plant. 12, 175–182.

Section editor: Z Rengel

Recent Advances of Plant Root Structure and Function, 151–158
Edited by O. Gašparíková, M. Čiamporová, I. Mistrík and F. Baluška
© 2001 *Kluwer Academic Publishers.*

Effect of water stress upon root meristems of pea seedlings: The role of quantitative and qualitative changes in protein patterns

Donato Chiatante[1,3], Lucia Maiuro[2], Gabriella Stefania Scippa[2] & Antonino Di Iorio[1]
[1]*Centro Servizi, Facoltà Disciente Matematiche, Fisiche, Naturali Università Degli Studi del Molise, Via Mazzini 8, 86170 Isernia, Italia*
[2]*Dipartimento cis Facoltà di Agraria, Università Degli Studi del Molise, Via de Sanctis, 86100 Campobasso, Italia.*
[3]*Corresponding author**

Abstract

To investigate the effect of water stress on the activity of meristems we used an hydroponics culture of pea seedlings to which polyethilene-glicole (PEG) has been gradually added to give the external water potential of −1.7 MPa. We focused our attention on primary root meristem although the meristematic tissue of plumules has been considered as well. Our data showed that under water stress conditions, meristematic cells from primary root arrest their activity irreversibly, while shoot meristems rapidly recover their activity when the seedlings are placed in a normal water conditions. However, the inhibition of cell division of primary root meristems does not interfere with the development of a complete root apparatus; in fact, lateral roots are still formed during water stress from initials situated in the pericycle zone, which give origin to a meristematic tip. The different sensitivity to water stress shown by meristematic cells situated in primary roots, lateral roots or shoot plumules suggests that the tolerance to water stress in meristematic tissue may depend upon factors related to its developmental stage or to the type of 'mother initial' responsible for forming this tissue. The quantitative and qualitative differences in the content of specific proteins showed between meristems during water stress may support this idea. Some of the proteins examined in our samples have shown a striking immuno-affinity for an antibody raised against a conserved sequence common to all the Dehydrins. The proteins recognised by the anti-Dehydrins antibody accumulated in a considerable amount in meristematic cells of plumules during water stress conditions. Since Dehydrins have been proposed to play a protective role during water stress, it may be speculated that the better tolerance shown by shoot or lateral-root meristems is correlated with the accumulation of these specific proteins.

Introduction

All the nursery operations (handling, lifting, transportation) with seedlings, although carried out with extreme care (Chiatante, pers. obs.), seem to induce the desiccation of the root while do not affect the shoot (McKay, 1997; Persson, 1996). Root tip tissues are less protected against water loss and, therefore, particularly suffer from desiccation. Damages occurring to root tissues may jeopardise the development of a normal root apparatus and then the overall plant survival after replanting in open fields (McKay, 1997). Consequently, it is very important to investigate how plant tissues may tolerate water stress conditions especially during nursery operations.

It has been shown that root tip meristem, particularly sensitive to water stress, in some plants may tolerate desiccation for many hours without consequences for the their vitality (Zabielsking, 1969). Furthermore, root tip meristem is able to tolerate severe desiccation (5–7% of water content) in orthodox seeds (Chiatante and Brusa, 1994), but it is very sensitive in recalcitrant seeds. The confusion about tolerance characteristics of this tissue is increased by the evidence that in orthodox seeds (legumes and cereals), the same meristems desiccation-tolerant at embryonic stage give origin to a meristem which is sensitive during the vegetative stage. From these observations, it may be suggested that tolerance trait(s) in meristems is active only during the embryonic stage.

* FAX No: 00-39-0874-404678. TEL No: 00-39-0874-404666.
E-mail: chiatant@hpsrv.unimol.it

Nevertheless, this hypothesis does not explain how in pohychilohydric plants the meristem may tolerate repetitive desiccation events at any stage of development.

Different strategies have been used over the time to study water stress tolerance in plants. As results of these studies several genes (for a review, see Close and Bray, 1993) which may be involved in water stress tolerance have been isolated (Michel et al., 1993; Schneider et al., 1993) and the majority of them require the plant hormone Abscissic Acid to be induced. From homology between the deduced amino-acid sequences, some of these water-stress induced genes have been grouped in families and possible functions have been proposed. The Dehydrins (dehydration-induced) for example are a big family of water stress induced proteins which function seems to be related with the protection of cellular structures (Close et al., 1993). Most of the experimental works reported in the literature, involve the use of differentiated tissues of leaves and shoots. Much less it is known, at a molecular level, about the response of root tissues and in particular the meristems to water deficit.

The response of plants to water deficit is complex (Boyer, 1982; Bohnert et al., 1995; Bartels and Nelson, 1994) and is dependent from the organ, the tissue, and the stage of the development (Bohnert et al., 1995; Goday et al., 1994; Plant, 1995; Senaratua et al., 1995). It is not surprising that shoot and root of the same plant may have different responses to water deficit (reviewed in Close and Bray, 1993; Sharp et al., 1993) and it becomes reasonable to think that knowledge achieved in the shoot may be not useful to explain events taking place in the root.

From these considerations, work presented here aims to understand how meristematic cells respond to water stress conditions when the plant (*Pisum sativum*) is in a vegetative state giving particular attention to the effects at physiological and biochemical level of water stress in meristematic cells from primary root, lateral roots and shoot plumules.

Materials and methods

Preparation of seedlings and stress treatments

Dry seeds of *Pisum sativum* L. cv Lincoln were sterilised with 15% (v/v) sodium hypochlorite for 30 min. After been rinsed in running tap water for 1 h and left at room temperature for 4 h, seed were kept on moist agriperlite at 24 °C in the dark. Three days later, the seedlings were rinsed in tap water and their roots immersed in Hoagland's medium or alternatively in Hoagland's medium containing different concentrations of polyethylene glycol (PEG)according to Michel and Kaufmann (1973). For the long-term water stress experiment, seedlings were transferred every 24 h in solutions containing increasing concentration of PEG giving the solutions with water potential of -0.5, -0.8, -1.0, -1.5 and -1.7 MPa. For the short-term water stress experiment seedlings were transferred every 24 in solutions containing to obtain solutions with water potentials of -0.5, -1.0 and -1.7 MPa. For the recovery, after the PEG-treatment, the seedlings were rinsed several time with tap water then put in Hoagland's medium without PEG.

Root and shoot elongation

The elongation rate of root and shoot was measured at regular intervals of time during the entire experiment. The root length was measured from the tip to the cotyledons, while the shoot length was measured from the apex to the cotyledons.

The relative moisture content was determined by using this formula (FW-DW/FW) $\times$ 100, where FW is the fresh weight and DW is the dry weight of the root.

Protein fractionation

Primary-roots meristems (2 mm from the tip) or shoot terminal buds were excised and collected in extraction buffer (25 mM Tris/HCl pH 7.6, 10 mM $MgCl_2$, 10 mM β-mercaptoethanol and 30% (v/v) glycerol). The samples were homogenised in extraction buffer by using a glass potter. The homogenate was filtered with Miracloth, centrifuged for 30 min at 27 000 g. The supernatant recovered, saturated with ammonium sulphate and centrifuged to collect the precipitated proteins. The proteins were resuspended in 25 mM Tris/HCl pH 7.6, dialysed against the same buffer, freeze-dried to reduce the volume and used as a total extract preparation for electrophoresis and immunoblotting experiments. This protein preparation was named: soluble crude fraction.

Electrophoresis and immnuno-detection of Dehydrins

Electrophoresis and immuno-blotting experiments were performed according to Chiatante and Brusa (1994). A polyclonal antibody was prepared by injecting in rabbit a synthetic peptide of 17 amino acids representing the conserved domain of Dehydrins (Chiatante et al., 1997). This antibody was used for immuno-blotting experiments of nuclear proteins separated by SDS–PAGE electrophoresis. A secondary goat anti-rabbit antibody containing alkaline phosphatase was used to detect the proteins recognised by the anti-Dehydrins primary antibody.

Extraction and purification of nuclei

Nuclei were purified from root and shoot meristems according to the protocol from Chiatante & Onelli, 1993.

Mitotic index

Meristematic root tips were fixed in ethanol-glacial acetic acid (3:1), and then prepared according to the standard Feulgen procedure, with a hydrolysis time of 10 min in 1 N HCl at 60 °C. The stained roots were squashed between slide and cover-slip with a drop of acetic acid, frozen on dried ice then washed (15 min each washing) in: 100, 95, 70, 50, 35, 0 % of cold absolute ethanol. After drying, slides were mounted with immersion oil and xylene (1:1 v/v) and viewed with a Leitz Orthoplan Microscope to determine the mitotic index.

Histological preparation of root of pea seedlings

For histological preparation, roots of pea seedlings have been collected by hand scalpel, fixed and dehydrated. The pre-infiltration has been obtained by leaving the samples for 24 h in a solution of ethanol and paraffin (1:1 v/v) then transferred in a solution containing ethanol and paraffin in the ratio 1:3 (v/v) for another 24 h. For the infiltration, the samples were kept in liquid paraffin for 24 h before hardening in the oven. After the paraffin inclusion, 1–2 μm sections were cut by a Leith microtome (1512). The sections were mounted on a slide stained with eosin and hematoxylin and observed by a Leitz Aristoplan microscope.

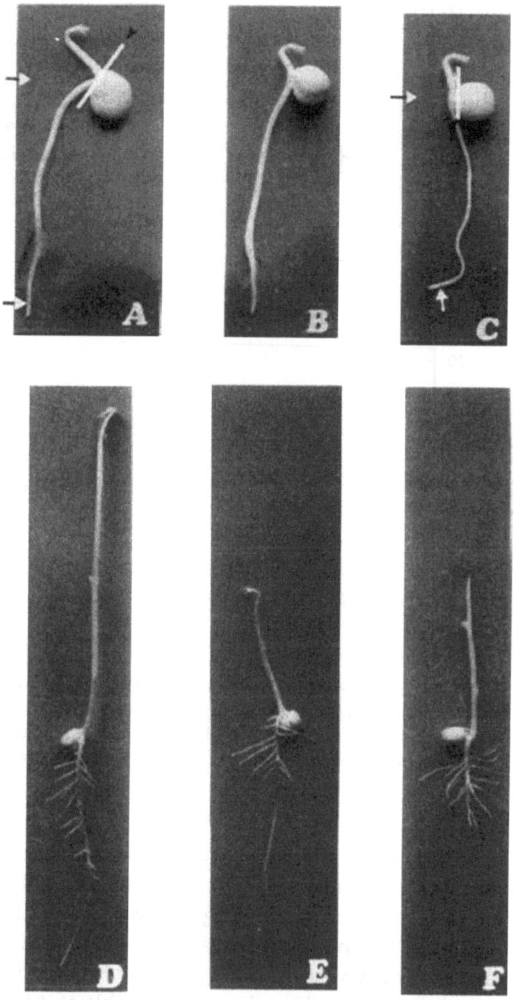

Figure 1. Effect of water stress on pea seedlings. (A): Control seedling. The arrows indicate the root axis which was measured during elongation; (B): pea seedling grown for 24h in a liquid medium containing PEG to produce a water potential of −0.5 MPa Pa; (C): pea seedling grown for 24h in a liquid medium containing PEG to produce a final water potential of −1.7 MPa. In this case, the root axis between the arrows shows clearly the reduction of diameter whereas the shoot seems not to be affected; (D): control seedling grown for 9 days; (E): pea seedling grown for 24h in a medium with PEG (−0.5 MPa) and recovered in a medium without PEG; (F): pea seedling grown for 24h in presence of PEG (−1.7 MPa) and recovered in a medium without PEG.

Results

The maximum external water-potential value of PEG (−1.7 MPa) to be applied in our experiments, was chosen for its similarity with the final value of water potential found in embryo tissues of pea seeds at the end of desiccation (Chiatante et al., 1997). Since it has been well established that the length and the intensity of the water stress are very important factors for plant survival, we made several attempts to establish the duration of the experiments for the stress treatment. Initially, pea seedlings grown for different lengths of time in agriperlite were transferred in hy-

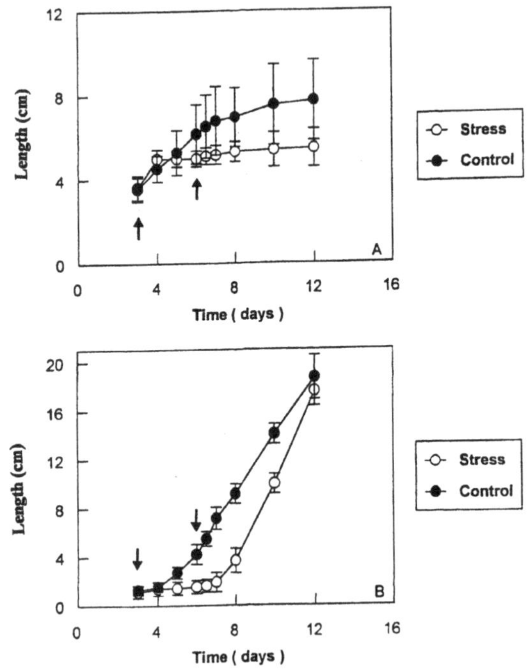

Figure 2. Effect of short-term water stress treatment on root and shoot elongation. (A): Root elongation; (B): Shoot elongation. The two arrows indicate the beginning and the end of the 3 day treatment of pea seedling with PEG. Values are the means of six replicates.

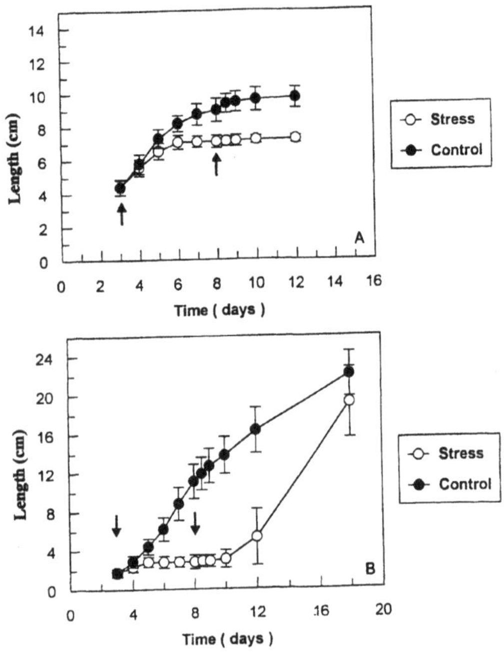

Figure 3. Effect of long term water stress treatment on root and shoot elongation. (A): root elongation; (B): shoot elongation. The two arrows indicate the beginning and the end of the 5 day treatment, of pea seedling with PEG. Values are the means of six replicates.

droponic culture in which PEG was added to reach a final water potential of −1.7 MPa. After a treatment of 24 h, root and shoot elongation was irreversibly inhibited. This result was most likely due to the fact that a very low value of water potential had been imposed too fast, making the tissues unable to adjust to the new water conditions. For this reason, we decided to decrease the water potential slowly and to select seedlings having the same initial length of the primary root and shoots to avoid the interference of a developmental-stage factor upon elongation measurements. In this experimental system, the seedlings conserved their green healthy plumules (Figure 1A–F) during the entire duration of stress treatments and they were able to recover completely at the end of the stress treatments.

A shot-time and long-time experimental procedure were used to apply the water stress to pea seedlings. In a short-time experiment (Figures 1D, E, F and Figure 2), the lowest value of water potential (−1.7 MPa) was achieved within 3 days, in a long-time experiment (Figure 3) it was achieved within 5 days. In both experimental systems seedlings were able to recover,

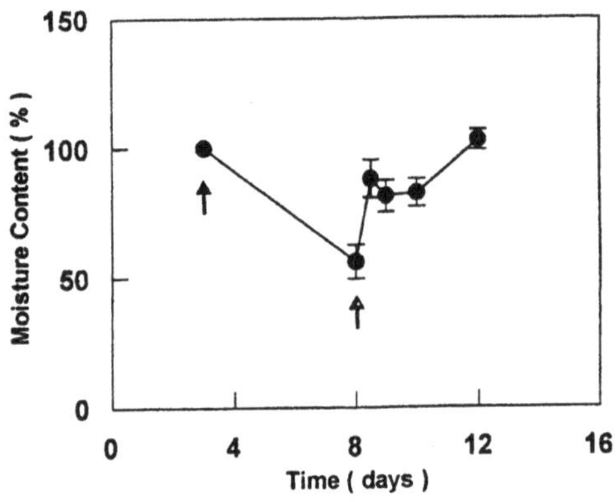

Figure 4. Loss of water from the primary root of pea seedlings during the water stress treatment. The two arrows indicate the beginning and the end of the 5 day water stress treatment. At the end of the treatment, seedlings were transferred in a fresh medium without PEG. Values are the means of six replicates.

although seedlings recover faster after a long-term treatment. Root diameters was reduced similarly after a short and long term treatment and it was probably induced by a loss of water from root tissues (Figure 1C). The measurement of moisture content has been used as indicator of loss of water from root tissues. Figure 4 shows that at the end of the water stress experiment the moisture content in the roots had dropped to 50% of the initial value and returned to its initial value during the recovery.

In the short-term experiment (Figure 2A), root elongation was arrested between 24 and 48 h from the beginning of water stress treatment when PEG concentration has been increased to give a water potential of −1.0 MPa. Similarly, in the long-term experiment (Figure 3A), root elongation was arrested after 3 days of water stress treatment when PEG concentration in the medium gave a water potential of −1.0 MPa. At the end of PEG treatment in both systems, we observed a recovery of root elongation, although the length in the primary-roots was 75% of the control. In *short-term* experiment (Figure 2B), the length of shoots increased slightly during the first 24 h of water stress treatment and remained unchanged for the whole period of water stress treatment. The recovery of shoot elongation (Figure 2B) started 2 days after the end of stress and reached the same value as the control after 6 days. In long-term experiments (Figure 3B), shoot elongation result inhibited after 3 days of water stress treatment and the recovery started only 3 days after PEG had been removed, and reached values similar to the control only after 10 days from the beginning of recovery (Figure 3B). Although seedlings respond similarly, we decided to adopt the long-term experiment as a routine treatment because it could leave more time for the seedlings to adjust to the new water potential conditions.

Beside the primary root, we have investigated the effect of water stress on initiation of laterals (Table 1). In long-term experiment, the number of new laterals decreased to 80.7% of control, whereas, their mean length was considerably shorter (Table 1). These data indicated the activity of meristems in the primary root tip was inhibited by water deficit but meristematic cells among initials produced in the pericycle layer were still active. Histological observation of primary root axis being under water stress showed (Figure 5) that the apex of new lateral was formed also during water stress treatment. It was evident that the initiation of laterals started at a shorter distance from the root apex in respect of control seedlings (not shown). At the end of stress treatments, we observed a complete recovery of elongation rate in laterals which doubled their length in few days (Table 1).

In contrast with the persistence of cell division activity in the meristematic apex of lateral roots, we observed the loss of cell division activity in the primary root meristem when water potential lower than −1.0 MPa was achieved (not shown). A consid-

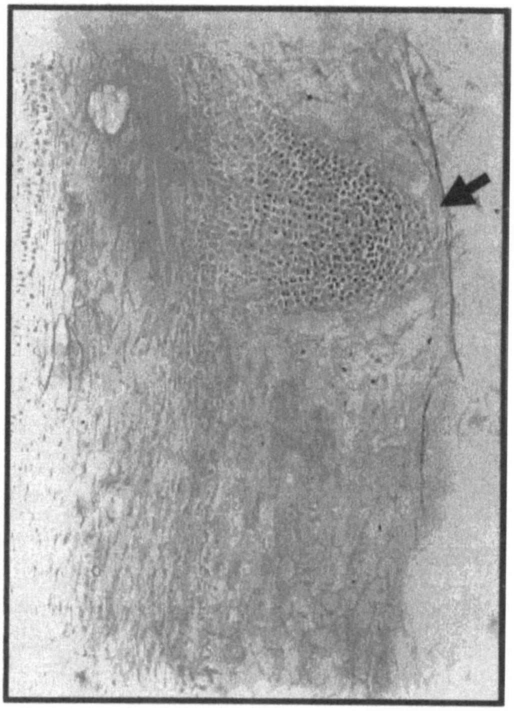

Figure 5. Longitudinal section of a primary root of pea seedling. The arrow indicates the meristematic tip of a lateral root which has been formed during stress treatment of pea seedlings by PEG.

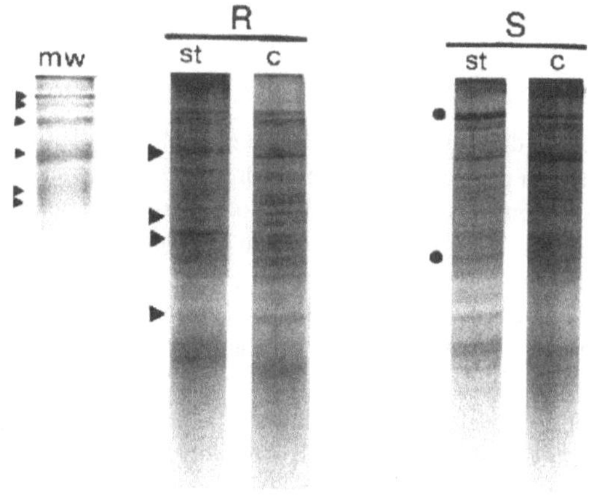

Figure 6. Biochemical analysis of proteins purified from root and shoot meristems. Protein purified from primary root meristems (R) or shoot plumules (S) were separated by SDS–PAGE electrophoresis and then stained by Blue Comassie. The arrows and dots point to the major proteins showing quantitative variations. MW: molecular weight markers (from the top: 123, 89, 67,50, 37.5, 34 kDa).

Table 1. Length (mm) of primary and lateral roots measured at the end of long-term treatment and after nine days of recovery. Seedlings used for the experiments had an initial length of 50±3 mm. Values are the means of six replicate standard deviation follows the signs ±

Stress treatment				Recovery			
Primary root		Lateral roots		Primary root		Lateral roots	
Control	Stress	Control	Stress	Control	Stress	Control	Stress
112±16	84±3.4	22±3.4	13±2	150±5	86±3	45±0.2	25±0.1

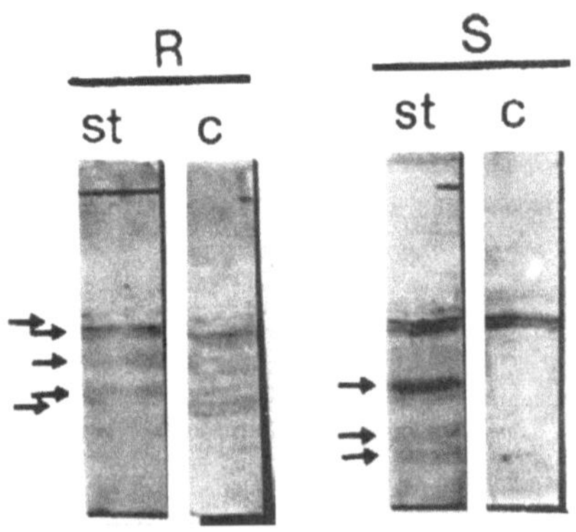

Figure 7. Immunoblotting of nuclear proteins by using the anti-Dehydrin antibody. Protein were extracted from nuclei of primary root (R) and shoots (S) under water stress (st) and non stress (c) conditions. The arrows indicate major proteins undergoing quantitative variations during water stress treatment of pea seedlings by PEG.

erable difference in behaviour between meristematic cells in primary roots, laterals or shoots during stress treatment was found. In order to understand whether these differences originated from changes at biochemical level, we compared the protein composition in a soluble crude fraction obtained from meristems of roots or shoots. For the biochemical analysis, the soluble crude protein fraction from shoots or primary-roots of water stressed seedlings were separated by SDS gel electrophoresis and compared with their relative non-stressed controls. Figure 6 shows quantitative changes of several proteins in the soluble crude fraction from primary root meristems under water stress. We have indicated only the major 6 proteins with molecular weight of 48, 32.5, 30, 28, 26.5 and 20 kDa, respectively (Figure 6-R-st arrows). Four proteins (48, 32.5, 28 and 20 kDa) showed a decrease of their content under stress condition, whereas other two major proteins with molecular weight of 30 and 26.5 kDa showed an increase of their content (Figure 6B-R-st dots). As well, we observed quantitative variations in protein patterns from water-stressed shoot.

To investigate if some of the proteins showing content variations have protective function, as suggested for Dehydrins, we used an antibody raised against a conserved Dehydrins domain in immuno-blotting experiment. Since we were particularly interested on the protection of the nucleus, we tested the antibody on proteins of purified nuclei from roots and shoots plumules (see Materials and Methods). As shown in Figure 7, the antibody recognised several proteins with an apparent molecular weight of 39 (doublet), 36, 30 (doubled), 26 kDa 22.5 kDa in nuclear extracts from root and shoot. The content of these proteins did not change in nuclei from primary root axis meristems under stress conditions (compare the proteins signed by the arrows in Figure 7R st-c). However, we found considerable changes between the content of the proteins with molecular weight of 39 (doublet), 36, 30 (doublet), 26, when they were extracted from shoot plumules of stressed or control seedlings (Figure 7S-st arrows).

Discussion

It is well established that water stress in plants evokes different responses, depending on tissue, organ and stage of development (reviewed by Bray,1993). Here we show how the same type of tissue localised in different organs may behave differently when subjected to water stress. Since very little has been studied about meristematic cells, and the effect of water stress on the regulation of the cell cycle has remained almost unexplored, we focused our attention to root and shoot meristems. Our investigations on meristems activity indicate that while the proliferation of meristematic cells from primary root is irreversibly inhibited by water stress, since never recover fully, meristematic cells from shoots and laterals roots are able to recover the activity when water condition comes to a normality. These differences are difficult to explain since meristematic cells of the primary root origin from the divisions of those initials which tolerate desiccation at embryonic stage (Chiatane & Onelli, 1993). Furthermore, it is unclear why in our experiments cell proliferation is arrested in meristematic cells of primary root, whereas it continues in meristematic cells derived from new initials localised in the pericycle zone which originate lateral roots. These results suggest that meristems (originally tolerant) become sensitive to water stress with age, while the new-formed meristems (in lateral roots) generated during water stress acquire water stress tolerance. However, tolerance persists in meristematic tissues of shoot plumules despite these tissues may be regarded as developmentally-old meristems. In our experiments, both meristematic tissues present similar sensitivity to the water potential (−1.0 MPa), but are characterised by different tolerance capabilities. These differences could be explained by the fact that shoots and primary roots perceive stress differently and the responses may be regulated by the expression of a different set of genes. A traditional approach to identify genes involved in water-stress tolerance is the biochemical analysis to identify changes in proteins levels (Pérez-Molpha-Balch et al., 1996). It is generally knowledge that water-limiting conditions results in a decrease or increase of many proteins-water stress related (Good and Zaplachinski, 1994; Leone et al., 1994). Recent experiments have demonstrated quantitative and qualitative changes in polypeptide patterns in the root regions in which elongation is inhibited at low water potential (Sharp et al., 1993). Our biochemical investigations by using SDS–PAGE electrophoresis show that meristematic cells in primary-roots and shoot plumules undergo quantitative changes in the content of several proteins obtained from a soluble crude fraction during water stress. From electrophoretic mobility, we are not able to establish whether proteins present in meristematic cell of primary-roots or shoot plumules are the same ones and studies are in progress to identify their amino acid sequence. However, it remains that the quantitative variations in the content of proteins observed in meristematic cells of primary-roots or shoot plumules might be the base of the differences observed in the physiological response of these tissues. If this is the case, it emerges from our experiment that water stress has affected the transcription in meristematic cells differently depending upon their organ-specificity. In the future, we want to expand this type of biochemical investigation also to the meristematic cells of laterals. The aim will be to understand whether the improved tolerance to water stress shown by meristematic cells of laterals derives from differences in the pattern of synthesis and degradation of specific proteins.

The quantitative variations of proteins such as LEA (Late Embryogenesis Abundant) and Dehydrins (known as LEA D-11 family) has been related to water stress tolerance (Close et al., 1993. In our extracts, a number of putative-dehydrin proteins have been identified by using an antibody raised against the conserved sequence of Dehydrins. Many putative-Dehydrins seem to be present in both shoot and primary root meristematic cells, but others have been found to be specific to only one of these. It is known that plant Dehydrins are encoded by multigene families and each dissected part of a seedling has a unique pattern of Dehydrin expression implying the presence of an extensive developmental control (Close et al., 1993). Our data confirm the difference in the organ distribution of Dehydrins in pea seedlings under water stress. To explain the difference of distribution observed in primary roots or shoots, we suggest that in meristematic cells, these proteins may be always present at a constitutive level, but they are accumulated during water stress when the intensity of the stress effect is not inhibitory to all the metabolic activities as probably happens in shoots. Therefore, the increased tolerance of shoots could be related to the accumulation of these Dehydrin-like proteins. The function of LEA and Dehydrins is still unknown (Xu et al., 1996 and references therein) although a protective role from damages deriving from water deficit in proteins or/and structures (membranes ?) has been suggested (reviewed in Bray, 1993; Dure et al., 1989; Dure, 1983). It is very interesting that nuclei from shoot meristematic cells increase the content of putative Dehydrins during the stress treatments as shown in extract obtained from immuno-blotting of the soluble fraction. It might be that these proteins are involved in the protection of nuclear structures from damages occurring during water loss. Data collected up to now are too preliminary and are still not enough to correlate the role of Dehydrins with the tolerance mechanisms (Bray, 1993; Cheng et al., 1993; Leone et al., 1994). However, works are in progress to better characterise the involvement of these proteins in the tolerance mechanism and to clarify if the tolerance of meristematic cells is regulated by positional and developmental factors.

Acknowledgements

The authors acknowledge MIRAAF (Ministero Italiano per le Risorse Agricole, Alimentari e Forestali) project: 'Biotecnologie Vegetali'., MURST (fondi 60%), and EC (Fair project no 1 CT95-0497) for supporting this research.

References

Bartels D and Nelson E D (1994): Approaches to improve stress tolerance using molecular genetics. *Plant, Cell Environ.* 17, 659–667.

Bohnert H J, Nelson D E, and Jensen R G (1995): Adaptations to environmental stresses. *Plant Cell* 7, 1099–1111.

Boyer J S 1982. Plant productivity and environment. *Science* 218, 443–448.

Bray E A (1993): Molecular responses to water deficit. *Plant Physiol.* 103, 1035–1040.

Chiatante D and Onelli E (1993): Nuclear proteins and the onset of cell proliferation in root meristem of *Pisum sativum*: QP47a novel acidic protein. *Seed Sci.Res.* 3, 35–42.

Chiatante D and Brusa P (1994): Increase of the content of QP47 (a desiccation-associated nuclear protein) in embryo cells during maturation of pea seeds. *Seed Sci.Res.* 4, 421–429.

Chiatante D, Scippa G S, Onelli E, Patrignani G and Maiuro L (1997): Nuclear and nucleolar organisation is affected by variations of moisture content in pea root meristems. *Plant Biosystem* 131(1), 13–24.

Close J T, Fenton D R and Moonan F (1993): A view of plant dehydrins using antibodies specific to the carboxy terminal peptide. *Plant Mol. Biol.* 23, 279–286.

Close J T and Bray E A (1993): *Plant responses to cellular dehydration during environmental stress*. The American Society of Plant Physiologists. 295 p.

Cheng Y, Weng J, Joshi C P, and Nguyen T H (1993): Dehydration stress-induced changes in translatable RNAs in *Sorghum*. *Crop Sc.* 33, 1397–1400.

Dure III L, Crouch M, Harade J, Ho T H, Mundy J, Quartano R, Thomas T, and Sung Z R (1989): Common aminoacid sequence domains among the LEA proteins of higher plants. *Plant Mol. Biol.* 12, 475–486.

Goday A, Jensen B A, Cullànez-Macià A, Mar Albà M, Figueraz M, Serratosan J, Torrent M, and Pagès M (1994): The maize Abscisic Acid-responsive protein *Rab*17 located in the nucleus and interacts with nuclear localisation signals. *Plant Cell* 6, 351–360.

Goday J A, Lunar R, Torres-Schuman S, Moreno J, Rodrigo R M, and Pintor-Toro J A (1994): Expression, tissue distribution and subcellular localisation of dehydrin TAS14 in salt-stressed tomato plants. *Plant Mol. Biol.* 26, 1921–1934.

Good A G and Zaplachinski S T (1994): The effects of drought stress on free aminoacid accumulation and protein synthesis in *Brassica napus*. *Physiol. Plant.* 90, 9–14.

Jones H G, Flowers T J and Jones M B (1996): Plants under stress. Cambridge: Cambridge University Press.

Jones J T and Mullet J E (1995): A salt and dehydration-inducible pea gene, Cyp15a, encodes a cell-wall protein with sequence similarity to cysteine proteases. *Plant Mol. Biol.* 28, 1055–1065.

Leone A, Costa A, Tucci M, and Grillo G (1994): Comparative analysis of short and long term changes in gene expression caused by low water potential in potato (*Solanum tuberosum)* cell suspension cultures. *Plant Physiol.* 106, 703–712.

McKay H M (1997). A rewiev of the effect of stresses between lifting and planting on nursery stock quality and performance. *New For.* 13, 369–399

Pèrez-Molphe-Balch E, Gidekel M, Sagura-Niets M, Honera-Estrella L, and Ochoa-Alejo N (1996): Effects of water stress on plant growth and root proteins in three cultivars of rice (*Oryza sativa*) with different levels of drought tolerance. *Physiol. Plant.* 96, 284–290.

Persson H (1996): Fine root dynamic in forest trees. *Acta Phytogeogr. Suec.* 81, 17–23.

Plant Z (1995): Sensitivity of crops plants to water stress at specific developmental stages: Revaluation of experimental findings. *Jsrael J.Plant Sci.* 43, 99–111.

Schneider K, Wells B, schmelzer F, Salamini F, and Bartles D (1993). Desiccation leads to the rapid accumulation of both cytosol and chloroplast proteins in the resurrection plant *Craterostigma plantagineum* Hochst. *Planta* 189, 120–131.

Senaratua T, Saxena P K, Rao M V, and Afele J (1995). Significance of the zygotic seed coat on quiescent and desiccation tolerance of *Medicago sativa*. *Phytochemestry* 40, 1383–1386.

Sharp R E, Voetbory G S, Saab I N, and Bernstein N (1993): Role of Abscisic Acid in the regulation of cell expression in roots at low water potential. *In*: Plant responses to cellular dehydration during environmental stress. Ed. Close T and Bray E A. The American Society of Plant Physiologists.

Verslues P E, Ober E S, and Sharp R E (1998): Root growth and oxygen relation at low water potentials. Impact of oxygen availability in Polyethylene Glycol solution. *Plant Physiol.* 116, 1403–1412.

Xu D, Duan X, Wany B, Hong B, Ho T H D, and Wu R (1996): Expression of a late embryogenesis abundant protein gene, HVA1, from barley confers tolerance to water deficit and salt stress in transgenic rice. *Plant Physiol.* 110, 249–257.

Zabielsking S (1969): The effect of various factors on the success of planting out in poplars. *Bul.Serv.Cult.Etud.Peuplier et Saul.* 33–53.

Recent Advances of Plant Root Structure and Function, 159–163
Edited by O. Gašparíková, M. Čiamporová, I. Mistrík and F. Baluška
© 2001 Kluwer Academic Publishers.

The distribution of lead in duckweed (*Lemna minor* L.) root tip

S. Samardakiewicz[1] and A. Woźny
Laboratory of General Botany, Faculty of Biology, Adam Mickiewicz University, Aleja Niepodległości 14, 61-713 Poznań, Poland. [1]*Corresponding author**

Key words: duckweed, Pb, transport, X-ray microanalysis

Abstract

While considerable information on lead distribution in the cells of terrestrial plants has been collected, little is known about lead localization in the cells of the aquatic plant. *Lemna minor* L. (duckweed) roots were examined using X-ray microanalysis. After 1-h treatment with lead, its concentration was the highest in small vacuoles. After 6 and 12 h, the lead content of cell walls gradually increased. The changes of lead level between vacuoles and cell walls may result from redistribution of this metal from symplast to cell walls or it may reflect increased apoplastic transport. Lead was not found in the ground cytoplasm of any variants of the experiments. This fact and presence of lead in small vesicles suggests that endocytosis may play the role in lead uptake in *Lemna*.

Abbreviations: app. conc. (%) – apparent concentration (%); cw – cell wall; EDS – energy dispersive spectrometry; v – vacuole

Introduction

Lead may be dangerous to living organisms. This results mostly from its high toxicity (Balsberg-Pahlsson, 1989; Bazzaz, et al.,1975; Borek et al., 1998; Davies, 1991; Foy et al., 1978; Obroucheva et al., 1998; Punz and Sieghardt, 1993; Woźny and Manikowska, 1990) and from the fact that it is one of the most widespread metals in the natural environment (Dörr et al., 1990; Moore and Ramamoorthy, 1984; Nriagu, 1979). There are still many unexplained questions about the relationship between lead and the organism, or between lead and the cell, e.g. the kinds of resistance employed by plants. These may include avoidance of the stress factor, namely limitation of its uptake. In turn, the entry of a stress factor into the protoplast intensifies the processes of tolerance, e.g. compartmentation in specific cellular structures (Punz and Sieghardt, 1993; Woźny and Krzesłowska, 1993). To determine the kind of the cell resistance strategy against lead, it is useful to identify the pathways of its transport in the organism and to localize lead deposits. Lead distribution within the cells of terrestrial plants is relatively well studied, but little is known about its distribution in aquatic vascular plants.

The aim of this study was to trace the successive stages of lead transport and deposition in *Lemna minor* root tip. Our investigations concentrated on this region because earlier studies utilizing the rhodizonate staining technique revealed that the root apex is an important site of lead intake in this plant (Kocjan et al., 1996).

Materials and methods

Duckweed (*Lemna minor* L.) was used for all experiments. Plants were grown hydroponically on liquid Wang medium (Wang, 1990). Stock cultures were incubated on Wang medium under a constant illumination of 66 μE m^{-2} s^{-2} at 23±1 °C. Plants grouped on the basis of morphological characteristics were treated with lead nitrate containing 15 μM Pb in a fifty-fold diluted Wang medium for 1, 6 and 12 h. Three plants per population were studied. The experiment was conducted in three series. A total of 9 plants were examined. Data were averaged and the standard deviations were calculated. The significance of differences was determined by Student's *t*-test (for $P < 0,05$).

* Phone: +4861 852-56-03. Fax No: +4861 852-36-15. E-mail: sas@main.amu.edu.pl. E-mail: adaw@main.amu.edu.pl

The dynamics of lead distribution in the root meristem region of *Lemna minor* was traced using X-ray microanalysis. Whole plants were prepared according to a previously described procedure (Kocjan et al., 1996). Osmium tetroxide, and in some cases also uranyl acetate were not used. The roots were divided into 2.5 mm length segments and cut into 100 nm sections which were later placed on copper grids and coated with carbon. The objects were then examined under a JEOL JEM 1200 EX transmission electron microscope supplied with a scanning attachment and a Link AN 10 000 Energy Dispersive Spectrometry (EDS) system for X-ray analysis of chemical elements.

Point analyses of 40 nm spot size as well as mapping analysis (DIGIPAD) were conducted at an accelerating voltage of 80 keV. The time of registration of signals forming the spectrum was 200 s for each point analysis and 120 min for the mapping analysis. Lead concentration was assessed on the basis of the peaks area for the emission line Lα (10 540 keV). Lead content data were processed using Link software for assessment of apparent concentration (%) [app. conc. (%)]. The term 'apparent concentration' in this context means the ratio of the number of counts under a Pb Lα peak (= the net integral) and the number of counts over the same channels from the control spectrum (without Pb). Maps of lead distribution were made in the DIGIPAD system. A region of the meristematic zone located at the apical part of the root cap was selected for examination. At the cross section of the root, a layer of protoderm, four layers of periblem and two layers of plerome were distinguished in this region. All these layers were analysed using the X-ray microanalysis.

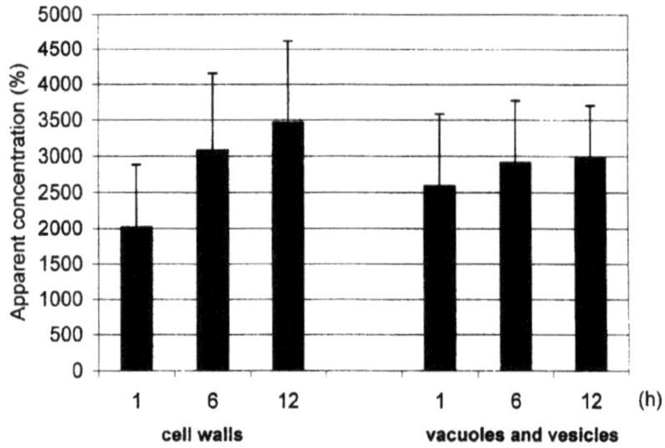

Figure 1. EDX analyses of apparent concentration [%] Pb in cell walls (cw) and vacuoles (v) in all the layers after 1, 6 and 12 h of treatment with lead (15 μM). Lead concentration was assessed on the basis of the window Pb La. n = 30, 9 assay 3 or 4 individuals each were done.

Results

After 1 h treatment with lead, the metal was localized in cell walls, vesicles and small vacuoles of all the distinguished cell layers. In each layer the lead content of vacuoles was significantly higher than that of cell walls (Figure 1). It was the highest in the outer layers of periblem cells, slightly lower in the protoderm and the lowest in the plerome (Figure 2). As far as cell walls are concerned, lead concentration in the protoderm (Figure 3) and plerome was very low. Larger amounts of lead were observed only in the walls of periblem.

After 6 h of incubation in lead solution, an increase in app. conc. (%) Pb was observed in the walls, vesicles and vacuoles of all the layers (though a difference of statistical significance was noted only in the walls), except for protoderm vacuoles, where a slight decrease was recorded (Figure 3). The rise in app. conc. (%) Pb in cell walls, in relation to 1 h, was higher than in vacuoles, (Figures 1 and 3). As a result, lead concentration in all the layers in general was higher in cell walls than in vacuoles after this period of time (Figure 1), but this difference was not significant. The lead content of cell walls was the highest in periblem cells, particularly in the walls neighbouring with intercellular spaces (Figure 2). The lowest lead level was detected in cell walls of the plerome. In vacuoles, a similar tendency was observed as in the case of cell walls, i.e. the lowest values were recorded in the plerome, while the highest in the periblem, particularly in the cells neighbouring with intercellular spaces.

After 12 h of treatment with lead, a further increase in app. conc. (%) Pb was detected (Figures 1 and 4), particularly in protoderm cell walls and vacuoles (Figures 2 and 3). For all the layers in general, app. conc. (%) Pb was higher in cell walls than in vacuoles (Figure 1), but this difference was not significant. However, lead distribution was different than in the earlier time variants. Lead concentration was the highest in the protoderm and the lowest in the deepest layers, i.e. in the plerome (Figures 2 and 3). In the protoderm (Figure 3) and periblem, lead was most abundant in the walls, whereas in the plerome lead was most abundant in the vacuole.

In all the studied variants lead was not recorded in the ground cytoplasm.

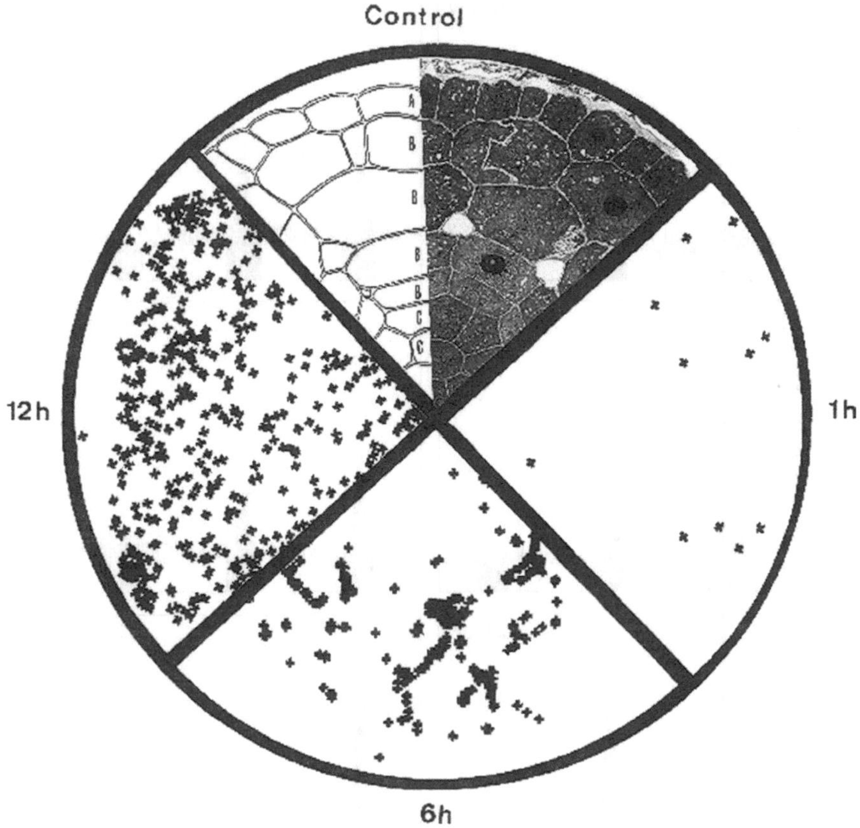

Figure 2. Mapping analysis of lead distribution (crosses) in the meristem after 1, 6 and 12 h of treatment with lead. In the control figure (without Pb) we submitted the anatomy scheme and part of meristem ultrastructure (×1600) in the *Lemna minor* root tip. Tissue distribution in each of 1/4 cross sections after Pb treatment is the same type as in the control. (A) protoderm, (B) periblem, (C) plerome.

Discussion

Literature data on lead uptake by terrestrial plants suggest that this element, after absorption by the root, is initially transported by the apoplast (Książek and Woźny, 1990; Lane and Martin, 1982; Punz and Sieghardt, 1993; Tung and Temple 1996; Wierzbicka, 1995). In the first stages of uptake, its presence was detected exclusively in the outer layers of the root. Only after some time it could be detected at the centre of the root (Książek and Woźny, 1990; Wierzbicka, 1995).

In this study, using the method of X-ray microanalysis, it was found that in the meristematic zone of roots of an aquatic plant (*Lemna minor*), this metal was present in the protoderm, periblem and plerome as early as after 1 h treatment with lead. Within the cell, lead was detected in small vacuoles, vesicles and cell walls. So fast a transport of lead across plant tissues was observed also in some terrestrial plants by other researchers (Książek and Woźny, 1990; Wierzbicka, 1995). For example, in the roots of *Allium cepa*, lead was transported to the successive layers of the root at a rate of 1 layer per 5 min, and reached the centre after 70–85 min (Wierzbicka, 1995).

When the lead dose was higher (longer time of incubation and/or higher concentration), this metal was also observed in the symplast of many terrestrial plants, usually several hours after the beginning of treatment (Krzesłowska and Woźny, 1996; Wierzbicka, 1995). It is also interesting that after 1 h treatment, high concentrations of this element were detected in the *Lemna minor* meristem region only in vacuoles. The question is how lead is taken up by *Lemna minor* cells: by means of diffusion and/or endocytosis? The fact that lead was not detected in the

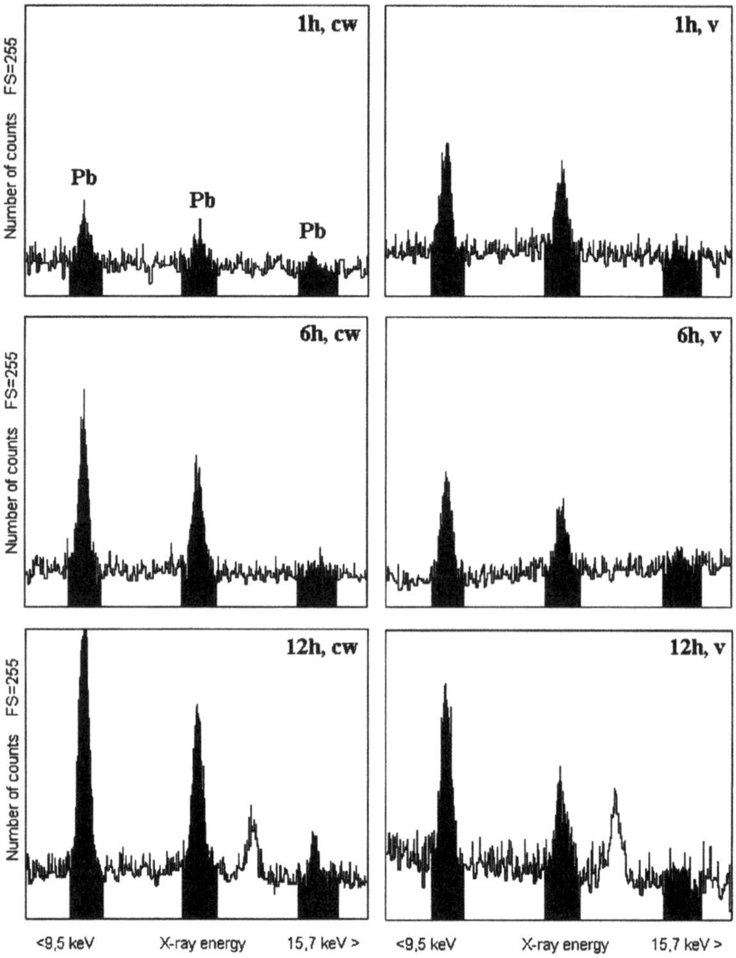

Figure 3. X- ray spectra (point analyses) from the cell walls (cw) and vacuoles (v) of the protoderm, after 1, 6, 12 h of treatment with lead. The spectrum, collected between 0 and 20 keV (only the part of the spectrum between 9,5 keV to 15,7 keV is shown), shows the L families of X-ray for Pb. The L family consists of the $L\alpha$ (10 540 keV), $L\beta 1$ (12 611 keV) and $L\gamma$ (14 762 keV) peaks which are black coloured. The y-axis is the number of counts in a given channel and the full scale deflection represents 255 counts in each spectrum.

ground cytoplasm but only in small vesicles (Figure 4) may suggest to us that also endocytosis plays the role in lead uptake in this species.

The changes in lead content proportions between cell walls and vacuoles after 6 and 12 h, in favour of cell walls, may result from redistribution of this metal from symplast to cell walls (Neumann et al., 1997), or it may reflect increased apoplastic transport after 12 h (Tung and Temple, 1996). Particularly high lead concentrations were detected in the walls of cells neighbouring with intercellular spaces. This seems to confirm the latter explanation.

Studies of lead distribution in the *Lemna minor* root meristem zone showed that, in contrast to the theoretical expectations, more lead accumulated in periblem cells than in the protoderm (after 1 and 6 h). Since in this region there are no mechanical barriers restricting apoplastic transport, explanation of this phenomenon requires further investigations. Moreover, a great deal of similarity to terrestrial plants was observed, including the rate of lead transport and participation of cell walls in apoplastic transport. At the same time, it was noted that in this plant, unlike in terrestrial plants, vesicles play a major role in lead accumulation after a short (1 h) treatment.

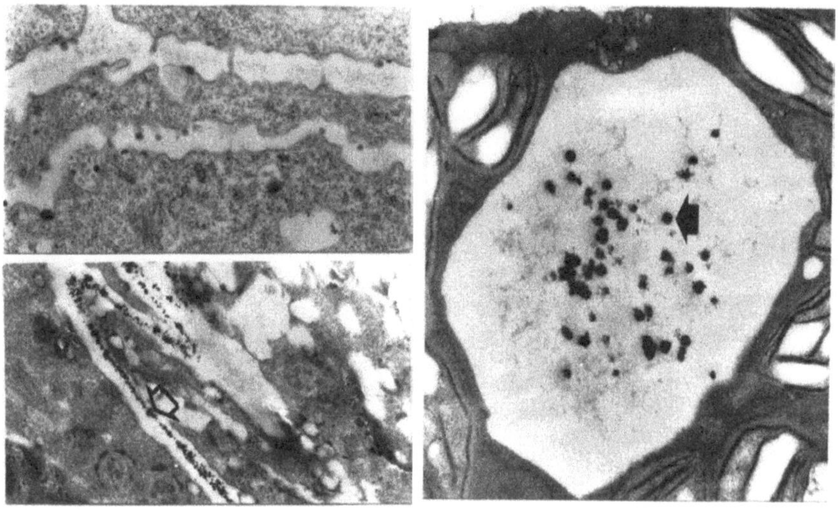

Figure 4. Lead localization in cross sections of meristem root cells (periblem) by transmission electron microscopy technique. Lead deposits in the vesicles and plasma membrane invaginations (1 h treatment, ×35000), cell wall (white arrow; 12 h treatment, ×25000) and vacuole (black arrow; 6 hours treatment, ×30000).

Acknowledgements

This research was supported by a grant from the State Committee for Scientific Research Nr 6PO4C 090 14. The X-ray microanalysis was carried out in the Electron Microscope Laboratory, Nencki Institute of Experimental Biology (Warsaw) under the supervision of Dr E. Wyroba. We wish to thank Mrs M. Samardakiewicz too for her technical assistance.

References

Balsberg-Pahlsson A M 1989 Toxicity of heavy metals (Zn, Cu, Cd, Pb) to vascular plants. Water Air Soil Pollut. 47, 287–319.

Bazzaz F A, Carlson R W and Rolfe G L 1975 Inhibition of corn and sunflower photosynthesis by lead. Physiol. Plant. 34, 326–329.

Borek S, Samardakiewicz S and Woźny A 1998 The effect of pH of water on lead toxicity in *Lemna minor* L. Biol. Bull. of Poznań 35, 19–24.

Davies M S 1991 Effects of toxic concentrations of metals on root growth and development. *In* Plant Root Growth: an Ecological Perspective. Ed. D Atkinson. pp 211–227. Blackwell Scient. Publ. Oxford.

Dörr H, Münnich K O, Mangini A and Schmitz W 1990 Gasoline lead in west German soils. Naturwissenschaften 77, 428–430.

Foy C D, Chaney R L and White M C 1978 The physiology of metal toxicity in plants. Ann. Rev. Plant Physiol. 29, 511–566.

Kocjan G, Samardakiewicz S and Woźny A 1996 Regions of lead uptake in *Lemna minor* plants and localization of this metal within selected parts of the root. Biol. Plant. 38, 107–117.

Krzesłowska M and Woźny A 1996 Lead uptake, localization and changes in cell ultrastructure of *Funaria hygrometrica* protonemata. Biol. Plant. 38, 253–259.

Książek M and Woźny A 1990 Lead movement in poplar adventitious roots. Biol. Plant. 32, 54–57.

Lane S D and Martin E S 1982 An ultrastructural examination of lead localization in germinating seeds of *Raphanus sativus*. Z. Pflanzenphysiol. 107, 33–40.

Moore J W and Ramamoorthy S 1984 Lead. *In* Heavy Metals in Natural Waters. Applied monitoring and impact assessment. Eds. J W Moore and S Ramamoorthy. pp 114–139. Springer – Verlag, New York – Berlin – Heidelberg – Tokyo.

Neumann D, Zur Nieden U, Schwieger W, Leopold I and Lichtenberg O 1997 Heavy metal tolerance of *Minuartia verna*. J. Plant Physiol. 151, 101–108.

Nriagu J O 1979 Global inventory of natural and antropogenic emission of trace metals to the atmosphere. Nature 279, 409–411.

Obroucheva N V, Bystrova E I, Ivanov V B, Antipova O V and Seregin I V. Root growth responses to lead in young maize seedlings. Plant Soil 200, 55–61.

Punz W F, Sieghardt H 1993 The response of roots of herbaceous plant species to heavy metals. Environ. Exp. Bot. 33, 85–98.

Tung G and Temple P J 1996 Uptake and localization of lead in corn (*Zea mays* L.) seedlings, a study by histochemical and electron microscopy. Sci. Total Environ. 188, 71–85.

Wang W 1990 Literature review on duckweed toxicity testing. Environ. Res. 52, 7–22.

Wierzbicka M 1995 How lead loses its toxicity to plants. Acta Soc. Bot. Pol. 64, 81–90.

Woźny A and Krzesłowska M 1993 Plant cell responses to lead. Acta Soc. Bot. Pol. 62, 101–105.

Woźny A and Manikowska 1990 The toxic effect of lead on the *Lemna minor* L. Bull. Soc. Amis Sci. Lettr. Poznań 28, 59–63.

Section editor: L V Kochian

Recent Advances of Plant Root Structure and Function, 165–170
Edited by O. Gašparíková, M. Čiamporová, I. Mistrík and F. Baluška
© 2001 Kluwer Academic Publishers.

Lead effects on cereal roots in terms of cell growth, root architecture and metal accumulation

N. V. Obroucheva[1,5], V. B. Ivanov[1], M. Sobotik[2], H. Bergmann[3], O. V. Antipova[1], E. I. Bystrova[1], I. V. Seregin[1] & L. K. Shpigun[4]
[1]*Institute of plant physiology, Russian Academy of Sciences, Moscow, Russia.* [2]*Federal Research Institute for Agriculture in Alpine regions, Irdning, Austria.* [3]*Institute of nutrition and environment, Fr.-Schiller University, Jena, Germany.* [4]*Institute of general and inorganic chemistry, Moscow, Russia.* [5]*Corresponding author**

Key words: barrier role, cell division, cell elongation, growth, lead, root, root architecture, root branching, toxicity

Abstract

The influence of lead nitrate on root growth of maize and barley seedlings was studied in terms of cell growth analysis with respect to root system morphology, lead accumulation and distribution among plant organs. At growth-inhibiting 10^{-3} *M* concentration in Petri dishes, lead nitrate retarded by half root extension, but did not affect the number of lateral or seminal roots that resulted in the modified root architecture, i.e. shorter and more compact root system. Lead unspecifically inhibited cell division and cell elongation. It was accumulated by root mainly in slime, rhizodermis and cortex, but almost did not penetrate endodermis that prevents lead translocation to shoots and explains not only its low content in shoots and ears, but weak and delayed effect on shoot growth as well. This barrier role of root system is emphasized.

Introduction

Lead is commonly classified as a dangerous pollutant widely spread in the soil and vegetation all over the industrial countries. However, it can be considered a natural element in lead-enriched soils too, as in the case near Arnoldstein (Austria) where its concentration in the upper soil layer amounts to 8–10 g/kg due to lead emission from lead ore deposit (Kilian, 1982). In both cases, the roots are a primary target of this heavy metal. Root growth inhibition is the primary response of the plant exposed to heavy metals.

Although the root growth inhibition by lead was shown and its accumulation by some plants was measured (Breckle, 1991; Davies, 1991), the relationship between root growth in terms of cell division, elongation and branching patterns as affected by lead distribution and accumulation was not studied at all. Up to now, it is unclear what is the contribution of inhibition of either cell division or cell elongation to the retardation of the root extention and alteration of root system architecture. The relative sensitivity of each growth process to lead treatment has not been compared on the same plant.

Here we have compared the lead effects on these processes in the roots of two cereal plants.

Materials and methods

The seeds of maize (*Zea mays*, cv. Diamant) and barley (*Hordeum vulgare*, cv. Krona) germinated in enameled trays on filter paper moistened with distilled water in dark thermostat at 25 ° and 20 °C, respectively. Then the seedling roots were exposed to lead nitrate solutions in Petri dishes on filter paper for various time periods at 27 °C (maize) or 20–21 °C (barley) in the dark. In outdoors soil culture experiments with barley, the plants were grown in various mixtures of Pb-enriched soil with standard soil (Sobotik et al., 1998). The experiments were performed with lead nitrate (analytically pure grade).

The roots of maize and barley of 20–30 mm in average were used for cell growth analysis. Root length, length of root-branching zone and number of lateral roots were recorded. Mature cell length, meristem

* FAX No: 7(095)9778018. E-mail: ivanov@ippras.ru

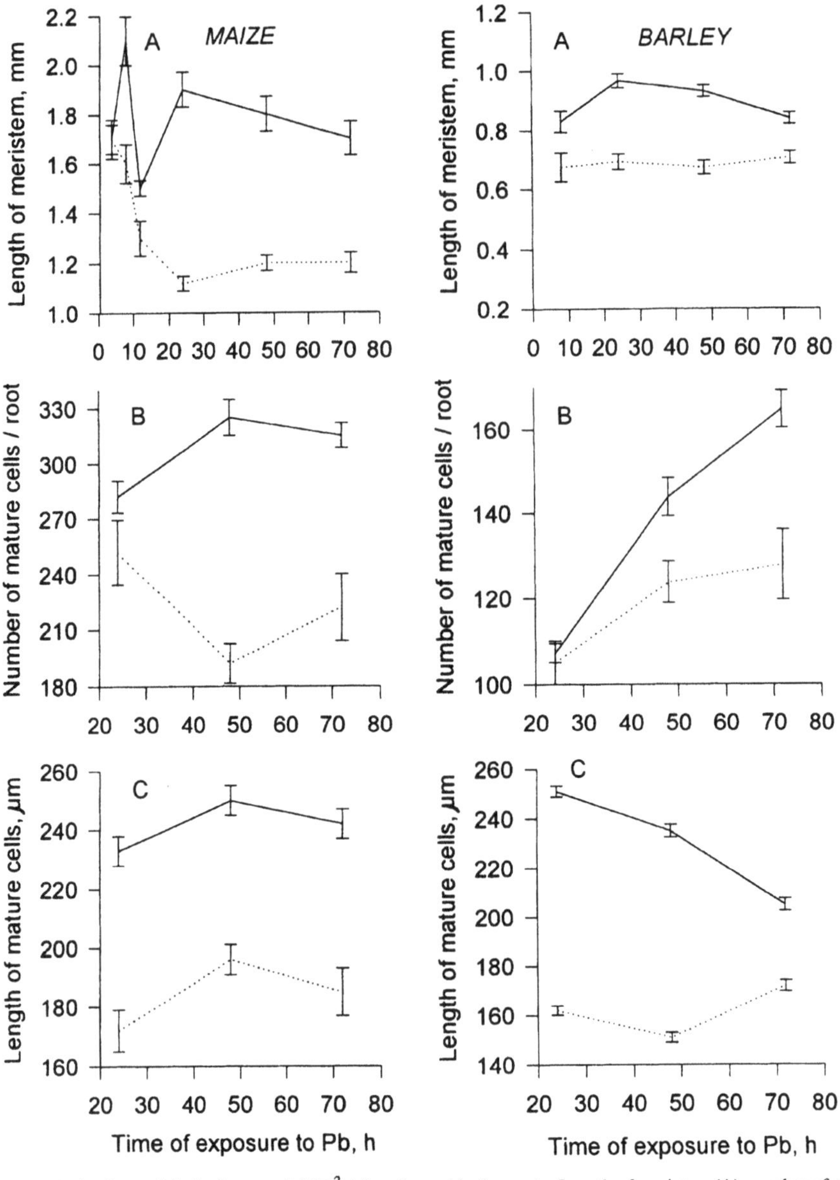

Figure 1. Cellular analysis of growth in lead-exposed (10^{-3} *M*) maize and barley roots. Length of meristem (A), number of cells along daily root increment in length (B) and length of fully-elongated cells (C) were measured in each of 20 roots along the 3rd cortical file. Control roots are shown with solid line, Pb-treated roots with dashed line. The results are represented as means ± se.

length and number of cells along the daily length increment were measured under microscope along the files of cortical cells on the hand-made sections of living roots (Ivanov et al., 1998; Obroucheva et al., 1998).

To reveal the location and distribution of absorbed lead histochemically, hand-made transversal sections were incubated either in dithizone (diphenylthiocarbazone) solution predissolved in the mixture of acetone:water (3:1) or in 4-(2-pyridylazo)resorcinol solution in 50 mm borax (Seregin and Ivanov, 1997). Both reagents gave similar results.

For measuring lead content in the roots and shoots, dry plant material (100 mg) was digested by 1 ml sub-boiling nitric acid followed for 10 min in a pressure-controlled microwave oven (PMD, FRG). With an AAS4 atomic absorbtion spectrometer (Carl Zeiss, DDR), an area of absorbance peak at 283.3 nm was measured (Martin, 1993). The modifier consisting of 1% palladium nitrate, saturated boric acid and 10% citric acid (1:1:1) was applied. In other experiments, after digestion and dissolving dry material, the lead content was measured by flow-injection spectrophotometry (Novikov et al., 1990).

Results

At 10^{-3} *M* $Pb(NO_3)_2$, the root growth was inhibited by half in 3 days. Higher (10^{-2} *M*) concentration was lethal for roots.

Cellular analysis of the growth inhibition by 10^{-3} *M* lead nitrate has shown (Figure 1) that root exposure to lead resulted from the inhibition of both cell division and cell elongation by about 20%. Inhibition of cell proliferation follows from the decrease in cell number in daily length increment. The shorter lengths of fully-elongated cells indicated the inhibition of cell elongation. Thus, the contribution of both processes to root shortening was almost equal, and they do not differ in their tolerance to lead. It is essential that cell elongation was curtailed during 12 h, that is when cell division remained almost unchanged. Along with metabolic disturbances, lead is known to decrease cell wall extensibility (Burzynski and Jakob, 1983). Therefore, cell elongation is directly damaged by lead independent of its effect on cell proliferation. The partial suppression of both processes indicates not a selective but toxic effect of lead on cell growth. Owing to inhibition of both cell growth processes in main, seminal and lateral roots, the length of root system was shorter in lead-treated maize and barley roots.

Another component of root architecture – root branching – was estimated only in maize roots, because, in barley seedlings, lateral roots appear much later. Number of seminal roots was unaffected by lead due to a very low rate of lead penetration through seed testa (Obroucheva et al., 1998). We compared here the length of branching zone with total length of the main root on the third day of root exposure to lead (Figure 2). The branching zone became shorter because of shorter length of mature cells situated along the main root axis. As a result of more significant reduction in root length due to inhibition of both cell division and elongation, the branching zone in treated

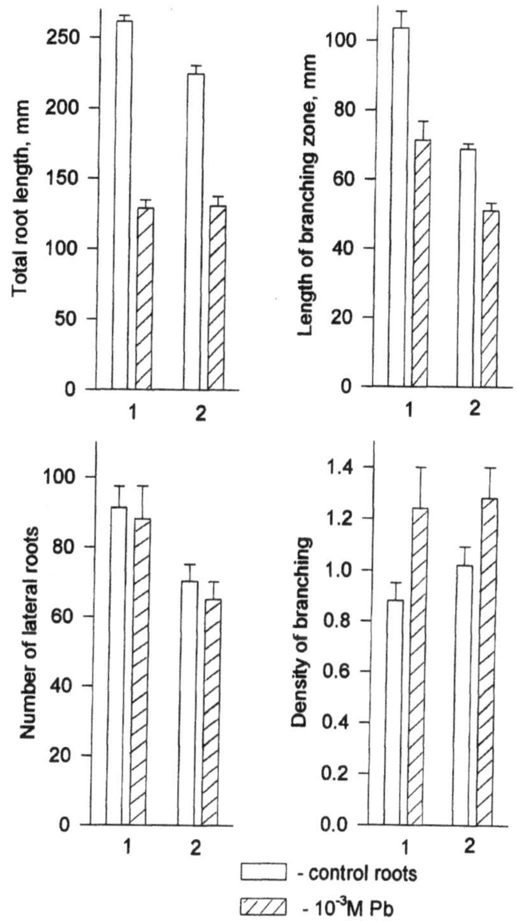

Figure 2. Root branching characteristics in control and lead-treated (10^{-3} *M*) maize roots (1,2 - two experiments). Data are given as means ± se.

roots comprised 55±5% (1st experiment) and 39±3% (2nd experiment) of total root length as compared to 40±2% (1st experiment) and 31±1% (2nd experiment) in control roots. The number of lateral roots was similar. Lead nitrate (10^{-3} *M*) did not affect the duration of lateral root development from the first division in pericycle to the appearance on root surface as calculated according to the method suggested by Ivanov et al. (1997). When branching density was calculated (number of lateral roots per branching zone length), root system of lead-exposed roots turned out to be shorter and denser than in control seedlings. Its greater compactness resulted from shorter distance between lateral roots because of shorter size of cells of main root along the branching zone. Thus, inhibition of both

Figure 3. Root systems of *Arrhenatherum elatius* plants grown in Pb-enriched soil near Arnoldstein (left, 50 cm) and Pb-depleted soil near Villach, Austria (right, 110 cm long). Photo made by Drs M. Sobotik and E. Lichtenegger.

cell division and elongation by lead ions causes the change in root system architecture: It becomes shorter and more compact. For example, such morphological modification may be vizualized on the photos of root system of *Arrhenatherum elatius* (Figure 3) grown on Pb-enriched (left) and Pb-depleted (right) soils.

Histochemical reaction for lead localization after maize root exposure for 1, 24 or 48 h to lead nitrate has shown that Pb was accumulated in the slime covering root cap and root itself and penetrated through rhizodermis to cortex, being located mainly in the cell walls. Endodermis is a barrier to further lead penetration to the stele (Figure 4). This explains the lack of injurous lead effect on the number of lateral roots. Lead simply does not penetrate through endodermis to their primordia in pericycle and cannot interfere with cell proliferation there. These results clearly indicate also that only a small fraction of root-absorbed lead reaches the xylem vessels and can be translocated upwards.

To evaluate the critical amount of lead to start the inhibition of root growth, the maize roots were exposed to lead nitrate solutions of various concentrations for various time periods, and lead content was measured in roots exhibiting 15–25% growth inhibition. The critical amount of absorbed lead to initiate the inhibition comprises 3.1–4.3 μg per root. At these lead levels in roots, no growth inhibition in shoots occurs, and lead amount per shoot is within the range of 0.2–0.4 μg, i.e. does not exceed 5–10% of lead accumulated by root. Only traces of lead were detected in shoots histochemically. Therefore, these data clearly indicate that roots play a role of barrier to lead transport to shoots.

The conclusion was confirmed by direct measurements of lead content in maize and barley roots and shoots after seed germination in water (H_2O - Pb series) or in Pb solution (Pb - Pb series) and further exposure to lead for 3 days (Table 1). At non-inhibitory 10^{-4}-10^{-5} M concentrations, lead content in roots was low, being much higher than in shoots.

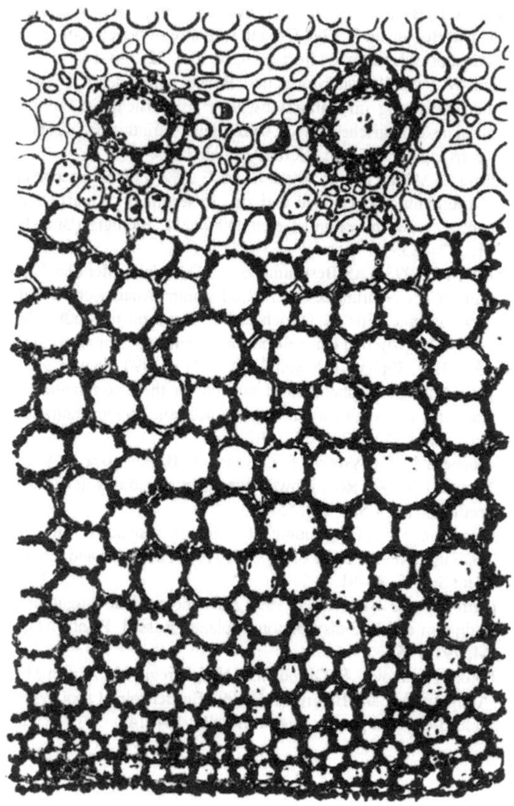

Figure 4. Schematic representation of lead distribution on the cross-section of lead-treated (10^{-3} *M*) maize root. Lead location after histochemical staining is shown by black colour.

Table 1. Lead content in roots and shoots of maize and barley seedlings exposed, after radicle emergence in water or lead, to various lead nitrate solutions (μg/g dry wt)

Treatment	Lead nitrate concentration (M)	Maize		Barley	
		Roots	Shoots	Roots	Shoots
H_2O - H_2O		3.7	0.9	36.9	5.7
H_2O - Pb	10^{-5}	9.9	5.7	39.1	9.9
Pb - Pb	10^{-5}	41.9	1.1	44.1	2.6
H_2O - Pb	10^{-4}	36.7	1.7	162.0	12.9
Pb - Pb	10^{-4}	41.2	2.8	195.0	11.7
H_2O - Pb	10^{-3}	369.0	11.2	1461.0	32.2
Pb - Pb	10^{-3}	369.0	11.3	2274.0	38.1
Pb - Pb	10^{-2}	5702.0	650.0	20525.0	869.0

At growth-inhibiting 10^{-3} *M* concentration, lead content markedly increased in roots of both plants while lead content in shoots remained at low, non-inhibitory level. Similar results were obtained by Leblova et al. (1986) with maize roots. In our experiments with toxic 10^{-2} *M* lead solution, roots absorbed very much lead and became unable to prevent its transport to shoots. At this concentration, lead penetration through endodermis to xylem vessels was also demonstrated histochemically. Higher lead accumulation by barley roots explains its greater sensitivity to lead as compared to maize.

Table 2. Lead content in barley plants grown in soil culture at various lead levels (μg/g dry wt)

94 days after planting			123 days after planting		
Soil	Leaves + stems	Ears	Soil	Leaves + stems	Ears
7371	133.0	5.9	8503	320.0	8.6
8968	82.2	7.4	10580	180.0	8.0
8300	85.0	6.9	10233	93.0	6.1
6795	58.2	4.6	10233	120.0	6.4
217	4.1	0.5	7544	62.0	4.6
193	2.5	0.0	2345	35.0	3.8
130	2.3	0.3	222	7.9	0.3
114	2.2	0.1	199	5.4	0.7
			247	5.8	0.9
			72	22.0	1.0

The barrier role of root system to lead influx may be also well illustrated by the experiments with barley plants grown in soil culture till grain filling (Table 2). Pb-enriched soil taken from Arnoldstein (Austria) was mixed in various proportions with standard humus soil. Leaves and stems contain rather low lead levels even in Pb-rich soils, up to the end of growing season. In the ears, lead content was the lowest. Some tissues in vascular bundles in the root collar, stem nodes and ear axes appear to prevent further lead translocation along the plants to the developing seeds.

The aforementioned results taken together are in accordance with the hypothesis for barrier functioning of roots with respect to lead, a heavy metal. In many cases, a high content of Pb and other heavy metals in roots and their low content in shoots were demonstrated (Leblova et al.,1986; Rauser, 1986), at least in the cereal plants. Apparently, this barrier role of root system at non-toxic concentrations of heavy metals is due not only to the presence of Casparian strips in the endodermis, but to binding of heavy metals along their way to endodermis to carbohydrates, phytochelatins, proteins and others.

Such barrier role of root system must be also taken into account when the functions of root system are discussed.

Acknowledgments

This work was supported by INTAS grant No 93-2432 (ext.) and Russian Foundation for Basic Research grant No 97-04-49219.

References

Breckle S-W 1991 Growth under stress: heavy metals. *In* Plant Roots, the Hidden Half. Eds. Y Waisel, A Eshel and U Kafkafi. pp 351–373. Marcel Dekker, New York.

Burzynski M and Jakob M 1983 Influence of lead on auxin-induced cell elongation. Acta Soc. Bot. Polon. 52, 231–239.

Davies M S 1991 Effect of toxic concentrations of metals on root growth and development. *In* Plant Root Growth: An Ecological Perspective. Ed. D Atkinson. pp 211–227. Blackwell Sci. Publ., Oxford.

Ivanov V B, Bystrova E I, Dubrovsky J G and Ploshinskaya M E 1998 Duration of lateral root formation in maize seedlings as affected by diverse factors. *In* Root Demographics and their Efficiencies in Sustainable Agriculture, Grasslands and Forest Ecosystems. Ed. JE Box. pp 777–787. Kluwer Academic Publishers, Dordrecht, The Netherlands.

Ivanov V B, Bystrova E I, Obroucheva N V, Antipova O V, Sobotik M and Bergmann H 1998 Growth response of barley roots as an indicator of lead-toxic effects. Angew. Bot. 72, 140–143.

Kilian W 1982 Waldstandorte und Böden sowie deren Schwermetallbelastung im Immissionsgebiet Arnoldstein. *In* Das Immissionsökologische Project Arnoldstein. Carinthia. Ed. G Halbwachs. pp 325–351. 11, Heft 39 (Sonderheft).

Leblova S, Mucha A and Spirhanslova E 1986 Compartmentation of cadmium, copper, lead and zinc in seedlings of maize (*Zea mays* L.) and induction of metallothionein. Biologia (Bratisl.). 41, 777–785.

Martin M-L 1993 AAS-Bestimmung von Cadmium, Blei und Thallium in der Nahrungskette. *In* Colloquium Analytische Atomspektroskopie. Eds. K Ditrich and B Wetz. pp 603–608. Universität Leipzig und Umweltforschungszentrum Leipzig-Halle.

Novikov E A, Shpigun L K and Zolotov Yu A 1990 Flow-injection extraction spectrophotometric method for the determination of lead and its combination with minicolumn preconcentration. Anal. Chim. Acta. 230, 157–162.

Obroucheva N V, Bystrova E I, Ivanov V B, Antipova O V and Seregin I V 1998 Root growth responses to lead in young maize seedlings. Plant Soil 200, 55–61.

Rauser W E 1986 The amount of cadmium associated with Cd-binding protein in roots of *Agrostis gigantea*, maize and tomato. Plant Sci. 43, 85–91.

Seregin I V and Ivanov V B 1997 Histochemical investigation of cadmium and lead distribution in plants. Russ. J. Plant Physiol. 44, 791–796.

Sobotik M, Ivanov V B, Obroucheva N V, Seregin I V, Martin M L, Antipova O V and Bergmann H 1998 Barrier role of root system in lead-exposed plants. Angew. Bot. 72, 144–147.

Recent Advances of Plant Root Structure and Function, 171–174
Edited by O. Gašparíková, M. Čiamporová, I. Mistrík and F. Baluška
© 2001 *Kluwer Academic Publishers.*

Diverse responses of root cell structure to aluminium stress

M. Čiamporová
Institute of Botany, Slovak Academy of Sciences, Dúbravská cesta 14, 842 23 Bratislava, Slovak Republic

Key words: aluminium, root cells, stress adaptation, structural response, *Zea mays*

Abstract

Plant cells respond to a certain stress factor in different ways depending on their developmental stage and type of tissue. Structural damage may be severe or even lethal in individual cells within a tissue that exhibits moderate or no effects of stress. In the case of aluminium toxicity, detailed observations of root tips of 3 day old *Zea mays* L., cv. TO360 seedlings revealed differences in the response of some cells. Two different structural changes appeared within root epidermis just behind the root cap. Cells with dark and shrunken cytoplasm occurred next to swollen cells with preserved cellular compartments. Within the root cortex, individual cells or a few cells of a file have severely damaged cytoplasm, in contrast to almost undisturbed cytoplasm of adjacent cells. Such extremely sensitive cells appear irregularly within the root apex. Their structural similarity with cells that are observed after a hypersensitive response in infected plant tissues suggests a role to accumulate aluminium, in order to allow the surrounding tissue to survive the stress.

Introduction

Stress-induced alterations of cell ultrastructure depend on both developmental and functional cell features. There are numerous examples of differences in structural responses to a certain stress factor, between cells of different age as well as between the cells belonging to different tissues. Cells of maize root epidermis were severely damaged while the neighbouring hypodermal cells revealed undisturbed structural integrity under copper stress (Ouzounidou et al., 1995). Similarly, structure of the cells of epidermis and peripheral cortex layers suffered more from aluminium than the cells in the central cylinder (Budíková et al., 1997). However, these responses could be ascribed to the location of the tissues on the root periphery and their immediate contact with the stress source, rather than to the abilities of the tissue to respond specifically. Along the root axis of aluminium stressed maize seedlings, the younger epidermal cells at about 2–20 mm from the root tip undergo serious structural damage while the older cells more proximally located do not accumulate aluminium or callose, and their structure remains unaffected (Budíková, 1999).

In our previous experiments, individual severely damaged cells with prominent deposits in their cell walls appeared within the cortex and stelar parenchyma of maize roots exposed to copper stress (Ouzounidou et al., 1995). Increased concentration of Al^{3+} in the root medium induced severe damage to the ultrastructure of peripheral root cap cells, cells of epidermis and of peripheral cortical layers within the meristematic and elongation root zones, while the cells of more centripetally located tissues revealed only higher vacuolation and irregularly thickened cell walls (Bennet et al., 1985; Ikeda and Tadano, 1993; Marienfeld et al., 1995). The aim of this paper is to point to the phenomenon of diverse stress responses in individual cells within the tissues of aluminium stressed maize roots.

Material and methods

Seeds of *Zea mays* L., cv. TO360 germinated on filter paper at 25 °C for 72 h in darkness. Seedlings with primary roots 40–50 mm long were transferred into the solution of 0.1 mM $CaCl_2$, (control) or exposed to

* Fax No: 00421-7-547 71948. E-mail: botuciam@savba.savba.sk. Tel: 00421-7-5941 2286 or 00421-7-547 75829

the solution of 50 μM $AlCl_3$ in 0.1 mM $CaCl_2$, pH 4.5 (Budíková et al., 1997) for 24 h. Five apical, 1 mm long segments of the primary roots were fixed with 3% glutaraldehyde and postfixed with 1% osmium tetroxide (both in 0.1 M Na-cacodylate buffer, pH 7.2), dehydrated in ethanol series followed by propylene oxide and embedded in Epon-Araldite resin. Semi-thin longitudinal and cross sections were stained with Toluidine blue. Ultra-thin sections stained with uranyl acetate and lead citrate were viewed with an electron microscope Tesla BS500.

Results and discussion

The microscopical observations were performed with the material characterised in our previous work (Budíková and Čiamporová, 1998). Primary root elongation of intact maize seedlings was 1.25 mm h^{-1}. The elongation of roots exposed to 50 μM $AlCl_3$ was reduced to 61% and 43% of the control plants during the 24 h and 48 h of treatment, respectively. Both in the control and in treated plants, the elongation zone was restricted to the apical 7 mm. The highest degree of the reduction of elongation rate occurred between the 2nd and 4th mm behind the tip of the root cap.

In accord with our previous results (Budíková et al., 1997), the exposure of maize primary roots to aluminium for 24 h resulted in an increased vacuolation of the cells, particularly in the apical region of the primary root. Epidermal and cortical cells were mostly shorter and wider than the cells in control roots (Figure 1 A,B). This could indicate that next to decreased mitotic activity (Clarkson, 1965), the change in orientation of the cell enlargement contributes to the inhibition of the root elongation growth. In the more proximal parts of the roots (1.5–20 mm behind the root tip), seriously damaged and dead cells of epidermis and peripheral cortex layers appeared (Budíková and Čiamporová 1998).

Detailed investigation of the root tissues revealed some diversity of structural responses within epidermis and cortex. As has already been shown (Budíková et al., 1997), two kinds of structural responses occurred in maize root epidermis. There were cells with extremely dense cytoplasm in which cellular organelles could not be distinguished (Figure 1C). Next to these cells, swollen epidermal cells with preserved cytoplasmic compartments were found in cross sections, as well as in longitudinal files (Figure 1 B,C). These diverse structural responses occur in very young epidermal cells close to the root apex. Swollen or dense cells can be discerned as soon as they cease to be covered by root cap cells (less than 1 mm behind the root tip). Structural observations are not sufficient to explain the reason for such diverse responses of the epidermal cells. The differentiation of epidermis into trichoblasts and atrichoblasts as a possible reason is probably irrelevant, because in maize roots all epidermal cells are potential trichoblasts. At the distance of about 1.5 mm behind the root tip, some of the epidermal cells undergo complete disintegration, while the neighbouring cells still retain their structural integrity (Figure 1D). More distant from root apex, up to about 25 mm, all epidermal cells undergo complete disintegration and sometimes also outer cortical cells are dead. Recent results of Blancaflor et al. (1998) provide a possible explanation, namely that of distortion of the dynamics of microtubules by aluminium in epidermal and outer cortical cells, but not in the tissues located more centripetally in the root. The morphology of epidermal cells occurring proximally to the distance of about 25 mm remained undisturbed (Budíková et al., 1997).

Individual cells being damaged, senescing or dead were not observed in the roots of control plants. This is in accord with the results of Wenzel and McCully (1991) who did not find enucleated and early senescing cells in longitudinal sections of intact maize, wheat, barley and oat roots. Within the cortical tissue of aluminium treated roots, some individual cells or a few cells within a file (Figure 1E) revealed high sensitivity to aluminium toxicity. Their cytoplasm was severely damaged, while the cytoplasm of adjacent cells did not show significant alterations (Figure 1F). Such cells or cell files seemed to appear irregularly within the root cortex. They appeared in very young cortical tissue, within the meristematic region of the root apex. Aluminium induced drastic changes in morphology of some cells were mentioned in barley root cortex (Figure 16, Ikeda and Tadano, 1993). Appearance of non-uniform responses of cortical cells in Al-treated wheat roots was reported by Votrubová et al. (1997). Individual cells with highly sensitive response were found in the cortex and stelar parenchyma of maize roots exposed to copper stress (Ouzounidou et al., 1995). Moreover, the copper stressed cells had extremely thickened cell walls and large deposits of a dense material between plasmalemma and the cell wall. The walls of cortical cells treated with aluminium were thickened to a lesser extent (Budíková et al., 1997).

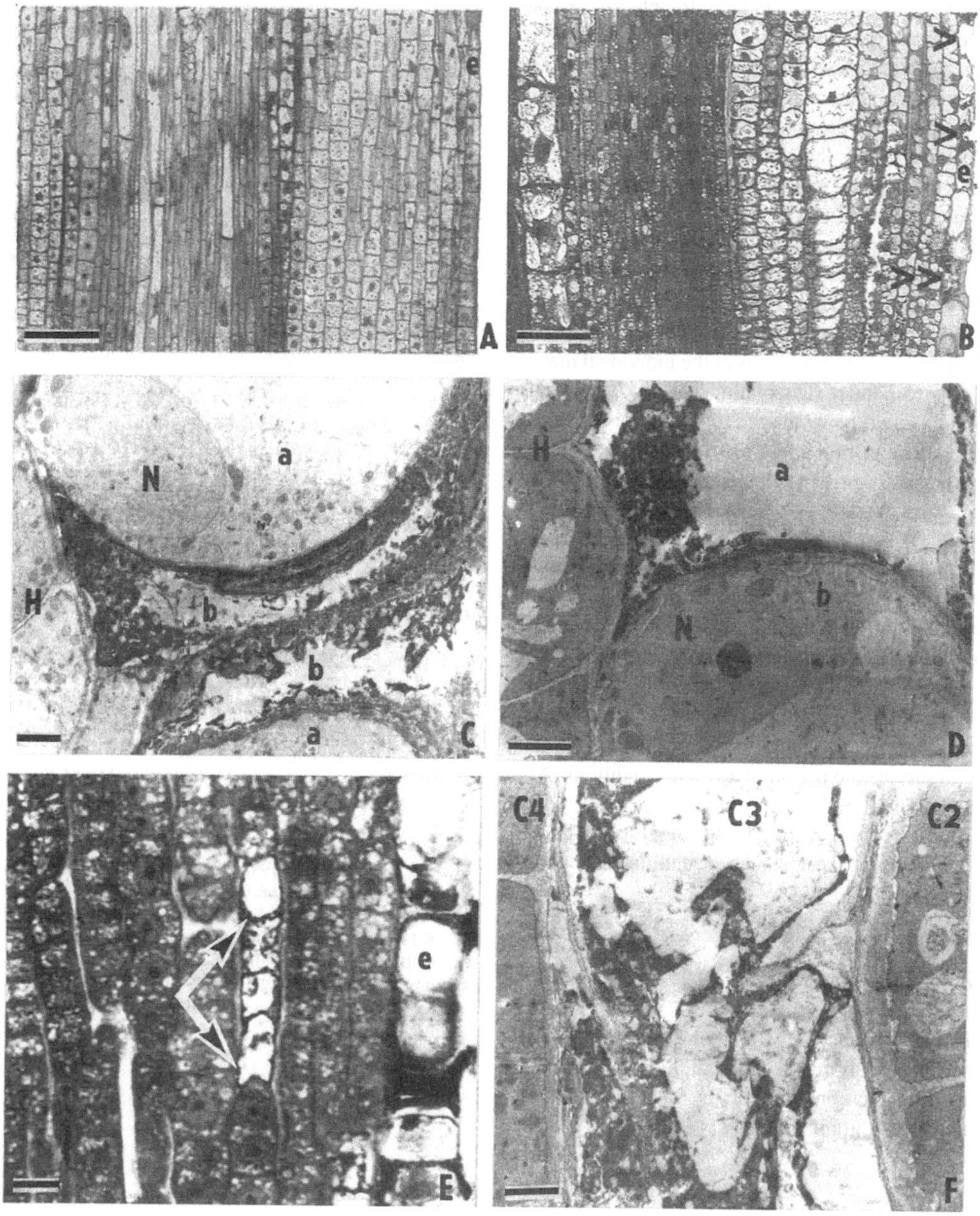

Figure 1. Light and electron microscopic documentation of maize root cells between 1 and 2 mm behind the root tip. Control (A) and aluminium treated (B) root segments. In the root epidermis (e) of the stressed roots disrupted cells (>) and cells with extremely dense cytoplasm (≫) occurred. Bars represent 100 μm. (C) Swollen epidermal cells (a) with preserved cytoplasm and organelles and two cells (b) with dense cytoplasm. (D) Complete distortion of the protoplast (a) and well preserved cytoplasm in the next epidermal cell (b). N=nucleus, H=hypodermal cells. Bars represent 3 μm. (E) Four seriously damaged cells in a file (arrows) of the root cortex surrounded by relatively unaffected cortical tissue. Cells in the epidermal layer (e) reveal diverse structural responses. Bar represents 20 μm. (F) Higher magnification of the damaged cells (C3) in the cortex while the cytoplasm integrity of the cells in the neighbouring hypodermal (C2) and fourth cortical layers (C4) remain preserved. Bar represents 2 μm.

It is difficult to explain the reason for the occurrence of the individually responding cells of similar age within a certain tissue. In *Vicia faba* root cortical tissue, the cells differed in the stage of their cell cycle and in enzyme characteristics, revealing an irregular 'mosaic pattern' of enzyme localisation (Beneš, 1983). This could indicate that the cells of the same tissue may not be identical and they may respond to a stress stimulus in different ways.

The individual cells of Al treated maize roots usually undergo drastic structural changes leading to distortion of their compartmentation and apparently also of their functions. With respect to ultrastructural appearance, the responses of the individual maize root cells to copper (Ouzounidou et al., 1995) and aluminium could be compared to a hypersensitive response (HR) e.g. in *Arabidopsis* leaves infected with *Pseudomonas syringae* (Pennel and Lamb, 1995). The authors consider HR as a form of programmed cell death (PCD). Salt stress could also trigger cell death with some apoptosis-related features in an immature region of barley roots. The specific TUNEL reaction for proving the DNA fragmentation characteristic of PCD was positive in most, but not all, nuclei in these root tissues (Katsuhara, 1997). Possibly, some cells undergo rapid disintegration associated with preferential ion accumulation, thus enabling the rest of the tissue to survive the period of adverse conditions. Application of specific tests to Al-stressed maize roots would be necessary to support the suggestion of such kind of protection mechanism at cellular and tissue levels.

Acknowledgement

I thank M. Ovečka, F. Čiampor jr. and V. Polák for their help with micro-photography and computer-aided documentation. Support by the Slovak Grant Agency VEGA, grant No. 5047 is acknowledged.

References

Beneš K 1983 Cell cycle aspects of differential enzyme localization in root tips. *In* Progress in Cell Cycle Controls. Proc. 6th Eur. Cell Cycle Workshop. Eds. J. Chaloupka, A. Kotyk and E. Streiblová. pp. 131-132. Institute of Microbiology, Czechoslovak Academy of Sciences, Praha.

Bennet R J, Breen C M and Fey M V 1985 The primary site of aluminium injury in the root of *Zea mays*. S. Afr. J. Plant Soil 2, 8-17.

Blancaflor E B, Jones D L and Gilroy S 1998 Alterations in the cytoskeleton accompany aluminium-induced growth inhibition and morphological changes in primary roots of maize. Plant Physiol. 118, 159-172.

Budíková S 1999 Structural changes and aluminium distribution in maize root tissues. Biologia Plantarum 42, 259-266.

Budíková S and Čiamporová M 1998 Growth and structural responses of maize primary roots to aluminium stress. In Progress in Botanical Research. Eds. I Tsekos and M Moustakas. pp 419-422. Kluwer Academic Publishers, Dordrecht, The Netherlands.

Budíková S, Čiamporová M, Ovečka M and Polónyi J 1997 Structural characterisation of maize root tip cells under aluminium stress. Acta FRN Univ. Comen. 29, 47-52.

Clarkson D T 1965 Effects of aluminium and some other trivalent metal cations on cell division in the root apices of *Allium cepa*. Ann. Bot. 29, 309-316.

Ikeda H and Tadano T 1993 Ultrastructural changes of the root tip cells in barley induced by a comparatively low concentration of aluminium. Soil Sci. Plant Nutr. 39, 109-117.

Katsuhara M 1997 Apoptosis-like cell death in barley roots under salt stress. Plant Cell Physiol. 38: 1091-1093.

Marienfeld S, Lehmann H and Stelzer R 1995 Ultrastructural investigations and EDX-analyses of Al-treated oat (*Avena sativa*) roots. Plant and Soil 171: 167-173.

Ouzounidou G, Čiamporová M, Moustakas M and Karataglis S 1995 Responses of maize (*Zea mays* L.) plants to copper stress I. Growth, mineral content and ultrastructure of roots. Environ. Exp. Bot. **35**, 167-176.

Pennel R I and Lamb C 1997 Programmed cell death in plants. The Plant Cell 9, 1157-1168.

Votrubová O, Černohorská J, Hlídková R, Krobová J and Eliášová K 1997 The effect of aluminum on growth and structure of wheat root. J. Appl. Genet. 38B, 277-281.

Wenzel C L and McCully M E 1991 Early senescence of cortical cells in the roots of cereals. How good is the evidence? Am. J. Bot. 78:1528-1541.

Section editor: H Lambers

Recent Advances of Plant Root Structure and Function, 175–186
Edited by O. Gašparíková, M. Čiamporová, I. Mistrík and F. Baluška

© 2001 Kluwer Academic Publishers.

Relationship between microtubule orientation and patterns of cell expansion in developing cortical cells of *Lemna minor* roots

S. Inada and S. Sato[1]
Department of Biology and Earth Sciences, Faculty of Science, Ehime University, Bunkyo-cho 2-5, Matsuyama 790-8577, Japan. [1]*Corresponding author**

Key words: cell wall, cortical cells, cortical microtubules, immunofluorescence, *Lemna minor*

Abstract

In actively growing cortical cells in the elongation zone of *Lemna minor* L. roots, both longitudinal (radial and tangential) and transverse walls expand in both length and width. The longitudinal walls of the three types of cortical cells in the root (i.e. outer, middle and inner) showed the largest expansion in the longitudinal axis. In contrast, the inner cortical cells exhibited the least expansion in width, whereas the middle cortical cells displayed the largest expansion in width. Thus, the profiles of the expansion of longitudinal walls were characteristic for the three types of cortical cells. In this study, both the orientation of cortical microtubule (MT) arrays and their dynamic reorientation, and the density of cortical MTs, were documented and correlated to the patterns of cell wall expansion. Significantly, transverse arrays of cortical MTs were most prominent in the radial walls of the inner cortical cells, and least so in those of the middle cortical cells. Toward the base of roots, beyond the elongation zone, the orientation of cortical MTs shifted continuously from transverse to oblique and then to longitudinal. In this case, the rate of shift in the orientation of cortical MTs along the root axis was appreciably faster in the middle cortical cells than in the other two types of cortical cells. Interestingly, the continuous change in cortical MT orientation was not confirmed in the transverse walls which showed much smaller two-dimensional expansion than the radial walls. Additionally, the presence of fragmented or shortened cortical MTs rapidly increased concomitantly with the decrease of transversely oriented cortical MTs. This relationship was especially prominent in the transverse walls of the inner cortical cells, which displayed the least expansion among the three types of cortical cells investigated. In the root elongation zone, the density of cortical MTs in the inner cortical cells was about three times higher than that in the other two cortical cell types. These results indicate that in the early stage of cell expansion, the orientation of cortical MTs determines a preferential direction of cell expansion and both the shifting orientation and density of cortical MTs affect the magnitude of expansion in width of the cell wall.

Introduction

Evidence from numerous studies has demonstrated that cortical MTs are involved in the control of plant cell morphogenesis. In rapidly elongating roots of *Raphanus sativus*, MTs in cortical cells of the elongation zone are oriented perpendicular to the root axis, whereas in non-elongating cells these cortical MTs are predominantly oblique (Traas et al., 1984). Similar observations have also been reported in other plants such as pea (Hogetsu and Oshima, 1986), *Nitella* (Wasteneys and Williamson, 1987) and maize (Baluska et al., 1992). Furthermore, cell growth is isodiametric when cortical MTs are either randomly oriented or disrupted with drugs (Hogetsu and Shibaoka, 1978; Melan, 1990; Shibaoka, 1991). Collectively, these data suggest that the orientation of cortical MTs plays an important role in deciding the direction of cell expansion because cell expansion is typically perpendicular to the orientation of cortical MTs.

In roots of *Azolla pinnata*, the interpolation of cortical MTs diminished or ceased in cortical cells which were expanding in all dimensions (Hardham and Gunning, 1979). In contrast, the density of cortical MTs increased rapidly in elongating sieve elements but diminished when these cells ceased to elongate

* FAX No: 899279630. E-mail: ssato@sci.ehime-u.ac.jp

(Hardham and Gunning, 1979). Traas et al. (1984) also reported that the density of cortical MTs in meristematic cells was high compared to that in cortical cells that had ceased expansion. These results parallel the observations of Seagull (1986) who showed that cortical MTs in elongating cotton fibres are predominantly transverse but change their orientation to oblique and eventually to longitudinal when the elongation rate of the cotton fibers decreases.

In contrast, cortical MTs are oriented parallel to the long axis in tip-growing root hairs which nevertheless grow longitudinally (Emons, 1982, 1989; Emons and Mulder, 1998; Traas et al., 1985). This observation is not consistent with the widely accepted idea that the cortical MT array controls the arrangement of cellulose microfibrils, thus limiting the direction of cell elongation by the tensile strength of the microfibrils. It is interesting to note that cortical MTs change orientation at right angles to the direction of externally applied mechanical stress (Fischer and Cyr, 1998; Fischer and Schopfer, 1998). This finding implies that mechanical stress can induce a new orientation of the cortical MT array. Baskin et al. (1994) indicated that the MT-stabilizing drug, taxol, allowed some radial expansion of cells even though cortical MTs were transversely oriented. In view of these results, the orientation of cortical MTs may not be primarily responsible for determining the pattern of cell expansion through controlling the orientation of cellulose microfibrils. Therefore, various aspects of the cortical MT array still need to be studied and correlated with the patterns of cell expansion. The roots of duckweed are appropriate for this purpose because they display a simple organization of tissues. In particular, the three morphologically-distinct layers of cells which constitute the cortex (inner, middle and outer cortical cells), and the endodermis, show characteristic profiles of cell expansion. In the present study, we examined relationships between the patterns of cell expansion in these cortical cell types and aspects of the cortical MT array, such as its orientation and dynamic reorientation, and density. The results are discussed in light of how cortical MT arrangements may be involved in the process of cell expansion.

Materials and methods

Plant culture

Plants of duckweed, *Lemna minor* L., were collected from a pond in the greenhouse of the Faculty of Science, Ehime University. The plants were surface sterilized and cultured at 23–25 °C under 25–30 μmol m^{-2} S^{-1}continuous light (white day-light fluorescent lumps) in a mineral medium (Hillman, 1961) containing 10 g l^{-1} sucrose but lacking tartaric acid. The number of plants increased linearly with time to reach 150–200 individuals after 20 days. At this time, roots had attained 4.5–5.0 cm in length and were mostly curved. In this study, 1.0–1.5 cm young straight roots were used.

Light microscopy

Root tips of 3 mm in length were excised and cut into 1 mm segments. These segments were fixed overnight at 4 °C in 40 g l^{-1} glutaraldehyde in 67 mM phosphate buffer (pH 7.17). The fixed segments were rinsed in distilled water, briefly pre-stained with 5 mg ml^{-1} toluidine blue to contrast cytoplasm, rinsed in distilled water again, and embedded in Spurr's resin (Spurr, 1969). Longitudinal and cross sections about 1 μm thick were made with an LKB Ultramicrotome and the sections transferred onto clean glass slides. The sections were then post-stained with 100 mg ml^{-1} methyl violet to contrast cell walls. Cell length and width, and distance from the root apex, were measured with an Olympus VM-30 video micrometer. To avoid underestimating cell length and width, serial sections were made and a section showing cell files with maximum cell width was selected for these measurements.

Immunofluorescence microscopy

The protocol for embedding with polyethylene glycol (PEG) 1540 was basically the method of Brown et al. (1989). Briefly, 4 mm-long root tips were fixed overnight at 4 °C in 40 g l^{-1} paraformaldehyde in 67 mM phosphate buffer (pH 7.17) containing 1 g l^{-1} Nonidet P-40 (NP-40) and 100 g l^{-1} dimethyl sulfoxide (DMSO), rinsed in buffer, then dehydrated through a graded ethanol series (0.12 0.24, 0.59, 0.75 g ml^{-1}, and absolute ethanol). They were incubated in 1:3, 1:1, and 3:1 (v/v) PEG:ethanol, then in absolute PEG for 30 min each, and kept in fresh absolute PEG in a constant-temperature incubator at 60 °C for 30 min. The PEG was then cooled to room temperature to harden. Ten μm-thick longitudinal sections were cut with a rotary microtome and transferred onto glass slides using forceps. The sections were rinsed in three changes of distilled water for 5 min each to remove the PEG and then air dried. The

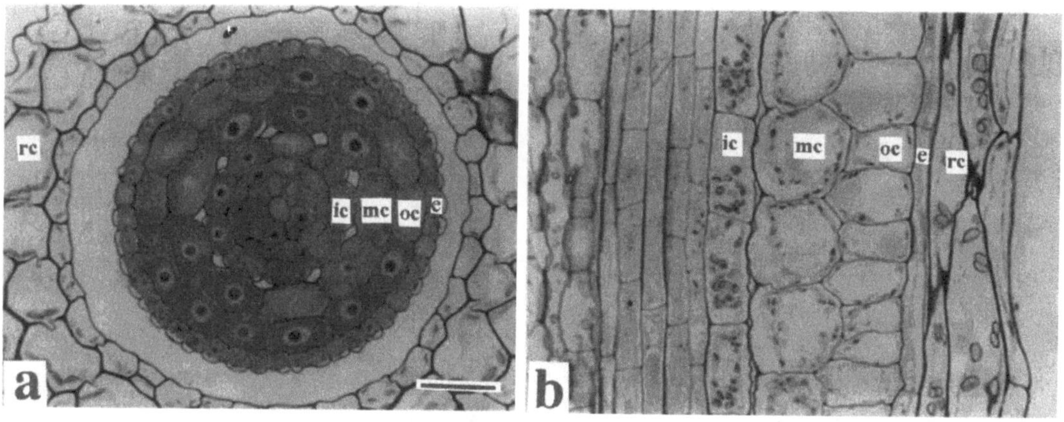

Figure 1. Light micrographs showing sections of *Lemna minor* roots. (a) Cross-section approximately 250 μm from the root apex. The root cap (rc) encloses the root. The root consists of epidermis (e), three layers of cortical cells, endodermis and central cylinder. Bar, 20 μm. (b) Longitudinal section of the elongation zone. The three types of cortical cells have distinctive morphologies. e–epidermis, oc–outer cortical cell, mc–middle cortical cell, ic–inner cortical cell, rc–root cap.

sections were then processed for immunostaining as follows. First, the sections were treated with detergent solution consisting of 10 g l^{-1} NP-40, 5 mM ethylene glycol-bis [β-aminoethyl ether]-N,N,N',N'-tetraacetic acid (EGTA), 0.5 mM phenylmethylsulfonyl fluoride (PMSF) and 1 mg ml^{-1} leupeptin in 67 mM phosphate buffer (pH 7.0), and then with blocking solution containing 0.5 g ml^{-1} normal goat serum, 0.5 mM PMSF, 1 mg ml^{-1} leupeptin and 1 g l^{-1} sodium azide in 67 mM phosphate buffer (pH 7.0). The sections were rinsed in PBS (136 mM NaCl, 6 mM Na_2HPO_4, 2.7 mM KCl, 1.5 mM KH_2PO_4, pH 7.0) containing 0.01% NP-40, and incubated overnight at 37 °C in a primary antibody (mouse monoclonal antibody raised against chick brain β-tubulin, Amersham) diluted 1:50 in PBS (pH 7.0). After washing in PBS, the sections were incubated in darkness for 1 h at 37 °C with FITC-conjugated secondary antibody (goat anti-mouse IgG, Sigma) diluted 1:20 in PBS. The sections were rinsed three times in PBS and mounted in a solution consisting of 9 volumes of glycerol: volume of PBS containing 100 mg ml^{-1} 1, 4-diazabicyclo [2,2,2] octane (DABCO) to suppress fading. The preparations were examined using a Nikon epi-fluorescence microscope, and photographs were taken with Kodak Tri-X pan 400 film. The orientations of cortical MTs were classified into three categories according to their angle relative to the long axis of the cell; transverse (T, 60-120 °), oblique (O, 30-60 ° and 120-150 °), and longitudinal (L, 0-30 ° and 150-180 °).

Some specimens were embedded in methacrylate resin for immunostaining according to a slightly modified method of Baskin et al. (1992). Apical 4 mm segments of roots were fixed overnight at 4 °C in 40 g l^{-1} paraformaldehyde in 50 mM piperazine-N,N'-bis[2-ethanesulfonic acid] (PIPES), pH 7.0, containing 5 mM EGTA and 2 mM $MgSO_4$. After rinsing in 100 mM PIPES containing 5 mM EGTA, 2 mM $MgSO_4$ and 1 mM dithiothreitol (DTT), the root segments were dehydrated through a graded ethanol series (10, 30, 50, 70, 90 g ml^{-1} ethanol containing 1 mM DTT, and absolute ethanol containing 10 mM DTT). The following steps for embedding were carried out at 4 °C. The methacrylate mixture was made by mixing four volumes of buthylmethacrylate, one volume of methylmethacrylate, 5 mg ml^{-1} benzoin ethyl ether and 10 mM DTT. The samples were passed through an ethanol-methacrylate mixture series, 2:1, 1:1, 1:2 (v/v) ethanol:methacrylate for 30 min each and then through three changes of absolute methacrylate mixture for 1 h each, and kept in the final change for at least 24 h. Polymerization was done overnight by ultraviolet irradiation using a TUV-200 ultraviolet polymerizer (Dosaka EM Co., Japan). Longitudinal and cross sections were made with a rotary microtome and placed on glass slides coated with 1 mg ml^{-1} poly-L-lysine. The embedding resin was removed from the sections through three changes of absolute acetone and then 50% acetone for 5 min each. The procedures for immunostaining of microtubules in these sections were as described above, except that in this case the primary antibody was a 1:1 mixture of mouse monoclonal an-

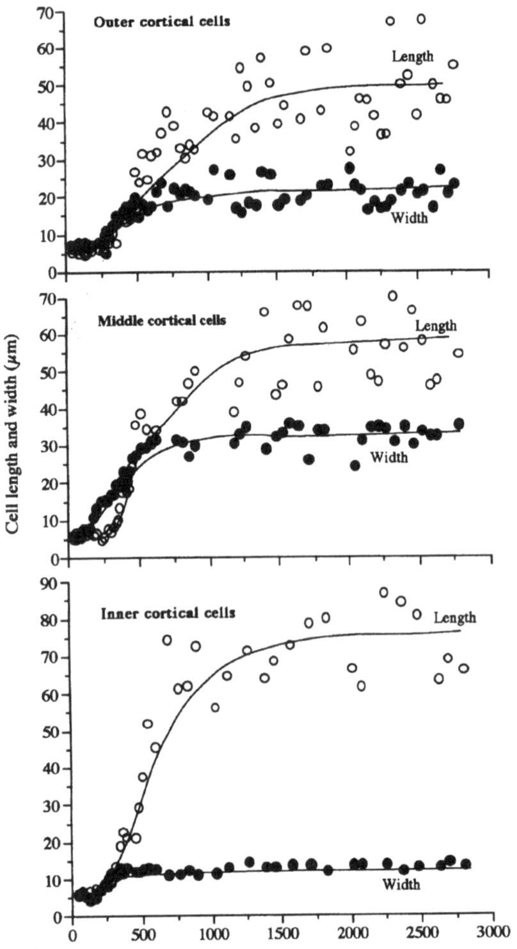

Figure 2. Relationship of cell length and width of radial walls of the three types of cortical cells relative to their position along the longitudinal axis of the root. The length and width of cells measured from longitudinal sections of roots are plotted against distance from the root apex. Data are from three plants, and the solid curves were fitted by eye.

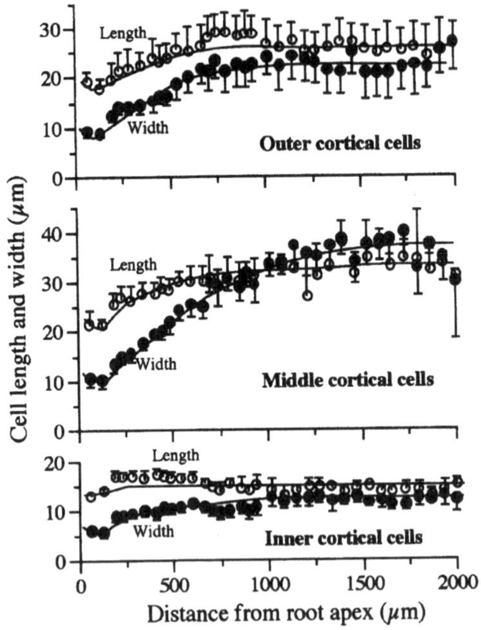

Figure 3. Relationship of cell length and width of transverse walls of the three types of cortical cells relative to their position along the longitudinal axis of the root. Each cross-section exhibited 15, 9 and 9 cells, respectively, for the outer, middle and inner cortical cells. Hence, data is represented by an average value and standard deviation. Data are from three plants, and the solid curves were fitted by eye.

tibodies raised against chick brain α- and ß- tubulin (Amersham) and used at a final dilution of 1/20 in PBS.

Transmission electron microscopy

Apical 4 mm segments of roots were fixed in 40 g ml^{-1} glutaraldehyde in 67 mM phosphate buffer (pH 7.17) and then post-fixed in 10 g l^{-1} osmium tetroxide in 33 mM phosphate buffer (pH 7.17). The specimens were dehydrated and embedded in Spurr's resin (Spurr, 1969). Ultrathin sections were prepared with an LKB Ultramicrotome, double-stained with uranyl acetate and lead citrate, and examined using a JEOL JEM-2000EX transmission electron microscope operated at 80 kV. The number of cortical MTs in selected cells was counted by direct observation using the electron microscope and the density of cortical MTs (expressed per μm of cell membrane) was estimated.

Results

Root tissues

The root cap consists of three layers of cells and is connected only at the apex of a root enclosing the root 1–1.5 mm long. As reported by Melaragno and Walsh (1976), the root is composed of seven layers of cells; the epidermis, three cortical layers (outer, middle and inner cortical cells), the endodermis, sieve elements, and xylem (Figure 1a). The meristem close to the root apex is composed of small, flat cells. In contrast, cortical cells at least 600 μm from the root apex have a distinctive morphology specific to each tissue type

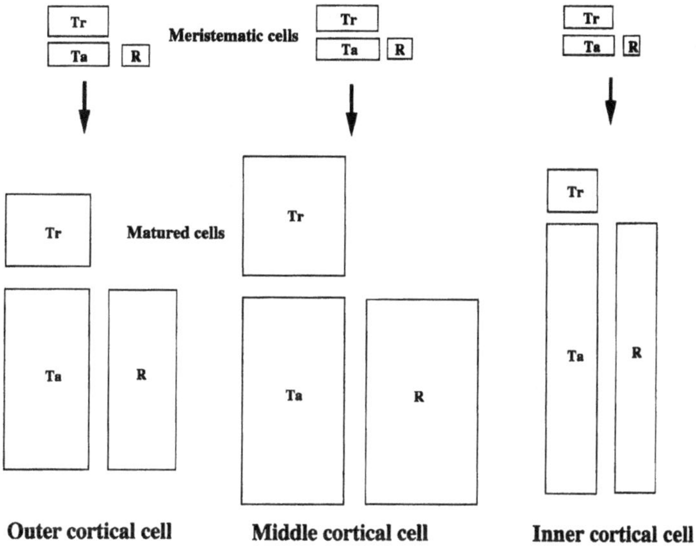

Figure 4. Schematic representation of growth of radial (R), tangential (Ta) and transverse (Tr) walls in meristematic and mature cells. See text for details.

(Figure 1b). The cortical cells were appreciably larger than other cell types, and hence a long span of cell files was relatively easy to obtain, and cells in each cortical layer have a characteristic morphology. In our study, therefore, we focused mainly on the cortical cells.

Measurements of cell expansion along the root axis

Both the length and width of cortical cells in longitudinal sections were measured along the root axis (Figure 2). The longitudinal cell walls of cortical cells are defined as two types, walls parallel with the surface of a root (tangential walls) and those perpendicular to the tangential walls (radial walls). As measurements were made from approximately median longitudinal sections of roots, often the radial walls of cells were exposed. Thus, we examined the radial walls. The length and width of cells were estimated as the values at the plateaus of the curves fitted by eye. Cell length rapidly increased within 1500 μm from the root apex (0–1500 μm region) for each category of cortical cells, but did not increase in length beyond 1500 μm from the root apex (Figure 2). Maximum cell length in mature cells was approximately 49, 58 and 75 μm for outer, middle and inner cortical cells, respectively. In contrast, expansion in cell width ceased at about 500 μm from the apex for outer and middle cortical cells, and at about 300 μm for the inner cortical cells (Figure 2). Thus, increases in cell width of all cortical cells only occurred in a short, distal region of the elongation zone, resulting in mature cells of greater length than width. It is noteworthy that the width of middle cortical cells increased prior to increases in cell length, whereas both length and width increased simultaneously in both the outer and inner cortical cells. The cell width of fully expanded cells was about 21, 32 and 12 μm for outer, middle and inner cortical cells, respectively (Figure 2).

Cross sectioning of roots exposes the transversal walls of cortical cells. The contour length of tangential and radial walls in these cross-sections was conventionally referred to here as cell length and width, respectively. Both the length and width of outer cortical cells increased slightly in the 0–750 μm region of the root and reached maximum values of approximately 25 μm in length and 22 μm in width, respectively (Figure 3). For middle cortical cells, cell width increased rapidly within the first 1500 μm from the root apex, reaching maximal cell width of approximately 35 μm, whereas cell length increased from about 21 μm at the root apex to approximately 32 μm at a distance of 1500 μm from root apex. The inner cortical cells displayed considerably less expansion both in length and width compared with the other two types of cortical cells. When fully expanded, the ratio of cell length to width (L/W ratio) was over 6 for the inner cortical cells, but less than 2 for the middle cortical cells. The outer cortical cells recorded an L/W ratio between these two values. These values indicate

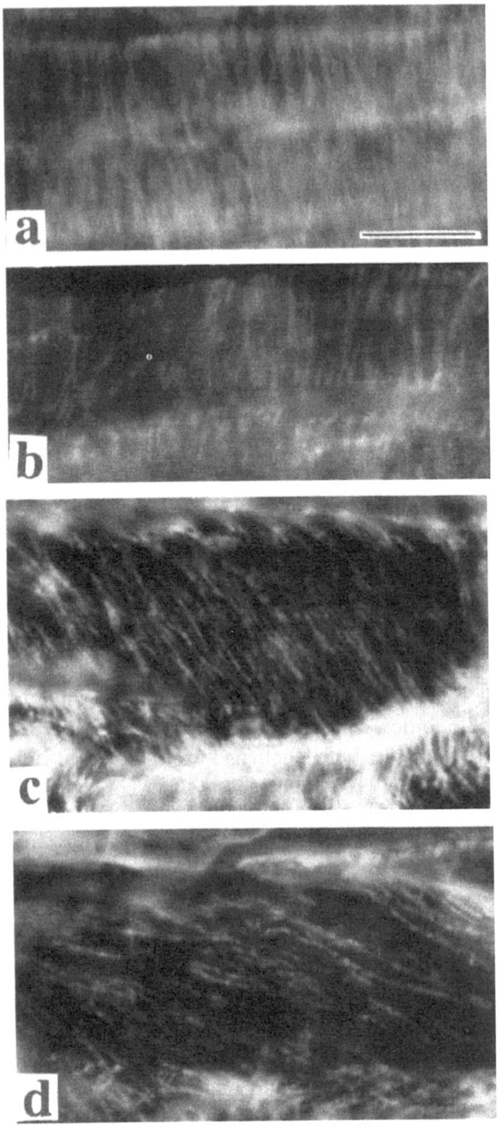

Figure 5. Immunofluorescence microscopy of methacrylate-embedded root tissue showing cortical MTs on radial walls of inner cortical cells. Representative images of cells in meristem (a), elongation zone (b) and post-elongation zone (c, d). The cortical MTs show transverse, oblique and finally longitudinal orientations in cells further along the longitudinal axis of the root. Bar, 10 μm.

that of the three types of cortical cells, the middle cortical cells exhibit the largest expansion in width, while the inner cortical cells display the least. Thus, a two-dimensional expansion of radial walls was characteristic to each type of cortical cells. In contrast, the transverse walls of each type of cortical cell displayed less expansion both in terms of cell length and cell width.

The shapes of the cell walls of cortical cells in both the meristem at the root apex and in the post-elongation zone, as determined by the data presented in Figures 2 and 3, are represented schematically in Figure 4. Although measurements of tangential walls were not performed directly, their profiles could be deduced from fixing those of both the radial and transverse walls. In general, the two-dimensional expansion of the tangential walls of each cortical cell type occurs in a very similar fashion to the radial walls, although the radial walls of outer and inner cortical cells appeared to expand less in width when compared with the tangential walls. In the outer and middle cortical cells, the transverse walls expanded considerably both in length and width, although this expansion was not as large as that of the longitudinal walls in these cells. In contrast, the transverse walls of the inner cortical cells showed only modest two-dimensional expansion.

Cortical MT arrays studied by immunofluorescence and TEM

Representative images of cortical MT arrays in longitudinal sections of roots embedded in methacrylate resin are shown in Figure 5. The orientation of cortical MTs shifted from transverse to oblique and then to predominantly longitudinal depending on cell position from the root apex. The obliquely- or longitudinally-oriented cortical MTs often appeared fragmented or shortened (Figure 5 c, d). Roots embedded in PEG were also processed for immunofluorescence localization and cortical MT arrays on radial walls were examined. Typically, and similar to the data shown for methacrylate-embedded tissue, cortical MTs were mostly oriented transverse to the root axis in the 0–1,000 μm region of the root, and mostly oblique in the region beyond 1000 μm from the root apex (data not shown). Thus, satisfactory images of cortical MTs could be obtained in roots embedded either in methacrylate resin or PEG.

The orientation of cortical MTs on radial walls of the three cortical cell types were examined in roots embedded in PEG (Figure 6). In the 0–1000 μm region, most cortical MTs of the inner cortical cells were oriented transverse to the root axis. In the same region of the root, the frequency of cortical MTs which had deviated from a transverse orientation was higher in the middle cortical cells compared to the inner cortical cells. Additionally, the frequency of transverse cortical

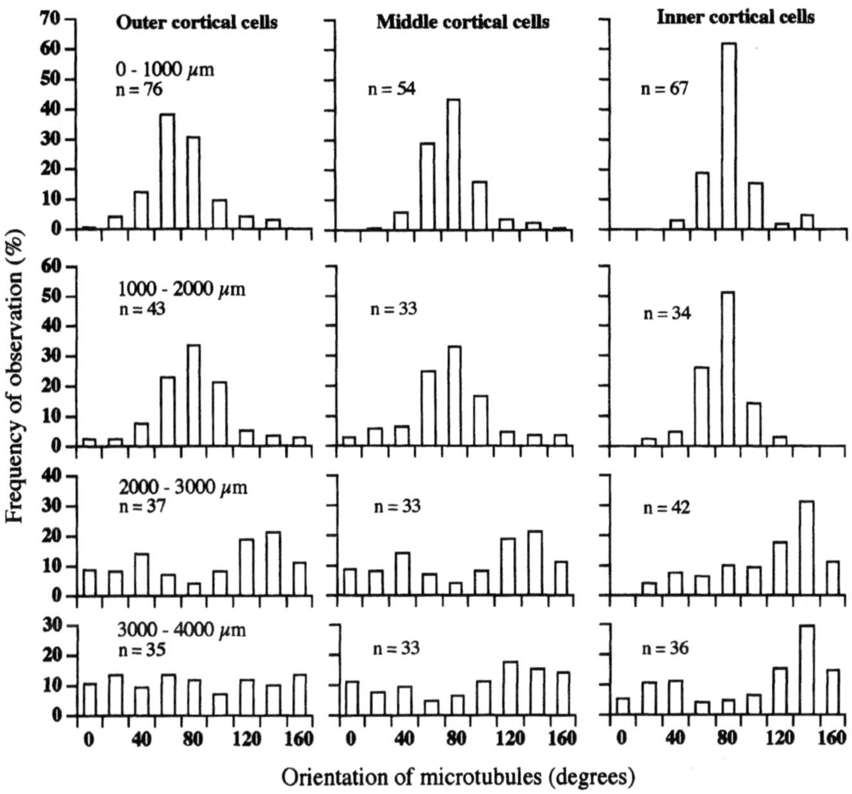

Figure 6. Frequency distribution of cortical MT orientation along the longitudinal axis of the root. Four regions of the root were examined; 0–1000, 1000–2000, 2000–3000, and 3000–4000 μm from the root apex. The orientation of cortical MTs gradually shifted from transverse to oblique, then to longitudinal, along the root axis. Data are from five plants, and the number of cells examined is indicated by 'n'.

MTs was lower in the outer cortical cells compared to the inner cortical cells. In the 1000–2000 μm region, the frequency of transversely-orientated cortical MTs was somewhat reduced compared to the more distal region for each of the three types of cortical cells, although a net transverse orientation of the cortical MTs predominated. In basipetal locations beyond 2000 μm from the root apex, the frequency of obliquely- and longitudinally-oriented cortical MTs increased in each of the cortical cell types, concomitantly with a marked decrease in the frequency of transversely-oriented cortical MTs. Of some note, the orientation of cortical MTs centered somewhat around 140 ° in the 2000–3000 μm region, especially in the inner and outer cortical cells. This may be due to examining in most instances only one of the two radial walls. This suggestion is supported by the observation that cortical MTs in the upper focal plane intersected those observed in a lower focal plane (Figure 7). These cortical MTs usually were observed to be continuous around the edges of cells, suggesting a connection of these intersected cortical MTs. We examined the orientation of cortical MTs in some cell files. This orientation shifted continuously from transverse to longitudinal at a location approximately 1000 μm from the apex (data not shown).

The transverse walls were classified into several categories according to the orientation and appearance of cortical MTs; these being transverse (T)-, oblique (O)-, longitudinal (L)-, random (R)- and fragmented (F)-types (Figure 8). The T-type was predominant in the first 500 μm of the root in all cortical cell types, and interestingly, the R-type was frequently observed in the same region (Figure 9). In the 500-1000 μm region, the T-type was again predominant in both the outer and middle cortical cells, while it was substantially reduced in the inner cortical cells, where the frequency of the L-, R- and F-types increased appreciably. The inner cortical cells with T-type cortical MTs decreased dramatically to almost 0% with

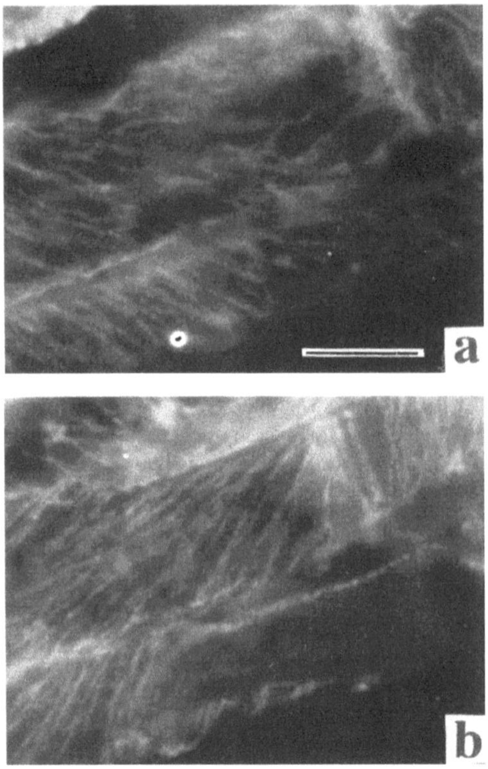

Figure 7. Immunofluorescence microscopy of methacrylate-embedded root tissue showing obliquely-oriented cortical MTs on opposite radial walls of the inner cortical cells. The cortical MTs on one side (a) intersect those on the other side (b) of the cell. Bar, 10 μm.

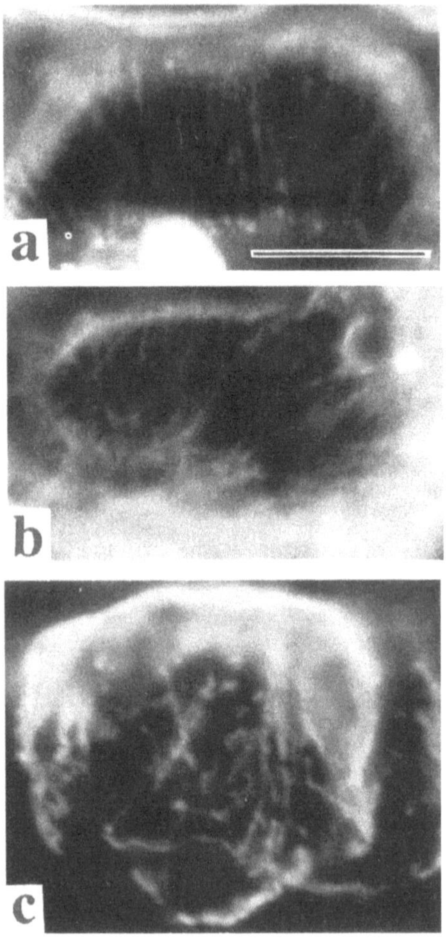

Figure 8. Immunofluorescence microscopy of methacrylate-embedded root tissue showing cortical MTs on transverse walls of the cortical cells. (a) Transverse cortical MTs in an outer cortical cell approximately 650 μm from the root apex. (b) Oblique cortical MTs in a middle cortical cell approximately 650 μm from the root apex. (c) Randomly-oriented and/or fragmented cortical MTs of a middle cortical cell approximately 400 μm from the root apex. Bar, 5 μm.

a corresponding large increase of the F-type in the 1000–1500 μm region of the root. In general, the cortical MTs of T- and O-types in the locations beyond 1000 μm from the apex became shorter and decreased in number. Most transverse walls of outer cortical cells in the 1500–2000 μm region were categorized as F-type. It is, therefore, of interest that the continuous shift in the orientation of cortical MTs as seen in the radial walls was not observed in the transverse walls.

Electron microscopy revealed profiles of cross segments of cortical MTs along the plasma membrane of the tangential walls (Figure 10 a, b). These cortical MTs were counted directly and the density of cortical MTs estimated for each type of cortical cells along the root axis (Figure 11). The density of cortical MTs in the outer cortical cells was similar to that in the middle cortical cells; density decreased almost linearly from a value of about 1.2 in the 0–1000 μm region of the root to about 0.4 in the region 3000–4000 μm from the apex. Notably, the density of cortical MTs in the inner cortical cells in the 0–1000 μm region was about three times that of the outer or middle cortical cells, and rapidly decreased toward the basipetal region. Interestingly, a high density of cortical MTs was also confirmed in the endodermal cells which increase in rate of expansion along the root axis similar to that documented for the inner cortical cells. In the 3000–4000 μm region of the root, cortical MTs of the inner cortical cells were orientated obliquely (Figure 10 c). In this case, the distance between cortical MTs was considerable compared to the spacing of MTs in the elongation zone. The density of cortical MTs on the transverse walls was about 0.3 in the 0–1000 μm region, and about 0.1–0.2 in the 1000–4000 μm region of the root.

Discussion

Some reports have suggested that cortical MTs are continuous along longitudinal walls (Cleary and Hardham 1988, Liang et al., 1996). Surveys with optical sectioning have displayed cortical MTs running continuously from one cell wall to the next in epidermal cells of *Datura stramodium* stems (Flanders et al., 1989). Using confocal scanning laser microscopy, cortical MTs visualized by microinjecting rhodamine-conjugated tubulin have been confirmed to be continuous along longitudinal walls in living epidermal cells of pea stems (Yuan et al., 1995). Furthermore, others have claimed that cortical MTs run in reverse orientation between opposite radial walls and turn along the corners of adjoining walls (Lloyd, 1984). Together, these observations suggest that cortical MTs are positioned in cells like springs running along longitudinal cell walls. This conclusion is also in parallel with the present observation that the arrays of cortical MT on the radial wall intersected those on the opposite wall.

The elongation zone of *L. minor* roots spanned between about 250 and 1500 μm from the root apex as judged by the measurements of cell length. In this zone, most cortical MTs on the radial walls of all the cortical cells were transversely oriented. Transverse cortical MT arrays have been reported in elongating cells of many plant species (Akashi and Shibaoka, 1987; Baluška et al., 1992; Hardham and Gunning, 1978; Hogetsu and Oshima, 1986; Ledbetter and Porter, 1963; Traas et al., 1984; Wasteneys and Williamson, 1987; Williamson, 1991). In the outer and inner cortical cells of *L. minor* roots, the frequency of transversely-oriented cortical MTs decreased toward the post-elongation zone simultaneously with an increase in the frequency of obliquely- and longitudinally-oriented cortical MTs. This observation is consistent with previous reports showing that oblique and/or longitudinal cortical MT arrays are seen in cells that have ceased to elongate (Baluška et al., 1992; Hogetsu and Oshima, 1986; Iwata and Hogetsu, 1988; Sakiyama and Shibaoka, 1990; Wasteneys and Williamson, 1987; Williamson, 1991). However, the differences in cortical MT orientation along the root axis were characteristic for each of the three types of cortical cells. In the elongation zone, the frequency of obliquely- and longitudinally-oriented cortical MTs in the middle cortical cells was higher than in the outer and inner cortical cells. In contrast, the transverse cortical MT array was most prominent in the radial walls of the inner cortical cells in the elongation zone. These results indicate that the orientation of cortical MTs in the early stage of cell elongation principally determines a preferential direction of cell expansion. This situation may be due to transversely-oriented cortical MTs producing tension along the width of cells, thus restricting the expansion of cells in width. It should be emphasized that the cellulose microfibrils deposited during the course of cell elongation will also interfere with the expansion of cells. It is possible that the cortical MT array may affect the expansion of cells only in a very early stage of cell growth because the cellulose microfibrils will become heavily deposited in later stages, thus overwhelming any effects of the cortical MT array on cell expansion.

In the elongation zone, the density of cortical MTs on tangential walls was much higher in the inner cortical cells than in the other two types of cortical cells. A value of about 3 was estimated for the inner cortical cells, meaning that the distance between cortical MTs was about 330 nm. In contrast, a density for cortical MTs of about 1 was estimated for the outer and middle cortical cells, indicating that the distance between MTs in these cells was about 1000 nm. The expansion in width of the latter two types of cortical cells was larger than that of the inner cortical cells. A relatively dense population of cortical MTs was also seen in the endodermal cells which showed much greater expansion in length than in width. These data indicate that the density of cortical MTs is implicated in the degree of expansion perpendicular to the orientation of these MTs. This hypothesis is reasonable, since the dense cortical MT population will produce a strong tension along the orientation of cortical MTs, thus restricting cell expansion in the direction parallel to the cortical MTs. In contrast, a lower density of cortical MTs will allow cells to expand to some extent in the direction parallel to the cortical MTs. This scenario is not relevant, however, in the post-elongation zone where cellulose microfibrils are intensively deposited around the cell membrane in this region.

The dynamic helix model proposed by Lloyd (1984) has been successfully argued to explain the mechanism by which cortical MTs can change their

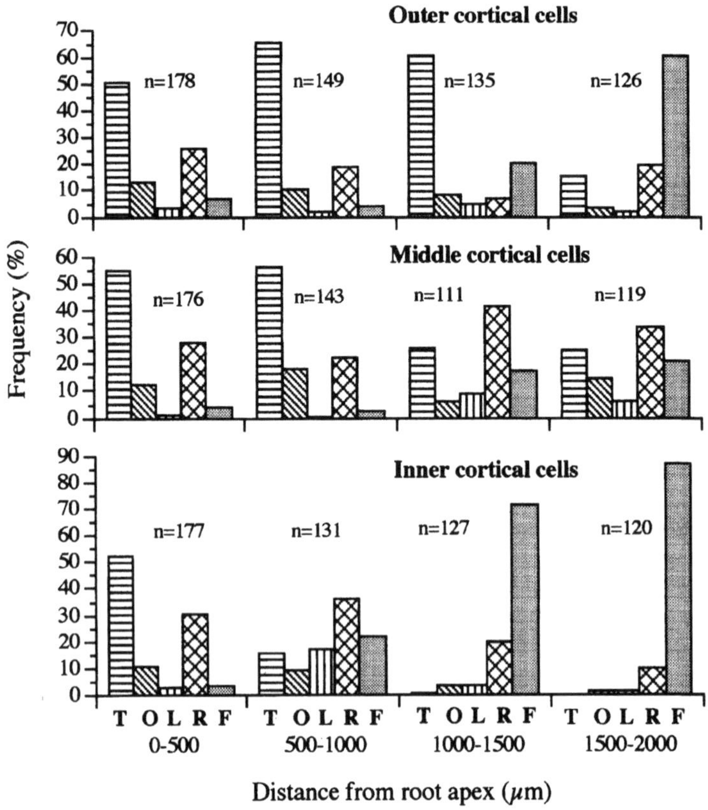

Figure 9. Frequency of cells with transverse (T), oblique (O), longitudinal (L), random (R) and fragmented (F) cortical MTs along the root axis. The T-type cells of the inner cortical cells dramatically decreased in the 500–1000 μm region of the root, whereas F-type cells increased substantially in the region greater than 1000 μm from the apex. Data are from five plants.

orientation. The helix model is partly based on the observation that the orientation of cortical MTs is reverse across the two opposing walls of a cell. The present study also confirmed the reverse orientation of cortical MTs between opposing radial walls. In this case, however, the pitch of each array was usually different. This observation is consistent with a report showing that the pitch of cortical MTs relative to the longitudinal axis of pea epidermal cells is different between the radial and tangential walls (Yuan et al., 1995). Thus, integrated arrays of cortical MTs appear to shift in pitch at the corners of cell walls. In general, the increase of the pitch of the cortical MT spiral along the cell membrane can explain the increase of steepness of cortical MT orientation relative to the longitudinal axis and simultaneously the decrease in cortical MT density. However, the decrease in cortical MT density reported in this study may not wholly depend on this explanation; it is possible that longitudinally-oriented cortical MTs were not counted, because only the longitudinal sections were examined by electron microscopy. Accordingly, the decrease in cortical MT density in the region beyond 1000 μm from the root apex may partly reflect the shift in cortical MT orientation from transverse to longitudinal.

Although the cortical MTs on transverse walls were oriented predominantly perpendicular to the longitudinal axis in the acropetal region of the elongation zone, the cells with obliquely- or longitudinally-oriented cortical MTs were seen in relatively high frequency compared with those along the longitudinal walls. In particular, the frequency of transversely-oriented cortical MTs on transverse walls of the inner cortical cells were remarkably reduced in the 500–1000 μm region when the expansion of transverse walls ceased. In addition, the fragmented cortical MTs were very frequently observed in the region beyond 1000 μm from the root apex. The continuous shift in

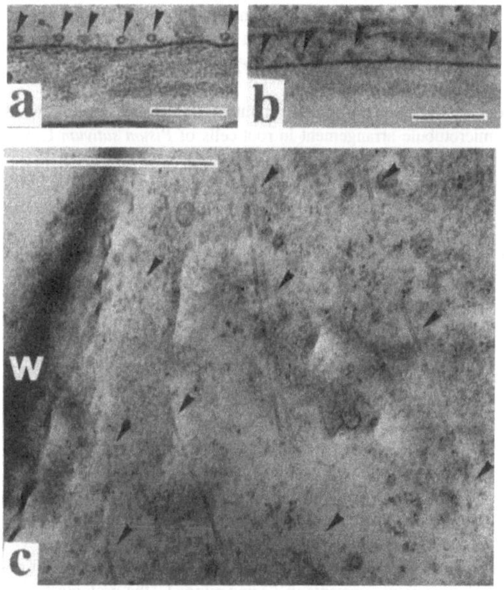

Figure 10. Electron micrographs of cortical MTs. The cross-sections of cortical MTs (arrow heads) are seen immediately adjacent to the tangential wall of an inner (a) and an outer (b) cortical cell in the elongation zone of the root. Bar, 0.2 μm. c. Oblique cortical MTs (arrow heads) of an inner cortical cell in the post-elongation zone. Bar, 1 μm. w–cell wall.

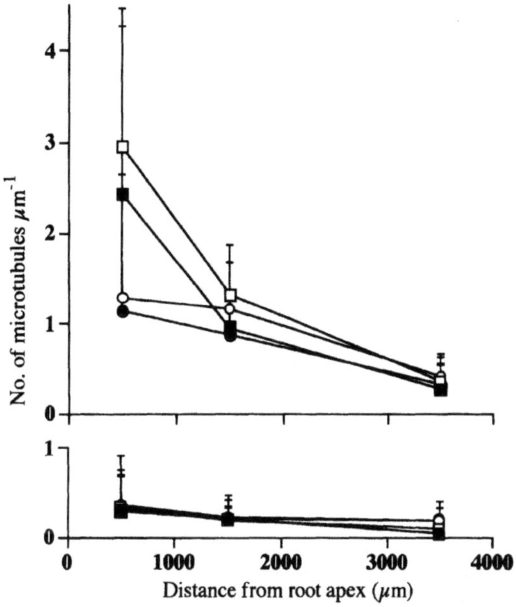

Figure 11. Density of the cortical MTs on tangential (top) and transverse walls (bottom). Density is defined as the number of cortical MT segments per μm of cell membrane. The number of cells examined were as follows: Inner cortical cells: 76, 38 and 30; Endodermis: 56, 36 and 26; Middle cortical cells: 52, 36 and 28; Outer cortical cells: 56, 36 and 26, respectively, for the 0–1000, 1000–2000 and 3000–4000 μm regions from the root apex. In the elongation zone, the density of cortical MTs on tangential walls was higher for the inner cortical cells (□) and endodermis (■) than for the outer (●) and middle cortical cells (○). * $P < 0.01$ versus outer and middle cortical cells. Data are from three plants.

the orientation of cortical MTs from transverse to longitudinal was not clearly seen on the transverse walls. These features of cortical MTs on cross walls may be implicated in the limited expansion of these walls. The expansion of the transverse walls in all cortical cells was much reduced compared to that of the radial and tangential walls. In particular, the inner cortical cells showed the least expansion of transverse walls both in length and width among the three types of cortical cells, and the net transverse orientation of cortical MTs rapidly diminished even in the region close to the root apex. Although the cortical MTs are first oriented perpendicular to the longitudinal axis of the cell, transverse cortical MT arrays may not be required to persist for any extended length of time because the transverse walls expand only marginally in length. An alternative explanation may be that the determination of shapes of both the radial and tangential walls inevitably leads to that of the transverse walls and therefore, the cortical MTs on the transverse walls are not heavily involved in controlling the shape of transverse walls. This hypothesis is supported by the low density of cortical MTs on transverse walls in all the cortical cells, even in the elongation zone. This low estimation may be partly due, however, to the examination being performed in median longitudinal sections, thus missing cortical MTs arrayed in parallel with this face of the cell.

Acknowledgements

We are very grateful to Prof. T. Beppu, Teikyo University of Science, for identifying the plant, and to Prof. S. Hasezawa and Dr. M. Fukuda, University of Tokyo, for their kind advice on improving the indirect immunofluorescence technique. We also express our sincere thanks to Prof. T. Shimmen, Himeji Institute of Technology, for his encouragement and support and to Dr. D. W. MacCurdy, University of Newcastle, Australia for reading the manuscript. We thank the Advanced Instrumentation Center for Chemical Analysis, Ehime University, for providing the opportunity to use an electron microscope. The research was partially supported by a Grant-in-Aid for Scientific Research (no.

09640793) from the Ministry of Education, Science, Sports and Culture.

References

Akashi T and Shibaoka H 1987 Effects of gibberellin on the arrangement and the cold-stability of cortical microtubules in epidermal cells of pea internodes. Plant Cell Physiol. 28, 339-348.

Baluška F, Parker J S and Barlow P W 1992 Specific patterns of cortical and endoplasmic microtubules associated with cell growth and tissue differentiation in roots of maize (*Zea mays* L.). J. Cell Sci. 103, 191-200.

Baskin T I, Busby C H, Fowke L C, Sammut M and Gubler F 1992 Improvements in immunostaining samples embedded in methacrylate: localizatio microtubules and other antigens throughout developing organs in plants of diverse taxa. Planta 187, 405-413.

Baskin T I, Wilson J E, Cork A and Williamson R E 1994 Morphology and microtubule organization in arabidopsis roots exposed to oryzalin or taxol. Plant Cell Physiol. 35, 935-942.

Brown R C, Lemmon B E and Mullinax J B 1989 Immunofluorescent staining of microtubules in plant techniques using polyethylene glycol (PEG) and Steedman's wax. Bot. Acta 102, 54-61.

Cleary A L and Hardham A R 1988 Depolymerization of microtubule arrays in root tip cells by oryzalin and their recovery with modified nucleation patterns. Can. J Bot. 66, 2353-2366.

Emons A M C 1982 Microtubules do not control microfibril orientation in a helical cell wall. Protoplasma 133, 85-87.

Emons A M C 1989 Helicoidal microfibril deposition in a tip-growing cell and microtubule alignment during tip morphogenesis: A dry-cleaving and freeze- substitution study. Can. J. Bot. 67, 2401-2408.

Emons A M C and Mulder M M 1998 The making of the architecture of the plant cell wall: How cells exploit geometry. Proc. Natl. Acad. Sci. USA 95, 7215-7219.

Fisher D D and Cyr R J 1998 Extending the microtubule/microfibril paradigm. Cellulose synthesis is required for normal cortical microtubule alignment in elongating cells. Plant Physiol. 116, 1043-1051.

Fischer K and Schopfer P 1998 Physical strain-mediated microtubule reorientation in the epidermis of gravitropically or phototropically stimulated maize coleoptiles. Plant J. 15, 119-123.

Flanders D J, Rawlins D J, Shaw P J and Lloyd C W 1989 Computed-aided 3-D reconstruction of interphase microtubules in epidermal cells of *Datura stramodium* reveals principles of array assembly. Development 106, 531-541.

Hardham A R and Gunning B E S 1978 Structure of cortical microtubule arrays in plant cells. J. Cell Biol. 77, 14-34.

Hardham A R and Gunning B E S 1979 Interpolation of Microtubules into cortical arrays during cell elongation and differentiation in roots of *Azolla pinnata*. J. Cell Sci. 37, 411-442.

Hillman W S 1961 Experimental control of flowering in *Lemna* . III. A relationship between medium composition and the opposite photoperiodic responses of *L. pepusilla* 6746 and *L. gibba* G3. Am. J. Bot. 48, 413-419.

Hogetsu T and Oshima Y 1986 Immunofluorescence microscopy of microtubule arrangement in root cells of *Pisum sativum* L. var. Alaska. Plant Cell Physiol. 27, 939-945.

Hogetsu T and Shibaoka H 1978 The change of pattern in microfibril arrangement on the inner surface of the cell wall of *Closterium acerosum* during cell growth. Planta 140, 7-14.

Iwata K and Hogetsu T 1988 Arrangement of cortical microtubules in *Avena* coleoptiles and mesocotyl and *Pisum* epicotyls. Plant Cell Physiol. 29, 807-815.

Ledbetter M C and Porter K R 1963 A 'microtubule' in plant cell fine structure. J. Cell Biol. 19, 239-250.

Liang B M, Dennings A M, Sharp R E and Baskin T I 1996 Consistent handedness of microtubule helical arrays in maize and Arabidopsis primary roots. Protoplasma 190, 8-15.

Lloyd C W 1987 The plant cytoskeleton: The impact of fluorescence microscopy. Annu. Rev. Plant Physiol. 38, 119-139.

Melan M A 1990 Taxol maintains organized microtubule patterns in protoplasts which lead to the resynthesis of organized cell wall microfibrils. Protoplasma 153, 169-177.

Melaragno J E and Walsh M A 1976 Ultrastructural features of developing sieve elements in *Lemna minor* L.-the protoplast. Am. J. Bot. 63, 1145-1157.

Seagull R W 1986 Changes in microtubule organization and wall microfibril orientation during *in vitro* cotton fibre development: an Immunofluorescent study. Can. J. Bot. 64, 1378-1381.

Shibaoka H 1991 The role of gibberellin in the formation of onion bulbs. In Gibberelins Eds. N Takahashi, BO Phinney and J MacMillan. pp 220-228. Springer-Verlag, New York

Sakiyama M and Shibaoka H 1990 Effects of abscisic acid on the orientation and cold stability of cortical microtubules in epicotyl cells of the dwarf pea. Protoplasma 157, 165-171.

Spurr A R 1969 A low-viscosity epoxy resin embedding medium for electron microscopy. J. Ultrastruct. Res. 26, 31-43.

Traas J A, Braat P and Derksen J W 1984 Changes in microtubule arrays during the differentiation of cortical root cells of *Raphanus sativus*. Eur. J. Cell Biol. 34, 229-238.

Traas J A, Braat P, Emons A M C, Meekes H and Derksen J 1985 Microtubules in root hairs. J. Cell Sci. 76, 303-320.

Wasteneys G O and Williamson R E 1987 Microtubule orientation in developing internodal cells of Nitella: A quantitative analysis. Eur. J. Cell Biol. 43, 14-22.

Williamson R E 1991 Orientation of cortical microtubules in interphase plant cells. Int. Rev. Cytol. 129, 135-206.

Yuan M, Warn R M, Shaw P J and Lloyd C W 1995 Dynamic microtubules under the radial and outer tangential walls of microinjected pea epidermal cells observed by computer reconstruction. Plant J. 7, 17-23.

Section editor: H Lambers

MIX
Papier aus verantwortungsvollen Quellen
Paper from responsible sources
FSC® C105338

If you have any concerns about our products,
you can contact us on
ProductSafety@springernature.com

In case Publisher is established outside the EU,
the EU authorized representative is:
Springer Nature Customer Service Center GmbH
Europaplatz 3, 69115 Heidelberg, Germany

Printed by Libri Plureos GmbH
in Hamburg, Germany